D1226521

Vacuum Manual

VACUUM MANUAL

Edited by

L. Holland

Formerly Director of Research
Edwards High Vacuum;
Associate Reader in Physics, Brunel University

W. Steckelmacher

Formerly Deputy Director of Research
Edwards High Vacuum

J. Yarwood

Professor of Physics
Central London Polytechnic

E. & F. N. SPON London

First published 1974
by E. & F. N. Spon Ltd.
11 New Fetter Lane, London EC4P 4EE

Printed in Great Britain by
William Clowes & Sons Ltd.
London, Colchester and Beccles

ISBN 0 419 10740 1

Distributed in the U.S.A.
by Halsted Press, a Division
of John Wiley & Sons, Inc., New York

Library of Congress Catalog Card Number 73-13378

Contents

Introduction

Vacuum apparatus is widely used in research and industrial establishments for providing and monitoring the working environments required for the operation of many kinds of scientific instruments and process plant. The vacuum conditions needed range from the relatively coarse vacuum requirements in applications covering diverse fields such as

food packaging, dentistry (investment casting), vacuum forming, vacuum metallurgical processes, vacuum impregnation, molecular distillation, vacuum drying and freeze drying etc.

to the other extreme involving the highest possible vacuum as in

particle accelerators, space technology – both in simulation and outer space, and research studies of atomically clean surfaces and pure condensed metal films.

Vacua commence with the *rough vacuum* region, i.e. from atmosphere to 100 Pa* passing through *medium vacuum* of 100 Pa to 0·1 Pa and *high vacuum* of 0·1 Pa to 1 μPa (10^{-6} Pa) until *ultra high vacuum* is reached below 1 μPa to the limit of measurable pressure about 1 pPa (10^{-12} Pa).

Two decades ago it was possible for a vacuum scientist or engineer to have a comprehensive knowledge of the operating principles, characteristic and chief manufacturers of the pumps, instruments and plant used in his field. Since then a whole family of new types of vacuum pump and analytical and process monitoring instruments have been developed and passed into routine manufacture. Also the ultra high vacuum region has been entered and new vacuum techniques have appeared often requiring use of vacuum systems, on an industrial scale. There are now many companies concerned with the production of basic vacuum components, such as pumps and instruments, and their related accessories. If it has become difficult for the vacuum worker to maintain an overall view of his field then obviously it is even more so for those who regard vacuum apparatus as tools or a means to an end.

Fortunately reference works are available reviewing most aspects of vacuum technology and its applications but there are no major works cataloguing the types and makers of

* The pascal (Pa) is the SI unit of pressure: 1 Pa = 1 N/m^2, although currently the Torr is still the most widely used pressure unit in vacuum technology. 100 Pa = 3/4 Torr approximately (1 Pa = 7·50062 mTorr, 1 Torr = 133·322 Pa). Wherever possible the SI unit of pressure has been used in the handbook but conversion of many of the graphs and tables reproduced has not always been possible. Also, there are many fields where vacuum pressure is only an indicator of satisfactory or adverse working conditions and there is no immediate incentive to adopt the base unit.

available vacuum products except for occasional exhibition catalogues which are limited by participation. The handbook presented here corrects the foregoing deficiency by giving classified information about the performance and characteristics of the pumps, instruments and process plant available from manufacturers all over the world. It also relates their characteristics and performance to basic vacuum data which can be applied in system design and uses. The data sections give information on: gas flow under viscous and molecular flow conditions, gas permeation through and desorption from vacuum materials and the properties of pump fluids, sealing compounds and greases. Finally, it was considered desirable to present the reader with reviews on selected topics on vacuum attainment, instrumentation and application for which recent advances had been made. By this means parity is maintained between treatment of new products in these fields and knowledge of their function. The authors and publisher express their thanks to the many manufacturers and research centres who have kindly supplied information on which much of the product and basic vacuum data given here is based. Furthermore, comments, criticisms and suggestions for additions from manufacturers and users alike are welcomed to ensure the comprehensive nature of future editions of this handbook.

We would like to express our appreciation for the help given by Jean Charman, Jennifer Roberts and Norma Ward, in the preparation of the manuscript.

March 1973 L. Holland, W. Steckelmacher and J. Yarwood

1 – Basic Data

1.1 Ultimate pressure (p_u)

The ultimate pressure is defined (BSI 2951, 1969) as 'the limiting pressure approached in a vacuum chamber after pumping for sufficient time to establish that further reduction in pressure will be negligible'.

A pumping system which provides at the chamber at the pressure p_u an effective pumping speed (i.e. volume rate of flow) of S will withdraw a quantity of gas (or vapour) per second Q from the chamber given by

$$Q = p_u S. \quad (1)$$

At the ultimate, Q will also be the influx of gas into the chamber. This influx will be due to the sum of the following effects:

(a) Leakage into the chamber of gas from the surrounding atmosphere; a quantity of gas per second denoted by Q_{leak}.

(b) Evolution of gas from the chamber inner walls and from any materials or components within this chamber. This *outgassing* will be denoted by Q_{outgas}.

(c) Back-diffusion from the pumping system, i.e. due to gas which passes from the backing (fore) pressure side of the pump(s) to the high vacuum side. Denoted by $Q_{back\text{-}diff}$.

(d) Back-streaming from the pumping system, prevalent where a pump fluid or lubricating fluid is used and some of the molecules of this fluid are scattered so as to travel towards the chamber instead of towards the fore pressure region. Added to this is back migration due to evaporation of the pump or lubricating fluid from surfaces in the pumping system. This is denoted by $Q_{back\text{-}stream}$.

Effect (a), leakage from the atmosphere, due to imperfect seals, may be made negligible by correct design and procedure. There will, nevertheless, remain an influx of gas due to permeation through the chamber walls. Q_{leak} then becomes Q_{perm}. Ignoring also the effects of back-diffusion and back-streaming, both of which can be reduced to insignificant levels by satisfactory design of the pumping system in relation to the choice of pump(s), baffles and/or cold traps $Q_{back\text{-}diff}$ and $Q_{back\text{-}stream}$ can both be set equal to zero. Equation (1) can be written

$$p_u = (Q_{perm} + Q_{outgas})/S. \quad (2)$$

This equation (2) does not lead to straightforward calculations in practice because Q_{perm} and Q_{outgas} are not necessarily constant during the pump-down, both depend on the nature of the gas, the materials involved and their pre-treatment; moreover S depends on the nature of the gas, the pressure and the kind of pumping system used. Nevertheless, it is clear that a knowledge of the outgassing and gas permeation characteristics of materials are important in design calculations.

To minimize the evolution of gas from materials, *degassing* is practised. Whereas outgassing is defined as the spontaneous evolution of gas from materials, degassing is the deliberate removal (by heating or particle bombardment) of gas from a material.

Degassing involves bake-out (in an oven) of the chamber undergoing pumping and, to obtain pressures below 10^{-4} Pascal unit, vacuum stoving (heating within a separate vacuum furnace) of materials and components within the vacuum chamber.

1.2 The evolution of gas from materials

The factor Q_{outgas} is due to:

(a) The release from the surface of bonded gas and vapour films where such bonding is due to physisorption and chemisorption, the latter giving rise to the higher binding energies. Considerable data on such binding energies is given by Redhead, Hobson and Kornelsen (1968). Roughly, physisorption energies can be taken as less than 90 kJ/mole (1 eV/particle) and greater than this value for chemisorption energies.

(b) The release of gas or vapour occluded in pores or in solution in the body of the material.

(c) The vapour pressure of the material; in this case, the gas or vapour evolved will normally be of molecules of the material itself.

(a) and (b) are reduced by previous cleaning treatment and by degassing; (c) is insignificant at room temperature for the chamber and the usual constructional materials employed except when liquids and polymers are involved.

1.2.1 Outgassing

Outgassing will depend greatly on the cleanliness of the material and how it is cleaned (e.g. chemical degreasing agents used).

Materials which have *not* been subjected to bake-out *in vacuo* or vacuum stoving will evolve predominantly water vapour when under vacuum (presuming that all traces of any organic materials, such as cleaning agents, have been previously removed). If base pressures below 10^{-4} Pa are not required, the usual constructional materials for vacuum chambers are mild steel and glass, and bake-out is not needed if the pumping speed available is adequate. To attain pressures below 10^{-4} Pa, certainly below 10^{-6} Pa, and especially if the evacuated chamber is to be isolated from the pumps subsequent to processing, bake-out of the chamber is essential. The recommended chamber materials are then either borosilicate glass or stainless steel (300 series, e.g. type 18/8, containing by weight approximately 18% chromium and 8% nickel).

Unbaked glass *in vacuo* evolves above all water vapour from its surface. Water vapour is also released from the interior of the glass, but at a much smaller rate, together with carbon dioxide, oxygen, nitrogen and minute traces of other gases. Molecular gases from unbaked Pyrex glass in an ultra-high vacuum (uhv) system (pressure below 10^{-6} Pa) are evolved at a rate of about 10^{-5} Pa m^3 s^{-1} per m^2 of surface,* but this figure depends on the condition of storage of the glass since manufacture and increases with alkali content of the glass.

The evolution of gas from glass when it is baked-out under vacuum reaches a peak rate at a temperature of about 150°C for soft glasses (e.g. soda-lime glasses) and at about 210°C for hard glasses (e.g. borosilicate glasses). This peak is caused by the release of surface gas, chiefly water vapour. Subsequently, at higher temperatures, the gas evolution rate falls greatly until temperatures of 300°C (for soft glasses) to 400°C (for hard glasses) are attained. Beyond such temperatures, gas is evolved as the glass itself undergoes decomposition, so outgassing will persist indefinitely.

TABLE 1.1 Maximum degassing (bake-out) temperatures for common glasses

Glass	Temperature (°C)
Soda-lime	350
Lead	350
Borosilicate (Pyrex, Hysil)	450

The rate of outgassing from a borosilicate glass such as Pyrex is reduced by about 10^9 to 10^{10} times (i.e. to about 10^{-14} Pa m^3 s^{-1}/m^2) after long term (minimum 4 hours) bake-out *in vacuo* at 450°C.

Considerable evidence exists that, for uhv systems, bake-out of glass need only be practised once at 350°C or above; provided that air is not admitted to above 10^{-4} Pa to the system, in subsequent runs a bake-out temperature of 200°C is satisfactory. The same is true for uhv systems having a chamber constructed from stainless steel.

Unbaked stainless steel *in vacuo* evolves chiefly hydrogen, water vapour and carbon dioxide at a rate of about 10^{-6} to 10^{-4} Pa m^3 s^{-1}/m^2 of surface, the lower figure being for a very clean polished surface after several hours of pumping. After bake-out for some hours (the time being decided chiefly by the dimensions of the vacuum chamber, the thickness of its walls and the maximum temperature, usually 450°C) polished clean stainless steel under uhv will evolve primarily carbon monoxide and hydrogen at a rate of about 10^{-8} Pa m^3 s^{-1}/m^2. Much longer bake-out times (16 hours or more) can reduce this figure by a factor of about 100 times. Calder and Lewin (1967) report that bake-out of stainless steel under uhv to 1000°C for 1 hour reduced the hydrogen outgassing rate to $1{\cdot}3 \times 10^{-11}$ Pa m^3 s^{-1}/m^2, as compared with $1{\cdot}3 \times 10^{-9}$ after bake-out for 12 hours at 400°C.

Dayton (1961) has surveyed the most reliable data on the outgassing of materials. He proposes for *unbaked contaminated surfaces* (i.e. which have been exposed to air for some time) the empirical equation

$$Q_{\text{outgas}} = 10^{-3}/t^a, \quad (3)$$

* 1 Torr = 133 Pa; 1 Torr litre = $1{\cdot}33 \times 10^{-1}$ Pa m^3; 1 Torr litre s^{-1} per cm^2 of surface = $1{\cdot}33 \times 10^3$ Pa m^3 s^{-1} per m^2 of surface.

TABLE 1.2 Outgassing from various materials *in vacuo.* No distinction is made between desorption of gas, volatilization and permeation. Data compiled from Blears, Greer and Nightingale (1960), Santeler, Holkeboer, Jones and Pagano (1966), Dayton (1959), Flecken and Nöller (1961), Connor, Buritz and Von Zweck (1961), Steinherz (1963), Diels and Jaeckel (1962)

Material	Pre-treatment	Time *in vacuo* (hour)	Outgassing rate in Pa m^3 s^{-1} from a surface of 1 m^2 area
Aluminium	Cleaned in stergene	10	$1{\cdot}1 \times 10^{-5}$
	Anodized	10	$1{\cdot}3 \times 10^{-4}$
	None	1	$1{\cdot}7 \times 10^{-3}$
Araldite D[1]	None	10	$1{\cdot}3 \times 10^{-3}$
		100	$5{\cdot}3 \times 10^{-4}$
Brass	Cast, washed	10	4×10^{-4}
Butyl rubber	None	1	2×10^{-3}
Chrome-plated steel	Polished, vapour degreased	1	$1{\cdot}3 \times 10^{-5}$
Copper	None	1	3×10^{-3}
Copper at 450°C	None	1	2×10^{-3}
	Degreased, pickled	1	$3{\cdot}5 \times 10^{-4}$
	Degreased	1	2×10^{-3}
Epoxy resins	None	1	$1{\cdot}4 \times 10^{-2}$
No. 200	Degassed, then 24 h in dry nitrogen	1	$2{\cdot}6 \times 10^{-5}$
Eppon	None	1	5×10^{-3}
		4	$1{\cdot}1 \times 10^{-3}$
Kel F	None	1	$5{\cdot}3 \times 10^{-5}$
Mild steel	Shot-blasted	10	8×10^{-5}
Molybdenum	None	1	9×10^{-4}
Mylar	None	1	4×10^{-3}
	Degassed	1	$2{\cdot}7 \times 10^{-4}$
Neoprene	None	1	4×10^{-2}
		10	2×10^{-2}
Nickel	None	1	8×10^{-4}
Nickel-plated steel	Polished, vapour degreased	1	$7{\cdot}5 \times 10^{-4}$
Nylon[2]	None	1	$1{\cdot}6 \times 10^{-2}$
Plexiglass	Degassed	1	$1{\cdot}3 \times 10^{-3}$
Polyethylene	None	1	$3{\cdot}5 \times 10^{-4}$
Porcelain	Glazed	1	$8{\cdot}6 \times 10^{-4}$
PVC[3]	None		7×10^{-3}
Silicone rubber	None	1	4×10^{-2}
Stainless steel	None	1	3×10^{-4}
		10	$2{\cdot}6 \times 10^{-5}$
	Polished, vapour degreased	10	$1{\cdot}9 \times 10^{-6}$
Silver	None	1	8×10^{-4}
Steatite	None	1	$1{\cdot}2 \times 10^{-4}$
Tantalum	None	1	$1{\cdot}2 \times 10^{-3}$
Teflon[4]	None	1	$6{\cdot}5 \times 10^{-3}$
Tungsten	None	1	$2{\cdot}6 \times 10^{-4}$
Viton A[5]	None	1	$3{\cdot}0 \times 10^{-5}$
Zirconium	None	1	$1{\cdot}8 \times 10^{-3}$

1 Hot-setting Araldite 1 has a low outgassing rate ($< 5 \times 10^{-6}$ Pa m^3 s^{-1} per m^2), the gas evolved being chiefly water vapour

2 Nylon is very hygroscopic.

3 Results of other workers for PVC indicate an outgassing rate of 8×10^{-2} Pa m^3 s^{-1}/m^2 even after several weeks of pumping; it should be avoided in vacuum systems.

4 Polytetrafluoroethylene (PTFE) (same as Teflon, Fluon) after an initial pump-down of about 50 hours at 100°C has an outgassing rate below 4×10^{-6} Pa m^3 s^{-1}/m^2.

5 Value given by Farkass and Barry (1961) who, however, give outgassing rate figures some 100 times lower than those recorded in this table for other O-ring elastomers at room temperature.

where Q_{outgas} is the outgassing rate in Pa m^3 s^{-1}/m^2 after 1 hour of pumping, t is the time in hours since the start of outgassing, and a is between 0·7 and 2·0 but is most frequently unity. In the first hour of pumping, the rate of outgassing is considerably greater and much more variable depending on the material and its surface treatment. Assuming $a = 1$, this implies an outgassing rate after the first hour of pumping an unbaked material which is largely independent of the constructional material used. This is a reasonable assumption for glass and metal chambers of mild steel, steel alloys, copper, etc. which are unbaked. Though much experimental work has been done on the outgassing rates of unbaked metals, the results are so varied and so dependent on the experimental conditions, that a better choice than Dayton's equation is only really possible in practice if more precise evaluation data in the particular circumstances are made available from experimental trials. However, Table 1.2 gives somewhat more precise data, and is necessary on considering other material than glass and metals.

Schittko (1963) gives experimental data on the gas evolution rates from a wide range of solid surfaces including those of plastics and elastomers. He emphasizes that the common practice of measuring gas evolution with the aid of ionization gauges (where pressures are usually 'nitrogen-equivalent' values) demands a number of precautions because surface reactions with walls, thermal dissociation at hot cathodes and gas decomposition all introduce large errors. Hence, a mass spectrometer with a low temperature emitter filament is preferred. He also asserts that the pressure–temperature course of the gas evolution quantity follows for almost all samples a law of the form:

$$Q = Q_0 \exp(t/t_0),$$

where t is the temperature of the sample and t_0 is room temperature, Q is the outgassing rate at the temperature t and Q_0 is this rate at t_0.

TABLE 1.3 Colwell (1970) gives data on outgassing rates of over 80 untreated refractory and electrical insulating and other materials suitable for use in vacuum furnaces; the tests were performed over the pressure range 10^{-1} to 10^{-5} Pa.

Material	Outgassing rate in Pa m^3 s^{-1}/m^2 of surface after		
	1 hour	4 hour	20 hour
Araldite	$2{\cdot}6 \times 10^{-3}$	$9{\cdot}3 \times 10^{-4}$	4×10^{-4}
Epoxy resin	$3{\cdot}6 \times 10^{-3}$	2×10^{-3}	8×10^{-4}
Expanded quartz	$7{\cdot}2 \times 10^{-4}$	2×10^{-4}	7×10^{-5}
Fused alumina*	$6{\cdot}7 \times 10^{-4}$	$1{\cdot}6 \times 10^{-4}$	$1{\cdot}1 \times 10^{-4}$
Mullite	$8{\cdot}3 \times 10^{-4}$	$1{\cdot}9 \times 10^{-4}$	$1{\cdot}1 \times 10^{-4}$
Polyethylene	$2{\cdot}6 \times 10^{-3}$	6×10^{-4}	2×10^{-4}
Pyrophilite (unfired)	$7{\cdot}3 \times 10^{-2}$	$1{\cdot}1 \times 10^{-2}$	$1{\cdot}7 \times 10^{-3}$
(fired)	2×10^{-3}	$7{\cdot}2 \times 10^{-4}$	$2{\cdot}8 \times 10^{-4}$
Sillimanite	7×10^{-4}	2×10^{-4}	$1{\cdot}1 \times 10^{-4}$
Syndanyo	1×10^{-1}	$4{\cdot}5 \times 10^{-2}$	$1{\cdot}5 \times 10^{-2}$
Tufnol†	$4{\cdot}7 \times 10^{-3}$ to $1{\cdot}2 \times 10^{-1}$	3×10^{-3} to 6×10^{-2}	2×10^{-3} to $1{\cdot}7 \times 10^{-2}$
Zircon	$1{\cdot}6 \times 10^{-3}$	$3{\cdot}2 \times 10^{-4}$	$6{\cdot}6 \times 10^{-5}$

* Alumina (Al_2O_3) is virtually impermeable to gases and is unaffected at temperatures up to 1700°C by air, water vapour or permanent gases.

† Values range for 5 different brands.

Beryllia (BeO) is similar to alumina in its resistance to gases in heating to 1700°C but it volatilizes quickly in the presence of water vapour because beryllium hydroxide is formed. Magnesia (MgO) is not recommended because it has poor thermal stress resistance, poor mechanical strength and reacts with metals at elevated temperatures. Zircon (a combination of ZrO_2 and SiO_2) is a useful material for fabricating vacuum furnace trays.

Mica surfaces can be regarded as saturated with gas; mica contains up to about 18% by weight of combined water.

Several results have been given on the outgassing rates of elastomers which can be degassed by bake-out at temperatures above 120°C and are suitable for the manufacture of O-rings and other sealing gaskets. Viton A is bakeable to 150°C. Hait (1967) reports that polyimide withstands bake-out up to 300°C, and has a lower outgassing rate than Viton A, the principal gases evolved at 300°C being H_2O, CO, CO_2 and H_2 whereas, on returning to room temperature, the primary residual gases are CO and CO_2.

Chernatony (1966) has made extensive tests on Viton A (a copolymer of hexafluoropropylene and vinylidine fluoride). O-rings of this elastomer degassed at pressures below 10^{-4} Pa at temperatures up to 290°C over 3 weeks gave subsequently an outgassing rate less than $6{\cdot}6 \times 10^{-7}$ Pa m^3 s^{-1}/m^2 and the solubility of air in Viton A at 18°C and relative humidity 50% was found to be $4{\cdot}25 \times 10^5$ Pa m^3 per m^3 of elastomer and due mainly to water vapour and CO_2. Viton O-ring seals are alleged to permit pumping to 10^{-6} Pa and lower with cryopumping to liquid nitrogen temperatures (77 K). A double O-ring of Viton with an evacuated interspace is recommended for uhv systems in which pressures of a few 10^{-8} Pa (nitrogen equivalent) are obtainable by means of a liquid nitrogen trapped mercury diffusion pump. Baking the Viton at 150°C and the chamber at 300°C can eliminate completely water vapour from the system. Furthermore, cryopumping with liquid helium enables pressures of 10^{-9} Pa and below to be obtained.

Chernatony and Crawley (1967) find that with an O-ring in a groove the chamber pressure is typically half than when the O-ring is fully exposed within the chamber but after repeated bake-out (e.g. to 200°C for Viton A) this situation is reversed because the fully exposed O-ring becomes depleted of gas so that it contributes a pressure some 100 times lower than the same ring used in a groove.

Table 1.4 gives their results for Viton A, fully exposed in the vacuum chamber.

TABLE 1.4 Viton A outgassing

	Outgassing rate in Pa m^3 s^{-1}/m^2
After 200°C bake-out	$2{\cdot}7 \times 10^{-7}$ at 20°C
	$2{\cdot}7 \times 10^{-6}$ at 110°C
	2×10^{-3} at 180°C
Unbaked	$1{\cdot}3 \times 10^{-4}$ at 20°C

Outgassing rates for polymers are approximately two orders of magnitude higher than for stainless steel (Bailey, 1964).

Strausser (1968) reports that hydrogen evolution from 304 stainless steel is not observed prior to bake-out. During a pump-down in which a stainless steel sample temperature was increased gradually to 150°C, at first, near room temperature, 90% of the gas released is water vapour. Then the outgassing rates for nitrogen, propane and methane increase by three orders of magnitude; subsequently, hydrogen evolution predominates and, with the sample at 150°C, the hydrogen outgassing rate remains relatively constant. Oxygen evolution is small. As the sample cools after bake-out, all outgassing rates fall drastically with hydrogen the predominant gas. All other gases have final (room temperature) outgassing rates at least two orders of magnitude below that of hydrogen.

Strausser also reports that final outgassing rates after a prolonged 150°C bake-out are approximately the same as after a shorter 300°C bake-out, alleging that bake-out above 150°C is not necessary for uhv systems. Furthermore, a system that is vapour degreased, chemically cleaned and glass-bead blasted produces a lower outgassing rate than one only vapour degreased with trichlorethylene, but the overall difference is less than one order of magnitude.

Barton and Govier (1968) report that bake-out of new 18/8/1 stainless steel at 450°C *in vacuo* successfully removes all gases except for traces of hydrogen which diffuses out of the metal. Gas adsorbed at the surface on re-exposure to the atmosphere can be removed by bake-out of a vacuum chamber for two to three hours at 450°C. For many purposes, where a pressure of 10^{-7} Pa is adequate, sufficient gas can be removed by bake-out at 200°C to 300°C for 12 hours. They assert that the most efficient methods of cleaning are those which remove the surface of the sample. Electropolishing giving a high polish as well as removing the surface produced the best results, but after the original 'soil' is removed by bake-out, such electropolishing is only marginally better than vapour degreasing in conjunction with a good quality machine for finish.

Alpert (1959) provides useful data on the gas evolution within uhv systems, summarized in Table 1.5.

TABLE 1.5

uhv system	Gas evolution	Outgassing rate in Pa m^3 s^{-1}/m^2
Baked borosilicate	Water vapour at room temperature	4×10^{-15}
Glass	Helium at room temperature	$1{\cdot}3 \times 10^{-12}$
Unbaked with borosilicate glass and metal valves	Molecular gases from glass and metal valves	$1{\cdot}3 \times 10^{-5}$
Baked stainless steel	Mostly CO	$6{\cdot}7 \times 10^{-12}$
Unbaked stainless steel	Molecular gases	$1{\cdot}3 \times 10^{-4}$

1.2.2 Degassing

TABLE 1.6 Maximum degassing temperatures for metals and graphite.

Material	Temperature (°C)	Gases evolved in order of decreasing amounts	Material	Temperature (°C)	Gases evolved in order of decreasing amounts
Copper	500	CO_2, CO, H_2O, N_2, H_2	Platinum	1000	
Graphite*	1800	N_2, CO at 2150°C	Steels	1000	CO, H_2, N_2, CO_2
Iron	1000	CO, H_2O, N_2	Tantalum	1400	CO, H_2, N_2, H_2O
Molybdenum	950†	CO, H_2, N_2	Titanium	1100	H_2, H_2O, CH_4
Nickel	950	CO, H_2O, CO_2, H_2	Tungsten	1800	CO, CO_2, H_2

* It is virtually impossible to degas graphite completely.
† Higher temperatures can be used but the metal becomes brittle.

TABLE 1.7 Percentage composition of gases evolved from different materials at a temperature T°C (Wagner and Marth, 1959)

Material	T (°C)	Pressure (Pa)	Gas fraction (%)					
			CO_2	H_2O	CH_4	CO	N_2	H_2
Iron	1000	6×10^{-4}	–	3	–	95	2	–
Molybdenum	1800	5×10^{-4}	4	5	–	80	10	1
Nickel	1000	$2{\cdot}5 \times 10^{-4}$	2	4	–	92	–	2
Tantalum	2200	3×10^{-4}	–	3	–	50	6	40
Titanium	1100	2×10^{-5}	–	45	5	–	–	50
Barium getter (1·5 mg)	900	1×10^{-6}	5	40	5	–	–	50
Fluorescent screen	200	1×10^{-7}	25	5	–	50	–	20
Mica	200	1×10^{-7}	–	3	0·5	67	26	3
	300	4×10^{-4}	–	13	3	50	22	11
	350	8×10^{-4}	–	10	2	34	14	39

1.2.3 Vapour pressure

The saturated vapour pressure p_v of a pure material increases with the absolute temperature T K in accordance with the equation

$$\log p_v = A - B/T, \tag{4}$$

where A and B are constants depending on the material in question.

TABLE 1.8 Melting points, boiling points and saturated vapour pressures of important elements (based on Honig and Kramer, 1969). All temperatures in degree Kelvin (K); 0°C = 273·15 K; m.p. = melting point; b.p. = boiling point (values of m.p. and b.p. from Kaye and Laby (1966)

Element	m.p. (K)	b.p. (K)	Temperature (K) for saturated vapour pressure in Pa of								
			$1{\cdot}33 \times 10^{-9}$	$1{\cdot}33 \times 10^{-7}$	$1{\cdot}33 \times 10^{-5}$	$1{\cdot}33 \times 10^{-3}$	$1{\cdot}33 \times 10^{-1}$	133	$1{\cdot}33 \times 10^{3}$	$1{\cdot}33 \times 10^{4}$	$1{\cdot}33 \times 10^{5}$
Aluminium	933·3	2793	820	910	1020	1175	1370	1845	2090	2410	2850
Antimony*	904	1860	476	526	584	658	754	1030	1250	1535	1920
Arsenic	1090†	900	354	388	428	480	544	680	742	820	912
Barium	1002	2063	490	550	626	726	862	1220	1420	1705	2130
Beryllium	1560	2745	824	914	1025	1170	1360	1820	2050	2370	2810
Bismuth*	544·5	1837	512	570	644	738	868	1180	1350	1575	1875
Boron	2303	3973	1370	1515	1690	1910	2200	2860	3200	3610	4150
Cadmium	594·2	1040	295	328	367	418	490	664	760	884	1070
Caesium*	301·6	951·6	213	241	273	319	382	542	638	772	980
Calcium	1112	1757	471	526	592	678	796	1080	1250	1470	1820
Carbon	3773	4100	1700	1855	2040	2270	2570	3170	3440	3770	4150
Chromium	2130	2945	960	1065	1180	1335	1545	2010	2260	2580	3000
Cobalt	1768	3200	1025	1140	1275	1450	1670	2200	2460	2810	3280
Copper	1356·6	2840	856	944	1060	1205	1400	1890	2130	2460	2900
Gallium	302·9	2480	708	790	892	1025	1205	1630	1850	2140	2530
Germanium*	1210·4	3107	922	1015	1130	1300	1520	2050	2320	2680	3200
Gold	1336	3080	920	1025	1160	1325	1545	2060	2330	2670	3140
Indium	429·3	2343	644	720	812	936	1105	1505	1720	2000	2400
Iridium	2727	4660	1580	1760	1950	2210	2540	3290	3670	4130	4750
Iron	1809	3135	996	1100	1225	1385	1605	2120	2390	2720	3220
Lead	600·6	2023	515	580	656	758	900	1250	1440	1700	2100
Lithium	453·7	1620	426	478	544	624	738	1020	1165	1360	1650
Magnesium	922	1363	383	427	480	550	642	868	984	1150	1410
Manganese	1515	2373	708	778	864	980	1140	1520	1720	2000	2370
Molybdenum	2890	4885	1590	1755	1950	2210	2560	3360	3770	4300	5000
Nickel	1726	3185	1030	1145	1275	1440	1660	2180	2450	2790	3240
Palladium	1825	3237	952	1065	1200	1370	1600	2150	2430	2800	3330
Phosphorus	317·3	550	146	166	191	220	259	348	396	463	564
Platinum	2043	4095	1335	1480	1655	1885	2180	2860	3190	3610	4170

TABLE 1.8 (*continued*)

Plutonium	913	3505	928	1035	1180	1360	1615	2230	2560	3000	3600
Potassium	336·4	1037	248	277	315	364	434	618	718	852	1070
Radium	973	1800	436	488	552	638	756	1060	1225	1490	1840
Rhenium	3453	5960	1890	2090	2340	2650	3070	4040	4550	5200	6050
Rhodium	2239	4000	1330	1470	1640	1855	2130	2780	3110	3520	4070
Rubidium	312	967	232	257	290	335	400	570	660	794	990
Selenium*	490	958	286	317	357	407	474	638	722	832	980
Silicon*	1685	3490	1085	1200	1335	1505	1750	2330	2630	3020	3550
Silver	1234	2436	720	798	898	1020	1190	1600	1810	2090	2500
Sodium	371	1156	293	328	372	430	508	716	824	968	1180
Strontium	1043	1648	436	485	548	628	740	1010	1160	1380	1695
Sulphur*	388·4	717·8	232	253	279	311	352	465	532	620	732
Tantalum	3250	5640	1925	2120	2370	2680	3070	4000	4460	5020	5780
Thorium	2028	5060	1460	1615	1815	2070	2440	3300	3750	4350	5200
Tin	505	2896	800	898	1020	1170	1380	1885	2150	2500	2940
Titanium	1933	3560	1140	1260	1405	1590	1845	2440	2730	3120	3640
Tungsten	3680	5935	2060	2280	2520	2840	3260	4220	4700	5280	6010
Uranium	1405	4405	1275	1425	1600	1835	2150	2900	3300	3800	4500
Yttrium	1799	3610	1070	1180	1320	1500	1750	2360	2690	3110	3700
Zinc	692·7	1184	336	374	421	482	566	760	870	1010	1210
Zirconium	2125	4775	1525	1690	1890	2150	2490	3300	3700	4200	4870

* More than one gas species present.
† At 35·8 atm.

The *rate of evaporation* in a vacuum of a metal at an absolute temperature T K is related to the saturated vapour pressure p_v. For those pure metals which sublime, and where free evaporation is possible, a useful working equation is

$$m = 4{\cdot}4 \times 10^{-3}\sqrt{M}\, p_v / T^{1/2}, \qquad (5)$$

where p_v is in pascal, m is the sublimation rate in kg/m^2 and M is the atomic mass of the metal.

TABLE 1.9 Vapour pressures of water at various temperatures t in °C

t in °C	100	80	60	40	20	0	−10	−20	−30	−50	−60	−80
p_v in Pa	$1{\cdot}0 \times 10^{5}$	$4{\cdot}7 \times 10^{4}$	2×10^{4}	$7{\cdot}35 \times 10^{3}$	$2{\cdot}3 \times 10^{3}$	$6{\cdot}1 \times 10^{2}$	$2{\cdot}6 \times 10^{2}$	$1{\cdot}1 \times 10^{2}$	38	3·90	1·1	$5{\cdot}3 \times 10^{-2}$

TABLE 1.10 Vapour pressures of mercury at various temperatures t in °C

t in °C	200	150	100	80	60	40	20	10	0	−10	−20	−30	−78*	−196†
p_v in Pa	2300	370	36	12	3·3	8×10^{-1}	$1{\cdot}5 \times 10^{-1}$	$6{\cdot}5 \times 10^{-2}$	$2{\cdot}5 \times 10^{-2}$	8×10^{-3}	$2{\cdot}4 \times 10^{-3}$	$6{\cdot}4 \times 10^{-4}$	4×10^{-7}	negligible

* Temperature of solid carbon dioxide with acetone.
† Temperature of liquid nitrogen.

TABLE 1.11 Vapour pressures at 20°C of some organic liquids

Liquid	b.p. (°C)	Vapour pressure in Pa at 20°C	Liquid	b.p. (°C)	Vapour pressure in Pa at 20°C
Acetone	56·5	$2{\cdot}5 \times 10^{4}$	Ethyl ether	34·5	$5{\cdot}6 \times 10^{4}$
Benzene	80·1	1×10^{4}	Isopropyl alcohol	82	$5{\cdot}1 \times 10^{4}$
Carbon tetrachloride	76·7	$1{\cdot}2 \times 10^{4}$	Methyl alcohol	64·1	$1{\cdot}3 \times 10^{4}$
Chloroform	61	$2{\cdot}1 \times 10^{4}$	Toluene	110·6	3×10^{3}
Ethyl alcohol	78·5	6×10^{3}	Trichlorethylene	88	8×10^{3}
Ethyl bromide	38·4	$5{\cdot}1 \times 10^{4}$	Xylene	140	$6{\cdot}7 \times 10^{2}$

TABLE 1.12 Vapour pressures of water, mercury and carbon dioxide at the temperatures of various refrigerants (Grigorov and Kanev, 1970)

Refrigerant	Temperature in °C at 10^5 Pa	Vapour pressure in Pa		
		H_2O	Hg	CO_2
Ice	0	$6{\cdot}1 \times 10^2$	$2{\cdot}5 \times 10^{-2}$	
Ice and salt	−18	$1{\cdot}1 \times 10^2$	4×10^{-3}	
Freon 22 in 1-stage refrigerator	−35	40	$9{\cdot}3 \times 10^4$	
Freon 22 in 2-stage refrigerator	−70	$2{\cdot}6 \times 10^{-1}$	$2{\cdot}6 \times 10^{-6}$	
Solid CO_2 with acetone	−78	$6{\cdot}6 \times 10^{-2}$	4×10^{-7}	$8{\cdot}9 \times 10^4$
Freon 13 in 2- or 3-stage refrigerator	−120 (minimum possible)			$1{\cdot}3 \times 10^3$
Liquid air	−187	Negligible*	Negligible	$1{\cdot}1 \times 10^{-4}$
Liquid nitrogen	−196	Negligible	Negligible	$1{\cdot}3 \times 10^{-6}$
Liquid hydrogen	−258	Negligible	Negligible	
Liquid helium	−269	Negligible	Negligible	

* Negligible implies $< 10^{-18}$ Pa.

TABLE 1.13 Vapour pressures of some gases (Honig and Hock, 1960)

Gas	b.p. (K)	Temperature K for saturated vapour pressure in pascal of								
		10^{-10}	10^{-8}	10^{-6}	10^{-4}	10^{-2}	1	10^2	10^4	10^5*
Argon	87·3	21·2	23·6	26·7	30·5	35·8	43·1	54·3	73·2	89·7
Carbon monoxide	94·7	21·4	23·7	26·6	30·2	34·8	41·3	51·0	67·1	83·9
Carbon dioxide	81·7	62·1	68·3	76·0	85·6	98·0	114·3	137·3	172·8	197·7
Helium	4·22									
Hydrogen	20·4	2·87	3·20	3·70	4·39	5·37	6·88	9·53	14·9	21·2
Methane	111·7	25·2	28·1	31·9	36·8	43·3	52·7	67·1	91·5	114·7
Neon	27·1	5·78	6·46	7·33	8·47	10·04	12·28	15·6	21·9	27·2
Nitrogen	77·4	18·9	21·0	23·6	26·9	31·3	37·3	46·8	63·2	79·7
Oxygen	90·2	22·7	25·1	28·1	31·7	36·6	43·1	53·9	74·3	92·5
Xenon	165·1	40·4	45·0	50·7	58·1	68·0	214·8	255·8	139·3	167·7

* Atmospheric pressure (standard) = $1{\cdot}013 \times 10^5$ Pa.

TABLE 1.14 Vapour pressures of some outgassed plastics and elastomers (Jensen, 1956)

Material	$\log p_v = 133\ (A - BT^{-1})$		Saturated vapour pressure
	A	B	p_v at 25°C
Butyl rubber	11·4	4900	$1{\cdot}3 \times 10^{-3}$
Hycar rubber	19·4	7400	$5{\cdot}3 \times 10^{-4}$
Mylar	3·09	3000	$1{\cdot}3 \times 10^{-5}$
Nylon cloth	10·0	5600	$1{\cdot}3 \times 10^{-7}$
Polyethylene	7·4	4500	4×10^{-6}
Saran*	6·4	4200	4×10^{-6}
Teflon†	4·3	3400	$1{\cdot}3 \times 10^{-5}$
Vinyl rubber	11·5	5900	$1{\cdot}3 \times 10^{-6}$

* Polyvinylidene chloride.
† PTFE or fluon.

Some of these materials can be used in vacuum systems at lower pressures than indicated by the vapour pressures given in Table 1.14 (Bailey, 1964). Indeed, normal outgassing of most plastics masks the vapour pressure phenomenon proper.

1.3 Permeation of gases through solids

The passage of gas through a solid barrier – *permeation* – involves the diffusion of gas through the solid and surface phenomena of sorption and desorption. In the simplest case, the concern is a slab of homogeneous solid material of cross-sectional area A m^2 and uniform thickness d m. Let p_1 be the gas pressure on one (outer) surface of this slab (which is usually part of the wall of a vacuum chamber and p_1 is the partial pressure in the atmospheric air of the gas in question) and p_2 is the gas pressure at the other (inner) surface, where $p_1 > p_2$. Then

$$Q = PA(p_1^n - p_2^n)/d, \qquad (6)$$

where Q is the permeation throughput (in Pa m^3 s^{-1}), and P is the permeation constant of the material which depends on the gas, the solid material and the temperature.

n is the inverse of the number of atoms (or radicals) into which the gas molecule splits when sorbed in the solid. For gases in non-metals, $n = 1$; for the diatomic gases in metals, $n = \frac{1}{2}$.

The unit for P is Pa m^3 s^{-1}/m^2 m^{-1} Pa = m^2 s^{-1}.

P is small for many gas–solid systems but increases rapidly with temperature T K, approximately in accordance with the equation

$$P = \text{const.} \exp(-1/T). \qquad (7)$$

The gas permeation through various metals of the gases hydrogen, oxygen, nitrogen and carbon dioxide as a function of temperature is given in a number of texts. Fig. 1.1 is from data given by Espe (1966), based on Waldschmidt (1954). Calder and Lewin (1967) report that the permeation from the atmosphere of hydrogen through stainless steel is $1{\cdot}3 \times 10^{-14}$ m^2 s^{-1} and is very small relative to other sources of hydrogen in stainless steel uhv chambers at room temperature.

The permeation constants for some gases through some elastomers and polymers are given in Table 1.15. It is stressed that gas permeation depends on temperature, purity of the gas and the material and the surface condition of the material. The permeation of gases through elastomers increases by about two orders of magnitude on raising the temperature from 25°C to 150°C.

Bailey (1964) gives a nomogram in conjunction with a table of figures to enable gas permeations to be calculated at various temperatures.

A permeation coefficient greater than 10^{-11} m^2 s^{-1} is large for high vacuum applications. For example, an elastomer of surface area 10^{-1} m^2 and of thickness 5×10^{-3} m across which exists a pressure of 1 atm = 10^5 Pa suffers a throughput of gas for $P = 10^{-11}$ m^2 s^{-1} given by equation (6) to be

$$Q = 10^{-2} \times 10^{-11} \times 10^5/5 \times 10^{-3} = 2 \times 10^{-6} \text{ Pa m}^3 \text{ s}^{-1},$$

assuming that $n = 1$.

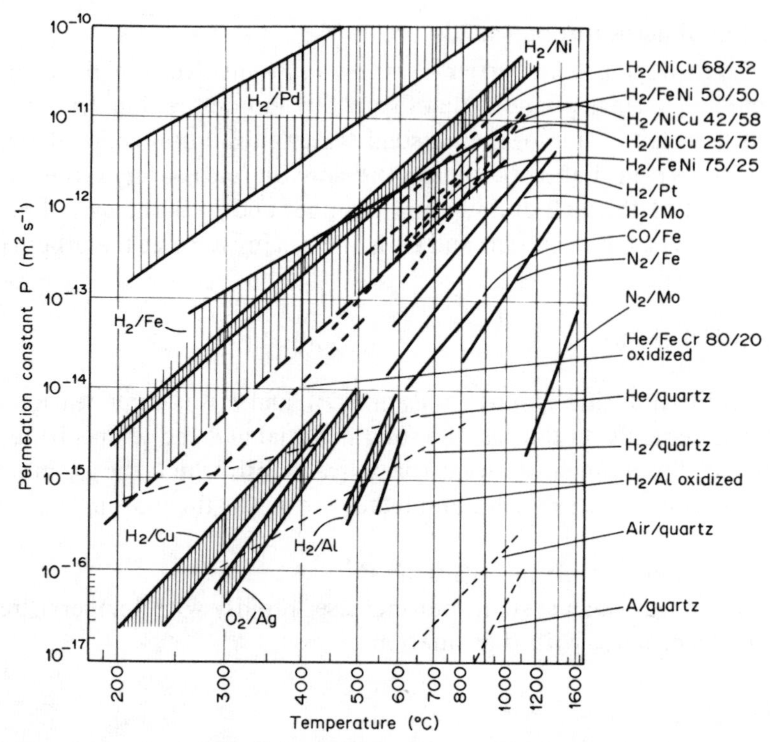

Fig. 1.1 Permeation constants ($m^2 s^{-1}$) for various gas–metal combinations as a function of temperature.

TABLE 1.15 Gas permeation constants (P) for some elastomers and polymers at about 20°C

	P in $m^2 s^{-1}$				
Material	Nitrogen	Oxygen	Hydrogen	Helium	Argon
Butyl[1]	–	–	$4{\cdot}5 \times 10^{-12}$	–	–
Hypalon (GR – S)[1]	$1{\cdot}6 \times 10^{-12}$	–	–	–	–
Neoprene (CS 2367)[2]	$6{\cdot}8 \times 10^{-13}$	$2{\cdot}3 \times 10^{-12}$	$1{\cdot}1 \times 10^{-11}$	$7{\cdot}5 \times 10^{-12}$	$2{\cdot}1 \times 10^{-12}$
Neoprene (CS 2368B)[2]	$1{\cdot}9 \times 10^{-13}$	$1{\cdot}4 \times 10^{-12}$	$7{\cdot}5 \times 10^{-12}$	$7{\cdot}2 \times 10^{-12}$	$1{\cdot}2 \times 10^{-12}$
Nygon[2]	$5{\cdot}2 \times 10^{-13}$	$1{\cdot}8 \times 10^{-12}$	$2{\cdot}1 \times 10^{-11}$	9×10^{-12}	2×10^{-12}
Nylon 31[3]	–	–	$1{\cdot}2 \times 10^{-13}$	5×10^{-11}	–
Nylon 51[3]	–	–	$4{\cdot}7 \times 10^{-13}$	$8{\cdot}2 \times 10^{-13}$	–
Perbunan (PB 60)*,[2]	–	6×10^{-13}	$7{\cdot}5 \times 10^{-12}$	$7{\cdot}5 \times 10^{-12}$	$1{\cdot}3 \times 10^{-12}$
Perspex[3]	–	–	$2{\cdot}5 \times 10^{-12}$	$5{\cdot}3 \times 10^{-12}$	–
Polystyrene[3]	–	$8{\cdot}4 \times 10^{-13}$	$1{\cdot}2 \times 10^{-11}$	$1{\cdot}4 \times 10^{-11}$	–
Polyethylene[3]	9×10^{-13}	$2{\cdot}7 \times 10^{-12}$	$7{\cdot}5 \times 10^{-12}$	$5{\cdot}3 \times 10^{-12}$	$2{\cdot}5 \times 10^{-12}$
Polyvinyltoluene (PVT)[3]	–	$5{\cdot}8 \times 10^{-13}$	$1{\cdot}5 \times 10^{-11}$	$1{\cdot}4 \times 10^{-11}$	–
PTFE[3]	$2{\cdot}3 \times 10^{-12}$	$7{\cdot}5 \times 10^{-12}$	$1{\cdot}8 \times 10^{-11}$	$5{\cdot}3 \times 10^{-10}$	$4{\cdot}4 \times 10^{-12}$
Rubber (natural 337)[2]	$5{\cdot}3 \times 10^{-12}$	2×10^{-11}	$3{\cdot}4 \times 10^{-11}$	$2{\cdot}2 \times 10^{-11}$	$1{\cdot}5 \times 10^{-11}$
Silastomer 80[2]	$2{\cdot}7 \times 10^{-10}$	5×10^{-10}	$9{\cdot}5 \times 10^{-10}$	$2{\cdot}9 \times 10^{-10}$	$5{\cdot}3 \times 10^{-10}$
Thiokol[1]	–	–	3×10^{-12}	–	–
Viton A[4]	–	–	$3{\cdot}5 \times 10^{-12}$	$7{\cdot}5 \times 10^{-12}$	–

* P60, PB60 and Gaco are all similar.

1 *Handbook of Chemistry and Physics* (Chemical Rubber Co., Cleveland, 1959).

2 Barton, R. S., *AERE Report* Z/M 210 (1958).

3 Barton, R. S., *AERE Report* M 599 (1960).

4 Barton, R. S., *private communication* (see Bailey, 1964).

The pumping speed S demanded to cope with this permeation at a pressure of 10^{-6} Pa is given by equation (1) to be

$$S = 2 \times 10^{-6}/10^{-6} = 2 \text{ m}^3 \text{ s}^{-1} = 2000 \text{ l s}^{-1}.$$

For uhv applications ($p_u < 10^{-6}$ Pa) several common elastomers are too permeable to gas. However, the calculation is more complex than indicated above because the pressure difference across the material for a given gas is the *partial* pressure of the gas in the atmosphere (Table 1.16), and a uniform pressure gradient across the sample has been assumed.

TABLE 1.16 Partial pressures of gases in the atmosphere at s.t.p. at sea-level (Smithsonian Physical Tables, 1954)

Gas	Partial pressure (Pa)	Gas	Partial pressure (Pa)	Gas	Partial pressure (Pa)
Nitrogen	$7{\cdot}85 \times 10^{4}$	Carbon dioxide	20 to 40	Hydrogen	$5{\cdot}05 \times 10^{-12}$
Oxygen	$2{\cdot}13 \times 10^{4}$	Neon	1·81	Xenon	$8{\cdot}1 \times 10^{-3}$
Argon	$9{\cdot}35 \times 10^{2}$	Helium	$5{\cdot}32 \times 10^{-1}$	Ozone	$2{\cdot}02 \times 10^{-3}$
Water vapour	$(6{\cdot}5 \text{ to } 40) \times 10^{2}$	Krypton	$1{\cdot}12 \times 10^{-1}$	Radon	7×10^{-15}

Several authors report on the permeation of gases through glasses. The most widely used glass is Pyrex, e.g. the borosilicate Corning glass 7740 for which the permation constant for air at room temperature is 6×10^{-15} m² s⁻¹. Altemose (1961) has measured the rate of permeation of helium through twenty different glasses. A plot of the reciprocal of the absolute temperature of the glass against the logarithm of the permeation rate yields a straight line. Values at 400°C range from $7{\cdot}5 \times 10^{-12}$ to $7{\cdot}5 \times 10^{-14}$ m² s⁻¹.

Table 1.17 gives results for some gas–glass and gas–silica combinations.

TABLE 1.17 Permeation constants for some gas–glass and gas–silica combinations*

Combination	Temperature (K)	P m² s⁻¹)	Author(s)
Helium/fused silica	193	$5{\cdot}3 \times 10^{-16}$	Braaten and Clark (1935)
	400	$4{\cdot}5 \times 10^{-13}$	
Ne/fused silica	773	$1{\cdot}0 \times 10^{-13}$	T'sai and Hogness (1932)
	1273	$1{\cdot}2 \times 10^{-12}$	
A/fused silica	1123	$1{\cdot}2 \times 10^{-14}$	Barrer (1951)
H_2/fused silica	473	$1{\cdot}65 \times 10^{-14}$	Barrer (1951)
	1273	$7{\cdot}5 \times 10^{-12}$	
N_2/fused silica	923	5×10^{-14}	Johnson and Burt (1922)
	1173	9×10^{-13}	
He/Pyrex	195	$3{\cdot}7 \times 10^{-17}$	Norton (1953, 1954, 1957)
	538	$1{\cdot}3 \times 10^{-12}$	
He/Corning 7900	298	$1{\cdot}1 \times 10^{-13}$	Altemose (1961)
	754	$1{\cdot}4 \times 10^{-11}$	
He/Corning 7740	374	$6{\cdot}8 \times 10^{-14}$	Altemose (1961)
	765	$6{\cdot}3 \times 10^{-12}$	
He/Corning 1723	612	4×10^{-15}	Altermose (1961)
	790	4×10^{-14}	

* See also Diels and Jaeckel (1962).

Aluminosilicate glasses have much lower helium permeation than the borosilicate glasses and withstand bake-out to 600°C but glass-blowing is much more difficult.

Mica (muscovite) is reported by Espe (1959) to have a permeation coefficient for helium of less than $0{\cdot}75 \times 10^{-19}$ m^2 s^{-1} at 100°C and below $0{\cdot}75 \times 10^{-17}$ m^2 s^{-1} at 415°C.

REFERENCES

ALPERT, D. (1959) *Vacuum*, **9**, 91.
ALTEMOSE, V. O. (1961) *J. Appl. Phys.*, **32**, 1309.
BAILEY, J. R. (1964) 'Handbook of vacuum physics' (Ed. Beck, A. H.), **3**, Part 4: 'Properties of miscellaneous materials' (Pergamon Press, Oxford).
BARRER, R. M. (1951) 'Diffusion in and through solids' (Cambridge University Press).
BARTON, R. S., and GORIER, R. P. (1968) *Proc. 4th Intl. Vac. Cong.*, Part 2, 775 (Institute of Physics, London).
BLEARS, J., GREER, E. J., and NIGHTINGALE, J. (1960) *Adv. Vac. Sci. and Tech.*, **2**, 473 (Pergamon Press, Oxford).
BRAATEN, E. O., and CLARK, C. G. (1935) *J. Amer. Chem. Soc.*, **57**, 2714.
B.S.I. 2951 (1969) 'Glossary of terms used in vacuum technology', Part 1: 'Terms of general application' (British Standards Institution, London).
CALDER, R., and LEWIN, G. (1967) *Brit. J. Appl. Phys.*, **18**, 1459.
CHERNATONY, L. de (1966) *Vacuum*, **16**, 13, 129, 247 and 427.
CHERNATONY, L. de, and CRAWLEY, D. J. (1967) *Vacuum*, **17**, 551.
COLWELL, B. H. (1970) *Vacuum*, **20**, 481.
CONNOR, R. J., BURITZ, R. S., and von ZWECK, T. (1961) *Trans. 8th Natl. Vac. Symp. and 2nd Intl. Cong.*, **2**, 1151 (Pergamon Press, Oxford).
DAYTON, B. B. (1959) *Trans. AVS Vac. Symp.*, **6**, 101.
DAYTON, B. B. (1961) *Trans. AVS Vac. Symp.*, **8**, 42.
DIELS, K., and JAECKEL, R. (1962) 'Leybold Vakuum-Taschenbuch' (Springer, Berlin).
ESPE, W. (1966) 'Materials of high vacuum technology' (Pergamon Press, Oxford); a translation from 'Werkstoffkunde der Hochvakuumtechnik' (VEB Deutscher Verlag, Berlin, 1959).
ESPE, W. (1959) *Vakuum-Technik*, **8**, 29.
FARKASS, I., and BARRY, E. J. (1960) *Trans. AVS. Vac. Symp.*, **7**, 35.
FLECKEN, F. A., and NOLLER, H. G. (1961) *Trans. 8th Natl. Vac. Symp. and 2nd Intl. Cong.*, **1**, 58 (Pergamon Press, Oxford).
GRIGOROV, G., and KANEV, V. (1970) 'Le vide poussé' (Masson, Paris).
HAIT, P. W. (1967) *Vacuum*, **17**, 547.
HONIG, R. E., and KRAMER, D. A. (1969) *R.C.A. Rev.*, **30**, 285.
HONIG, R. E., and HOCK, H. O. (1960) *R.C.A. Rev.*, **21**, 360.
JENSEN, N. (1956) *J. Appl. Phys.*, **27**, 1460.
JOHNSON, J., and BURT, R. (1922) *J. Opt. Soc. Amer.*, **6**, 734.
KAYE, G. W. C. and LABY, T. H. (1966) 'Tables of physical and chemical constants' (Longmans, London).
NORTON, F. J. (1953) *J. Amer. Ceram. Soc.*, **36**, 90.
NORTON, F. J. (1954) *Vac. Symp. Trans.*, 47.
NORTON, F. J. (1957) *J. Appl. Phys.*, **28**, 34.
REDHEAD, P. A., HOBSON, J. P., and KORNELSEN, E. V. (1968) 'The physical basis of ultrahigh vacuum' (Chapman and Hall, London).

SANTELER, D. J., HOLKEBOER, D. H., JONES, D. W., and PAGANO, F. (1966). 'Vacuum technology and space simulation' (NASA SP-105, STAR Index no. N66-36129).
SCHITTKO, F. J. (1963) *Vakuum-Technik*, **12**, 294.
STEINHERZ, H. A. (1963) 'Handbook of high vacuum engineering' (Reinhold, New York).
STRAUSSER, Y. E. (1968) *Proc. 4th Intl. Vac. Cong.*, Part 2, 469 (Institute of Physics, London).
T'SAI, L. S., and HOGNESS, T. (1932) *J. Phys. Chem.*, **36**, 2595.
WAGNER, J. S., and MARTH, P. T. (1957) *J. Appl. Phys.*, **28**, 1027.
WALDSCHMIDT, E. (1954) *Metall*, **8**, 749.

1.4 Gas flow in vacuum systems

1.4.1 Maxwell's distribution of molecular velocities

For a single component of velocity, v_x the number of molecules $dn\ (v_x)/n$ with velocities in the range between v_x and $v_x + dv_x$ is given by

$$dn(v_x) = n\left(\frac{M}{2\pi RT}\right)^{1/2} \exp(-Mv_x{}^2/2RT)\ dv_x.$$

For the resultant velocity v (where $v^2 = v_x{}^2 + v_y{}^2 + v_z{}^2$), the number $dn(v)$ of molecules with velocities in the range v and $v + dv$ is:

$$dn(v) = n\left(\frac{2M^3}{\pi R^3 T^3}\right)^{1/2} v^2 \exp(-Mv^2/2RT)\ dv.$$

The most probable velocity v_{max} (velocity at maximum of distribution curve),

$$v_{max} = \alpha = (2kT/m)^{1/2} = (2RT/M)^{1/2} = 129(T/M)^{1/2} \quad (m/s)$$

Mean velocity

$$\bar{v} = 4(kT/2\pi m)^{1/2} = 4(RT/2\pi M)^{1/2} = 145{\cdot}5(T/M)^{1/2} \quad (m/s)$$

Mean square velocity

$$\overline{v^2} = 3kT/m = 3RT/M = 2{\cdot}49 \times 10^4 T/M \quad (m^2\ s^{-2}).$$

Root mean square velocity

$$(\overline{v^2})^{1/2} = (3kT/m)^{1/2} = (3RT/M)^{1/2} = 158(T/M)^{1/2} \quad (m/s).$$

$\bar{v}$ at 20°C and 100°C for a number of gases.

	H_2	N_2	Air	He	A	Kr	Hg
20°C	1754	470·7	464·2	1245	394·2	272·3	175·9
100°C	1979	531·0	523·8	1405	444·7	307·2	198·4

$$\overline{v^2} = 3\pi(\bar{v})^2/8 = 3\alpha^2/2.$$

Impingement rate N of molecules on a surface in a chaotic gas (per unit area), with n = molecular density

$$N = \tfrac{1}{4}n\bar{v} = n(kT/2\pi \mathrm{m})^{1/2} = 36{\cdot}4n(T/\mathrm{M})^{1/2}.$$

In terms of pressure $p = nkT$

$$N = p/(2\pi \mathrm{m}kT)^{1/2} = 2{\cdot}653 \times 10^{24} p/(\mathrm{M}T)^{1/2}.$$

1.4.2 Mean free path, molecular diameter and viscosity

The mean free path of a particle is the average distance which it travels in the gas between successive collisions with other particles. This depends on the molecular density n and the diameter of the particles as well as that of the molecules in the gas. For identical molecules with δ = molecular diameter, the mean free path λ is given by

$$\lambda = 1/(\sqrt{2}\pi n\delta^2),$$

or in terms of pressure $p = nkT$,

$$\lambda = 3{\cdot}107 \times 10^{-18}\, T/p\delta^2 \quad \text{(m)}.$$

The viscosity η of the gas is related to δ by

$$\eta = \frac{0{\cdot}499\rho\bar{v}}{\sqrt{2}\pi n\delta^2}, \quad \text{(units: Ns/m}^2\text{; note: 1 centipoise (cP)} = 10^{-3}\,\text{Ns/m}^2\text{)}$$

where $\rho = p\rho_1$ = density of gas, $\bar{v} = 4(RT/2\pi \mathrm{M})^{1/2}$, $\rho_1 = \mathrm{M}/RT$. Therefore

$$\eta = 0{\cdot}499\rho\bar{v}\,\lambda \approx 0{\cdot}5\, \rho v\, \lambda,$$

i.e.

$$\eta = \left(\frac{2\mathrm{M}}{RT\pi}\right)^{1/2} \lambda p = 4\lambda p/\pi\bar{v},$$

$$\lambda = \left(\frac{\pi}{4}\right)\eta\frac{\bar{v}}{p} = \eta\frac{(R\pi/2\mathrm{M})^{1/2}}{p},$$

$$\lambda = 114{\cdot}51\,\frac{\eta}{p}\left(\frac{T}{\mathrm{M}}\right)^{1/2} \quad \text{(m)}.$$

For air at 20°C,

$$\eta = 1{\cdot}81 \times 10^{-5}\ \text{(Ns/m}^2\text{)},\ \delta = 3{\cdot}75 \times 10^{-10}\ \text{m},$$
$$\lambda = 0{\cdot}342/p \quad \text{(m)}.$$

Values of λ, p, δ, η for some common gases are listed in Table 1.26. Mean free path for ions

$$\lambda_i = \sqrt{2}\lambda,$$

for electrons

$$\lambda_e = 4\sqrt{2}\lambda.$$

Mean free path in binary mixture for molecules of Type A in a gas of type B,

$$(1/\lambda_A) = \sqrt{2}\pi n_A \delta_A{}^2 + \pi n_B \delta_{AB}^2 \sqrt{(1 + v_B{}^2/v_A{}^2)},$$

where v_A, v_B = average velocity of each type of molecule and $\delta_{AB} = \frac{1}{2}(\delta_A + \delta_B)$.

1.4.3 Quantity of gas (as amount of substance)

Mole (mol): the amount of substance of a system containing as many elementary entities as there are atoms in 0·012 kg of carbon-12.

In this definition, the elementary entities must be specified and may be atoms, molecules ions or electrons.

Molecular and molar weights

M' = molecular weight
= relative weight per mole to that of ^{12}C taken as 12·00.
M = molar weight (kg mol^{-1})
= weight (kg) per mole = $M' \times 10^{-3}$ (kg mol^{-1}).

Atomic mass unit u

u = 1/12 the mass of an atom of the ^{12}C nuclide
= $1{\cdot}6605 \times 10^{-27}$ kg.

Quantity of gas (as pressure-volume product)

Consider a volume V, filled with gas at a pressure p and a specified temperature, the quantity of gas q in pressure–volume units is given by

$$q = pV \text{ (Pa . m}^3\text{)}.$$

Equation of state

For an ideal gas:

$$pV = \frac{WR_0T}{M},$$

where W = mass of gas (kg), T = absolute temperature (K),
R_0 = universal gas constant per mole = 8·3143 J K^{-1} mol^{-1}. Therefore:

$$q = W\left(\frac{R_0}{M}\right) T.$$

Density and molecular density of gas

Density. The density ρ of a gas is defined by $\rho = W/V$ (kg/m^3). For an ideal gas, from the equation of state

$$\rho = \frac{W}{V} = p\left(\frac{M}{R_0 T}\right) = p\rho_1.$$

where $\rho_1 = M/R_0 T$ = density at unit pressure (kg m^{-3} Pa^{-1}).

Molecular density. The molecular density n in a volume V is the number of molecules in that volume divided by V.

The number of molecules in a mole of gas is Avogadro's number N_A,

$$N_A = 6{\cdot}0222 \times 10^{23}\ \text{mol}^{-1}.$$

The mass m of molecules of molar weight M is

$$m = M/N_A \quad \text{(kg)}.$$

The total mass of gas W in a volume V is

$$W = nVm = nVM/N_A \quad \text{(kg)}.$$

The equation of state becomes (using this expression for W)

$$p = nkT,$$

where $k = R_0/N_A$ = Boltzmann's constant = $1{\cdot}3806 \times 10^{-23}$ J K^{-1}.

Pumping speed or volume rate of flow

In any vacuum system, the rate of removal of a specified gas by a pump is expressed in terms of the pumping speed S or the volume rate of flow of the pump. This is defined as the volume V of gas removed per unit time t, at the pressure p at the inlet.

$$S = \mathrm{d}V/\mathrm{d}t,$$

at the pressure p.

In practice, S is determined from the rate of flow of a quantity of gas or throughput.

Throughput

The throughput expresses the rate of flow of a quantity of gas which passes through the vacuum system. The throughput Q of a gas passing through a cross-section A of a vacuum system is given by the quantity of gas q (in pressure volume units) at a specified temperature flowing in an interval of time divided by that time.

Hence if S is the pumping speed at the cross-section A, and under steady conditions, when p is constant, we have

$$Q = \mathrm{d}q/\mathrm{d}t = p\,\mathrm{d}V/\mathrm{d}t = pS.$$

Units

$$Q\ (\mathrm{Pa\ .\ m^3/s}), \qquad S\ (\mathrm{m^3/s}).$$

It is often convenient to express the pumping speed S at the cross-section A in terms of the throughput Q and pressure p at A by

$$S = Q/p \quad (\mathrm{m^3/s}).$$

Mass flow rate

The mass of gas flowing per unit time across the cross-section A, at a specified temperature T(K). The throughput is proportional to the mass flow rate $G = \mathrm{d}W/\mathrm{d}t$ for an ideal gas (when the equation of state is assumed to apply). Since in that case

$$Q = p\ \mathrm{d}V/\mathrm{d}t = (R_0 T/M)\ \mathrm{d}W/\mathrm{d}t = R_0 TG/\mathrm{M}.$$

Molecular flux

The molecular flux refers to the total number of molecules crossing a surface A (in a given direction) per unit time. The molecular flux F can be expressed in terms of the mass flow rate G, for molecules of mass m (or molecular weight M)

$$G = \mathrm{d}W/\mathrm{d}t = mF = (M/N_\mathrm{A})F \quad (\mathrm{kg\ s^{-1}}).$$

Again, for an ideal gas at a specified temperature T, the throughput is proportional to the molecular flux, since from relation between G and Q

$$Q = (R_0 T/N_\mathrm{A})F = kTF.$$

From the equation of state for an ideal gas, $p = nkT$, where n is the molecular density, so that

$$S = Q/p = F/n$$

The pumping speed for an ideal gas is therefore given by the ratio of the molecular flux passing through a cross-section A divided by the molecular density in that region.

These are useful relations in high vacuum and ultra high vacuum technology, where the molecular concepts of gas flow are important.

Conductance between two cross-sections

Gas flowing to the pump from the vessel being evacuated generally passes through a series of pipes and components which represent a resistance to flow. This results in a difference in pressure $(p_1 - p_2)$. The conductance U between two cross-sections is defined by the ratio:

$$U = Q/(p_1 - p_2) \quad (\mathrm{m^3/s}).$$

It is assumed that the throughput Q enters the component where the pressure is p_1, that there is no additional gas source (e.g. leak) or gas sink (pump or sorbent surface) so that the gas throughput at the outlet, where the pressure is p_2, is unchanged. In this case the equation of continuity applies, i.e. the throughput Q remains the same at each successive cross-section 1, 2, . . . n and the conductance for each is

$$U_i = Q/(p_i - p_j).$$

The effective pumping speeds S_1, S_2 . . . are given by

$$Q = p_1 S_1 = p_2 S_2 \dots$$

It follows that

$$\frac{1}{S_1} = \frac{1}{S_2} + \frac{1}{U}.$$

Formally one can draw an equivalent electrical network to such a vacuum system, but care must be taken not to drive the analogy too far. For conductances U_1 and U_2 . . . in series, the combined conductance U is given by

$$\frac{1}{U} = \frac{1}{U_1} + \frac{1}{U_2} + \dots.$$

Conductances in parallel add up as

$$U = U_1 + U_2 + \dots.$$

Conductance in terms of molecular flux and molecular densities

For an ideal gas, we have seen that $Q = kTF$, and with the molecular densities n_1 and n_2 corresponding to p_1 and p_2 where $p_1 = n_1 kT, p_2 = n_2 kT$. Hence

$$U = \frac{Q}{p_1 - p_2} = \frac{FkT}{(n_1 - n_2)kT} = \frac{F}{n_1 - n_2}.$$

This is mainly of interest for molecular flow considerations.

Transmission probability

Under molecular flow conditions, a useful alternative to the concept of the conductance of a component, is the related quantity, the transmission probability. Consider the molecular impingement rate N_1 (per unit area) at the entrance surface of area A_1, then the probability of transmission W_{12} for molecules through the component to emerge at A_2 is related to the flux of gas F_{12} (from A_1 to A_2) through the component simply by

$$F_{12} = A_1 N_1 W_{12}$$

(assuming no other sources and sinks of gas – i.e. no wall sorption effects, leaks etc.). The net flux

$$F = F_{12} - F_{21} = A_1 N_1 W_{12} - A_2 N_2 W_{21}.$$

At constant temperature, conservation of flow gives $F = 0$ when $N_1 = N_2$, i.e.

$$A_1 W_{12} = A_2 W_{21} = A_1 W \text{ say so that}$$

$$F = A_1 W (N_1 - N_2).$$

The conductance

$$U = \frac{F}{n_1 - n_2} = A_1 W \frac{(N_1 - N_2)}{n_1 - n_2}:$$

For an orifice in a thin wall, all the molecules impinging on A_1 are transmitted, and the transmission probability $W = W_0 = 1$ in this case. The conductance for a thin orifice is therefore

$$U_0 = A_1 \frac{(N_1 - N_2)}{n_1 - n_2}.$$

Inserting this into expression for U, we have $U = U_0 W$, where U_0 is the entrance conductance or equivalent orifice conductance of the entrance to the component.

In the case of chaotic gas (e.g. in a large vessel) with random impingement rates and a Maxwellian velocity distribution, the impingement rate $N = (1/4)n\bar{v}$ where $\bar{v}$ = mean molecular velocity. From the Maxwellian velocity distribution

$$\begin{aligned} \bar{v} &= 4(kT/2\pi m)^{1/2} \\ &= 4(R_0 T/2\pi M)^{1/2}, \\ &= 145{\cdot}5 \times (T/M)^{1/2} \quad \text{(m/s)}, \end{aligned}$$

so that from above the entrance or orifice conductance U_0 is given by

$$U_0 = \frac{1}{4}\bar{v}A_1 = \left(\frac{kT}{2\pi m}\right)^{1/2} A_1 = 36{\cdot}4 \left(\frac{T}{M}\right)^{1/2} A_1, \quad (\text{m}^3/\text{s}),$$

and

$$U = \frac{\bar{v}}{4} A_1 W = \left(\frac{kT}{2\pi m}\right)^{1/2} A_1 W = 36{\cdot}4 \times (T/M)^{1/2} A_1 W \quad (\text{m}^3/\text{s}).$$

The orifice conductance U_0 is simply given by this expression, with $W = 1$. For all other configurations, the conductance is determined by evaluating the corresponding transmission probability W.

Pumping speed in terms of molecular flux and molecular densities

If F is the net molecular flux passing through the area A of the pump mouth, the pump may be characterized by F in relation to the incident flux F_1.

If σ = probability of capture of gas by the pump (assumed averaged over area A)

$$F = \sigma F_1.$$

The gas flux F_2, backscattered (or not pumped) is similarly given by

$$F_2 = F_1 - F = (1 - \sigma)F_1.$$

The throughput Q of gas pumped is (for an ideal gas)

$$Q = kTF = kT\sigma F_1,$$

since $p = kTn$, (for an ideal gas with molecular density n)

$$S = Q/p = F/n = F_1\sigma/n,$$

where n is measured in a suitable region near the pump mouth (similarly to problem of measuring p) to obtain correct value of pump speed S. In terms of impact rate N (per unit area) at pump mouth (area A):

$$F_1 = AN,$$
$$F = AN\sigma.$$

Hence

$$S = A\sigma N/n \quad \text{or} \quad S = \sigma U_0,$$

where U_0 = entrance conductance of pump. Under molecular flow conditions, for a chaotic gas (or gas in a large chamber connected to pump with suitable gas admission to randomize flow)

$$N = n\bar{v}/4,$$

where $\bar{v}$ = mean molecular velocity. The 'intrinsic speed' of a pump (connected to a large chamber) is therefore

$$S = A\sigma\bar{v}/4.$$

These relations are valid when F, σ and S are independent of the directional distribution of the incident flux F_1. Errors could arise however for pumps with highly directional pumping characteristics when comparing σ or S obtained with tubulated (or beamed) gas inputs, with the σ or S for large chamber (or randomized) gas inputs respectively. In many cases the first component above the pump mouth is a baffle or trap which helps to randomize the input characteristics even for pumps with directional properties.

1.4.4 Flow regimes

At low pressures, flow is characterized by the molecular mean free path λ in relation to a typical dimension of the component – e.g. the diameter d of the tube or orifice. The *Knudsen number* $K = \lambda/d$ gives an indication of the type of flow at low pressures.* Since it is the dimensionless ratio K which counts and not just the mean free path as such, one cannot predict the type of flow from the pressure alone – for example molecular flow will take place at quite high pressures in a microporous solid, while it is laminar in a large size tube at the same pressure.

* In some literature, the inverse of this, i.e. d/λ is termed the Knudsen number.

The following flow regimes are distinguished at low pressures:

molecular flow	$K = \lambda/d > 3$,
transition flow	$3 > \lambda/d > 1/10$,
slip flow	$1/10 > \lambda/d > 1/100$,
continuum flow	$1/100 > \lambda/d$.

At higher pressures the further characterization of continuum flow by the Reynolds number R_e as turbulent or viscous laminar is discussed (under turbulent flow) below.

Turbulent flow in vacuum systems

The criterion that determines if the flow will be laminar or turbulent is the dimensionless Reynolds number, defined for flow in a tube as

$$R_e = ud\rho/\eta,$$

where u = flow velocity = volumetric flow rate/cross-sectional area = S/A,

d = diameter of tube,
η = viscosity of fluid,
ρ = density of fluid, from the equation of state,

$$\rho = \frac{W}{V} = p\frac{M}{R_0T},$$

$$R_e = Sp\frac{dM}{AR_0T\eta} = \frac{QdM}{AR_0T\eta} = \frac{Gd}{A\eta}.$$

In the case of a tube of circular cross-section $A = \pi d^2/4$.

$$R_e = \frac{4}{\pi R_0}\left(\frac{M}{\eta T}\right)\left(\frac{Q}{d}\right) = 0{\cdot}152\left(\frac{M}{\eta T}\right)\left(\frac{Q}{d}\right).$$

For air at 20°C, $\eta = 1{\cdot}829 \times 10^{-2}$ centipoise, $T = 293$ K, $M = 29$, $Q = 1{\cdot}21 \times 10^{-4}\,R_e d$,

$$R_e = 8250Q/d.$$

The flow becomes turbulent when R_e is in the range 1200 to 2200 and certainly when $R_e \geqslant 2200$

$$\text{turbulent flow: } Q \geqslant 1{\cdot}44 \times 10^4\left(\eta\frac{T}{M}\right)d \quad (\text{Pa m}^3/\text{s}).$$

$$\text{laminar flow: } Q \leqslant 7{\cdot}84 \times 10^3\left(\eta\frac{T}{M}\right)d \quad (\text{Pa m}^3/\text{s}).$$

For air at 20°C, flow is turbulent when $Q \geqslant 0{\cdot}267d$ and laminar when $Q \leqslant 0{\cdot}145d$.

Turbulent flow conductance in circular section tube

From Fanning's equation, the conductance of a 'long' tube is given by

$$U = \frac{\pi}{4}\sqrt{\frac{p_1 + p_2}{p_1 - p_2}\frac{d^5 R_0 T}{fML}} \quad (\mathrm{m^3/s}).$$

It should be noted that this covers both turbulent and laminar flows in long tubes, by taking the appropriate friction factor f (see Fig. 1.2) for $R_e \leqslant 1200$, the friction factor

$$f = \frac{64}{R_e} = 16\pi\eta\,\frac{d}{Q},$$

and then the conductance reduces to that of laminar flow:

$$U = \frac{\pi d^4}{128\eta L}(p_1 - p_2)\bar{p} \quad \text{where } \bar{p} = \frac{p_1 + p_2}{2}.$$

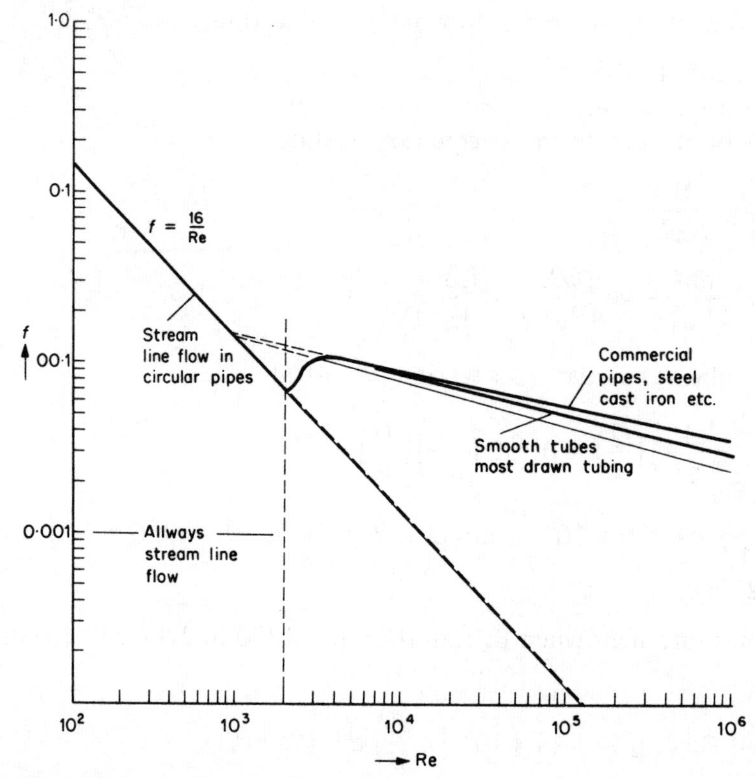

Fig. 1.2 Friction factor f for long tube, circular cross-section.

$Re = \dfrac{du}{\eta}\rho = \dfrac{4G}{\pi d\eta}$ d = diameter u = viscosity
u = velocity G = mass rate of flow
ρ = density

For very large ratios of d/wall roughness use $f = 0{\cdot}02$ for turbulent flow. (From W. H. McAdams (1954), *Heat Transmission* (McGraw Hill, New York).)

It should be noted that this only applies to straight tubes which are not too short. Flow through an orifice or short tubes at high flow rates (and relatively high pressures) is not friction controlled, but limited by the energy conservation laws.

High speed flow through short tubes and small apertures (also termed enthalpy controlled flow by Schumacher, 1961)

Assume p_1 is a relatively high pressure (say near atmospheric) on one side of a short tube (of area A, length $L \ll d$) and the pressure p_2 on the other side is decreased, then the flow velocity increases until it reaches the velocity of sound. Under this condition, the pressure ratio $r = p_2/p_1$ approaches a critical value r_c so that further reduction in p_2 produces no increase in flow velocity or rate of flow.

When $r = p_2/p_1 \geqslant r_c$

$$Q = Ap_1 r^{1/\gamma} \sqrt{\frac{2\gamma}{\gamma - 1} \frac{RT}{M} [1 - r^{(\gamma-1)/\gamma}]} \quad \text{(Pa m}^3\text{/s)}.$$

The conductance is given by

$$U = Q/(p_1 - p_2).$$

As expected this is independent of the viscosity of the gas. For air at 20°C, $\gamma = 1{\cdot}403$, $T = 293$°K, the conductance

$$U = 766 A r^{0{\cdot}712} \sqrt{1 - r^{0{\cdot}288}}/(1 - r) \quad \text{(m}^3\text{/s)}.$$

The critical pressure ratio is

$$r_c = \left(\frac{2}{\gamma + 1}\right)^{\gamma/(\gamma-1)}$$

$$\text{Note:} \quad \gamma = \frac{C_p}{C_v} = \begin{cases} 1{\cdot}66 \text{ for monatomic gases (e.g. inert gases)} \\ 1{\cdot}4 \text{ for diatomic gases (e.g. } N_2, O_2 \ldots) \\ 1{\cdot}3 \text{ for triatomic gases (e.g. } CO_2) \\ 1{\cdot}1 \text{ for many-atomic gases (e.g. organic vapours, oil vapours)} \end{cases}$$

the corresponding critical pressure ratios are:

$$r_c = \frac{p_2}{p_1} = \begin{cases} 0{\cdot}49 \text{ for } \gamma = 1{\cdot}66 \\ 0{\cdot}53 \quad \gamma = 1{\cdot}4 \\ 0{\cdot}55 \quad \gamma = 1{\cdot}3 \\ 0{\cdot}60 \quad \gamma = 1{\cdot}1 \end{cases}$$

for air at 20°C, $r_c = 0{\cdot}53$ at this and lower pressure ratios ($r \leqslant 0{\cdot}53$)

$$U = 200A/(1 - r) \quad \text{(m}^3\text{/s)}$$

$U = 200A$ approximately when $r \leqslant 0{\cdot}1$.

Gunther, Jaeckel and Oetjen (1955) also considered the case of a short cylindrical tube for which flow conditions were intermediate between the high speed orifice flow and laminar flow conditions and obtained experimental data for air and water vapour. Schumacher (1961) considered the application of vacuum apparatus with small openings to pass intense particle beams into the atmosphere.

1.4.5 Viscous laminar flow in tubes

For long tubes, and relatively small pressure gradients.

(*a*) *Circular cross-section*

R = radius, L = length of tube, $D = 2R$,

η = coefficient of viscosity (at temperature T).

From Poiseuille's law

$$U = \frac{\pi}{8} \frac{R^4}{\eta L} \bar{p} = \frac{\pi}{128\eta} \frac{D^4}{L} \bar{p} \quad (\text{m}^3/\text{s}),$$

where

$$\bar{p} = \text{mean pressure in tube} = (p_1 + p_2)/2 \quad (\text{Pa}).$$

For air at 20°C,

$$\eta = 1{\cdot}829 \times 10^{-5} \quad (\text{N s/m}^2)$$

(note: 10 poise = 1 N s/m^2) and then

$$U = 1{\cdot}34 \times 10^3\, D^4 \bar{p}/L \quad (\text{m}^3/\text{s}).$$

(*b*) *Flow between coaxial cylinders*

a = inner radius, b = outer radius, L = length of tube,

$$U = \frac{\pi}{8\eta} \frac{\bar{p}}{L} \left\{ b^4 - a^4 - \frac{(b^2 - a^2)^2}{\ln (b/a)} \right\} \quad (\text{m}^3/\text{s}).$$

(*c*) *Elliptical cross-section*

b = semi-minor, a = semi-major axis

$$U = \frac{\pi}{4\eta} \frac{\bar{p}}{L} \frac{a^3 b^3}{a^2 + b^2} \quad (\text{m}^3/\text{s}).$$

(*d*) *Triangular cross-section* (*equilateral*)

a = length of side, L = length of tube,

$$U = \frac{9\sqrt{3}}{20\eta} \ln \sqrt{3} \frac{\bar{p}}{L} a^4 \quad (\text{m}^3/\text{s}).$$

(*e*) *Rectangular cross-section*

a, b = length of sides, L = length,

$$U = \frac{1}{12\eta} \frac{a^2 b^2}{L} \bar{p} Y \quad (\mathrm{m^3/s}),$$

where

$$Y = \frac{a}{b}\left[1 - \frac{192}{\pi^5}\frac{a}{b}\left\{\tanh r + \frac{1}{3^5}\tanh 3r\right\}\right],$$

$$r = \frac{\pi}{2}\frac{b}{a}, \qquad b \geqslant a.$$

For $a \leqslant b < 3a$, a good approximation is:

$$Y = \frac{a}{b}\left[1 - \frac{192}{\pi^5}\frac{a}{b}(\tanh r + 0{\cdot}0045)\right].$$

For the whole range of a, b, a fairly good approximation (correct when $a = b$) is given by

$$Y = (a/b + b/a + n)^{-1}$$

with $n = 0{\cdot}371$.

For square section: $a = b$, $r = \pi/2$,

$$Y = 0{\cdot}422.$$

When $b > 3a$, a good approximation is

$$Y = \frac{a}{b}\left(1 - 0{\cdot}630\frac{a}{b}\right).$$

Thin slit, i.e. for $b \gg a$,

$$Y = a/b.$$

For air at 20°C,

$$U = 4{\cdot}56 \times 10^3 \frac{a^2 b^2}{L} \bar{p} Y \quad (\mathrm{m^3/s}).$$

In case of square section,

$$U_{\square} = 1{\cdot}92 \times 10^3 a^4 \bar{p}/L \quad (\mathrm{m^3/s}).$$

Transition and slip flow

The most important flow regimes are the extreme ones of molecular flow and continuum flow respectively since they cover the most extensive range under practical conditions. For many purposes it is sufficient to limit calculation of systems to those and consider the transition and slip flows as simple transitions and variations near these two limits. Theoretical and practical data on both slip and transition flows in tubes of various cross-sectional shapes were obtained by Dong (1956) and Dong and Bromley (1961).

Slip and transition flow for long circular section tubes

A useful empirical relation for the conductance, which bridges the regimes of laminar and molecular flow was obtained by Knudsen:

$$U = \alpha_1 \bar{p} + \alpha_2 \frac{1 + \beta_1 \bar{p}}{1 + \beta_2 \bar{p}} \quad (\mathrm{m^3/s}),$$

where

$$\alpha_1 = \frac{\pi}{8}\frac{R^4}{\eta L} = \frac{\pi}{128\eta}\frac{D^4}{L} = \frac{0{\cdot}3926}{\eta}\frac{R^4}{L},$$

$$\alpha_2 = \frac{2}{3}\cdot\frac{R}{L} A\bar{v} = 304{\cdot}76\frac{R^3}{L}\sqrt{\frac{T}{M}},$$

$$\beta_1 = (0{\cdot}2R\sqrt{MR_0/T})/\eta, \qquad \beta_2 = (0{\cdot}247R\sqrt{MR_0/T})/\eta,$$

and

$$\bar{p} = \text{mean pressure in tube} = (p_1 + p_2)/2.$$

At relatively high mean pressures $\bar{p}$ the first term soon becomes prominent compared with the second, so that at high pressures U corresponds entirely to the laminar flow conductance

$$U = \alpha_1 \bar{p}.$$

Similarly at low pressures, as the flow becomes molecular the conductance becomes independent of $\bar{p}$ and $U = \alpha_2$.

It is useful to write U in an alternative form involving the mean free path λ (corresponding to the mean pressure $\bar{p}$). Using the relationship with viscosity, $\eta = 0{\cdot}5\rho\bar{v}\lambda$, where ρ is the gas density corresponding to the mean pressure $\bar{p}$, $\rho = \bar{p}M/R_0T$, so that

$$\beta_1 \bar{p} = \sqrt{2\pi}R/\lambda = 2{\cdot}507R/\lambda = 1{\cdot}253D/\lambda,$$

$$\beta_2 \bar{p} = 2{\cdot}47\sqrt{2\pi}R/2\lambda = 3{\cdot}095R/\lambda = 1{\cdot}547D/\lambda.$$

In the first term, $\alpha_1 \bar{p}$ similarly, η can be written in terms of λ,

$$\alpha_1 \bar{p} = \pi R^4 \bar{p}/8\eta L = \pi^2 \bar{v} R^4/32L\lambda = (\bar{v}/4)(\pi R^2)(\pi R^2/8L\lambda).$$

This can be written in terms of the entrance conductance U_0 for a tube under molecular flow $U_0 = (\bar{v}/4)(\pi R^2)$ or in terms of the molecular flow conductance of a long tube: $U_m = U_0 8R/3L$, so that

$$\alpha_1 \bar{p} = U_0 \pi R^2/8L\lambda = U_m 3\pi R/64\lambda = U_m 3\pi D/128\lambda,$$

$$\alpha_1 \bar{p} = 0{\cdot}0736 U_m (D/\lambda),$$

and hence an alternative expression for the conductance of a long tube in the transition and slip flow is given by

$$U = U_m\left(0{\cdot}0736D/\lambda + \frac{1 + 1{\cdot}253D/\lambda}{1 + 1{\cdot}547D/\lambda}\right),$$

where λ is the mean free path corresponding to the mean pressure $\bar{p}$. Note that $D/\lambda = 1/K$ where K is the Knudsen number as discussed above. It is seen that when $K = 1$, $U = 0{\cdot}956U_m$, i.e. within 5% of molecular flow and certainly as K becomes large ($K > 3$ say) the terms in the bracket $\to 1$, so that $U \to U_m$ as expected.

Conductance of tubes and components under molecular flow conditions

Defining the intrinsic conductance of a duct or component as the conductance when the component is placed between two vessels large compared with the entrance or exist, one can assume that gas arrives at the entrance or exit in a random or chaotic manner. The conductance U is then given in terms of the transmission probability W by:

$$U = U_0W = \frac{\bar{v}}{4}A_1W = 36{\cdot}4(T/M)^{1/2}A_1W \quad (\text{m}^3/\text{s}),$$

where A_1 is the area at the entrance. For air at 20°C, $(T/M)^{1/2} = 3{\cdot}181$ and

$$U = U_0W = 116A_1W \quad (\text{m}^3/\text{s}).$$

Values of the transmission probabilities W and intrinsic conductance U for some simple geometrical shapes have been calculated and are tabulated below. For other shapes, and compound systems it is often possible to obtain approximate values, calculate the transmission probability using computer techniques (e.g. Monte Carlo type calculations) or by direct gas flow measurements (as used for pump speeds).

Long cylindrical tubes with uniform cross-sections

(*a*) *Arbitrary shape cross-section*

A = area of cross-section, L = length,
ρ = chord of cross-section making angle θ with normal to dS (on the perimeter)

$$I = \int_S \int_{-\pi/2}^{+\pi/2} \tfrac{1}{2}\rho^2 \cos\theta \, d\theta \, dS,$$

then

$$W = I/2LA.$$

Note: It has been suggested to use the formula:

$$W = \frac{8}{3}\frac{\pi A}{BL},$$

where B = perimeter, but this is only correct when the tube has a circular cross-section and in all other cases leads to errors.

In the following, the integral I has been evaluated to give W for the case of long tubes with circular, elliptical, coaxial, rectangular and triangular cross-sections.

(b) *Circular section, long tube*

R = radius (m), D = diameter (m) = $2R$, L = length (m), $L \gg D$,

$$A = \pi R^2 = \pi D^2/4 \quad (\text{m}^2),$$

$$W = 8R/3L = 4D/3L,$$

$$U_0 = 28{\cdot}6(T/M)^{1/2} D^2 \quad (\text{m}^3/\text{s}), \qquad U = U_0 W = \frac{2}{3}\frac{R}{L} A\bar{v} = \frac{2\pi}{3}\frac{R^3}{L}\bar{v}.$$

$$U = 38{\cdot}1(T/M)^{1/2} D^3/L \quad (\text{m}^3/\text{s}),$$

for air at 20°C, $(T/M)^{1/2} = 3{\cdot}181$,

$$U_0 = 91{\cdot}6D^2 \quad (\text{m}^3/\text{s}),$$

$$U = 121{\cdot}2D^3/L \quad (\text{m}^3/\text{s}).$$

(c) *Elliptical section, long tube*

a = semi – major axis, b = semi – minor axis,

$$A = \pi ab,$$

$$W = 8b/3L.$$

(d) *Coaxial tube (flow between concentric cylinders), long tube*

D_2 = outer diameter, D_1 = inner diameter (blocked), L = length, $L \gg D_2$,

$$A = \frac{\pi}{4}(D_2{}^2 - D_1{}^2) = \frac{\pi}{4} D_2{}^2(1 - e^2), \qquad e = D_1/D_2 = R_1/R_2,$$

$$W = D_2 X(e)/L,$$

$$X(e) = \frac{1}{1-e^2}\left[\frac{4}{3} - e + e^3 - \frac{2}{3}(1+e^2)E(e) + \frac{2}{3}(1-e^2)K(e)\right],$$

where $E(e)$ and $K(e)$ are the complete elliptical integrals as defined and tabulated for example in Jahnke and Emde (1933), pp. 78 and 80. $X(e)$ has been evaluated for a range of values of e.

$e = D_1/D_2$	0	0·1	0·2	0·3	0·4	
$X(e)$	1·333 = 4/3	1·232	1·126	1·014	0·896	
$e = D_1/D_2$	0·5	0·6	0·7	0·8	0·9	1·0
$X(e)$	0·770	0·637	0·500	0·355	0·198	0

(*e*) *Rectangular section, long tube*

a, b = length at sides, L = length,

$L \gg a, L \gg b,$

$A = ab,$ put $\delta = a/b,$

$$W = \frac{a}{L} Y(\delta),$$

where

$$Y(\delta) = \frac{1}{\delta} \ln(\delta + \sqrt{1+\delta^2}) + \ln\left(\frac{1+\sqrt{1+\delta^2}}{\delta}\right)$$

$$+ \frac{1}{3\delta^2} [1 + \delta^3 - (1+\delta^2)^{3/2}].$$

Note that $Y(1/\delta) = \delta Y(\delta)$. $Y(\delta)$ has been evaluated for a range of values of $\delta = a/b$.

$\delta = a/b$	0·1	0·125	0·2	0·333	0·5	0·667	1·0
$Y(\delta)$	3·42	3·32	2·88	2·40	2·05	1·80	1·48
$\delta = a/b$	10	8	5	3	2	1·5	
$Y(\delta)$	0·342	0·415	0·576	0·80	1·025	1·20	

(*f*) *Thin slit, long tube*

$L \gg a,$ $L \gg b,$ $a \gg b,$ $\delta = a/b,$

$$Y(\delta) = \frac{1}{2\delta}(1 + 2\ln 2\delta); \qquad W = \frac{b}{2L}\left(1 + 2\ln 2\frac{a}{b}\right).$$

Note: the thin slit in a short tube is discussed below.

(*g*) *Triangular section (equilateral), long tube*

a = length of sides, L = length, $L \gg a,$

$A = a^2\sqrt{3}/4 = 0{\cdot}433\, a^2$

$W = (\sqrt{3} \ln 3)\, a/2L = 0{\cdot}952\, a/L.$

Molecular flow conductance of short tubes (or tubes of any length)

In the case of short tubes, the conductance is very dependent on the manner in which gas is admitted and leaves the tube. The 'intrinsic' conductance, corresponding to the chaotic entry and exit of gas when the tube is connected between large chambers, is first considered for certain configurations. The conductance U is obtained from the transmission probability W and the entrance conductance U_0 as above.

(*a*) *Short circular cross-section cylindrical tube*

W was first calculated by Clausing for a range of values of R and L and has often subsequently been referred to as the *Clausing coefficient.* Values of W (as corrected by de Marcus (1956)) are listed in Table 1.18. Berman (1965) has obtained equations for the direct calculation of W for any value of L and R.

Defining

$$W = \omega_1 - \omega_2, \qquad L/R = a,$$

$$\omega_1 = 1 + a^2/4 - (a/4)(a^2 + 4)^{1/2}$$

$$\omega_2 = \frac{[(8 - a^2)(a^2 + 4)^{1/2} + a^3 - 16]^2}{72a(a^2 + 4)^{1/2} - 288 \ln(a + (a^2 + 4)^{1/2}) + 288 \ln 2}.$$

For small a, the series expansion may be useful (0·04% error at $a = 1$, less than 10^{-4}% at $a = 0{\cdot}5$).

$$W = 1 - (a/2) + (a^2/4) - (5a^3/48) + (a^4/32) - (13a^5/2560) - (a^6/3840) + \ldots.$$

For very small values of a, the first two terms of this series are often sufficiently accurate, i.e. $W = 1 - (a/2)$.

For $L/R \leqslant 0{\cdot}2$ errors are less than 1%, but rise to 4% at $L/R = 0{\cdot}4$. A better approximation for small a is obtained by putting

$$W = 1/(1 + 0{\cdot}5a) = 1/(1 + L/2R),$$

which agrees with the first three terms of/the series. This approximation is accurate to 10^{-2}% for $L/R \leqslant 0{\cdot}2$, better than 0·5% for $L/R \leqslant 0{\cdot}8$, better than 2% up to $L/R \leqslant 1{\cdot}4$.

For large a, the asymptotic equation may be used and is in error by about 1% at $a = 20$,

$$W \sim \frac{8}{3a} - \frac{2 \ln a}{a^2} - \frac{91}{18a^2} + \frac{32}{3}\frac{\ln a}{a^3} + \frac{8}{3a^3} - \frac{8(\ln a)^2}{a^4} + 0(1/a^4).$$

For very large a, the first term corresponds to the W for a long tube as given above.

TABLE 1.18 W for circular tubes (Clausing coefficients) of any length

L/R	W de Marcus (and Berman using series)	L/R	W de Marcus
0	1·0000	1·3	0·61425
0·1	0·95240	1·4	0·59736
0·2	0·90922	1·5	0·58148
0·3	0·86993	1·6	0·56651
0·4	0·83408	1·7	0·55236
0·5	0·801272	1·8	0·53898
0·6	0·77115	1·9	0·52628
0·7	0·74341	2·0	0·51423
0·8	0·71779	2·2	0·49185
0·9	0·69404	2·4	0·47150
1·0	0·67198	2·6	0·45289
1·1	0·65143	2·8	0·43581
1·2	0·63223	3·0	0·42006

TABLE 1.18 (*continued*)

L/R	W de Marcus (and Berman using series)	L/R	W de Marcus
3·2	0·40548	8	0·22530
3·4	0·39195	9	0·20669
3·6	0·37935	10	0·19099
3·8	0·36759	12	0·16596
4·0	0·35658	14	0·14684
4	0·35658	16	0·13175
5	0·31053	18	0·11951
6	0·27547	20	0·10938
7	0·24776		

(*b*) *Molecular flow in an annulus (between concentric cylinders) of any length*
The flow through a long annulus was calculated by Clausing (1926). Monte Carlo techniques were used by Davis (1960) to calculate the flow in tubes of arbitrary lengths, but covered a relatively limited range of values of lengths and radius. Berman (1969) has calculated the transmission probability over a large range using the variational method.

Defining

$$y = L/(R_2 - R_1),$$

where R_2 = outer radius, R_1 = inner radius of annulus, L = length,

$$r = R_1/R_2,$$
$$A = \text{area} = \pi(R_2{}^2 - R_1{}^2).$$

The transmission probability W as a function of y and r as obtained by Berman is tabulated (Table 1.19). The results found by Davis (1960) are shown in Table 1.20.

For practical use, an empirical equation was also obtained by Berman which shows a maximum error of 1·5% (for r near 1) valid for $0 \leqslant y \leqslant 100$.

$$W = \{1 + y[0{\cdot}5 - a \tan^{-1}(y/b)]\}^{-1},$$

where

$$a = \frac{0{\cdot}0741 - 0{\cdot}014r - 0{\cdot}037r^2}{1 - 0{\cdot}918r + 0{\cdot}05r^2},$$

$$b = \frac{5{\cdot}825 - 2{\cdot}86r - 1{\cdot}45r^2}{1 + 0{\cdot}56r - 1{\cdot}28r^2}, \quad 0 \leqslant r \leqslant 0{\cdot}9.$$

TABLE 1.19 Transmission probability ($W \times 10^4$) for an annulus, $y = L/(R_2 - R_1)$

y	$r = R_1/R_2$ 0·1	0·2	0·25	0·4	0·5	0·6	0·75	0·8	0·9	0·95
0·5	8017	8022		8030		8037		8043		8046
1·0	6737	6754		6783		6808		6829		6842
1·5	5842	5867		5915		5958		5997		6020
2·0	5175	5206		5266	5295	5324	5365	5378		5413
2·5	4655	4690		4758		4826		4894	4926	4940
3·0	4237	4274		4348		4423		4501		4558
3·5	3893	3931		4007		4087		4174		4241
4·0	3604	3642	3661	3720	3761	3804	3872	3896		3972
5·0	3123	3181		3260		3347		3448	3507	3538
6·0	2791	2828		2906	2948	2994	3071	3100		3201
7·0	2513	2548		2625		2712		2820		2929
8·0	2286	2321	2339	2395	2436	2481	2559	2589		2704
9·0	2099	2132		2204		2288		2496		2515
10·0	1914	1973		2042	2081	2124	2200	2230	2304	2352
12·0					1819		1933			
14·0					1617					
15·0	1414	1440		1499		1569		1666	1740	1792
16·0			1381		1456		1559			
18·0					1325					
20·0					1216		1310		1404	
25·0	921·7	941·1		984·5		1038		1116	1180	1230
30									1019	
35									897	
40									802	
50	496·0	507·6		533·9		567·4		618·2		700·4
100	258·9	265·4		280·1		299·2		328·9		380·5
200	132·7	136·1		144·0		154·3		170·6		200·1
500	53·97	55·40		58·69		63·04		70·00		82·99
1000	27·15	27·88		29·56		31·77		35·34		42·09
10^4	2·733	2·807		2·978		3·204		3·570		4·273
10^5	0·2735	0·2809		0·2980		0·3207		0·3575		0·4282

These results may be compared with those found for short coaxial tubes from Monte Carlo calculation of Davis (1960)

R_2 = outer tube, R_1 = inner tube, L = length and noting that $L/R_2 = y(1 - r)$

TABLE 1.20 W for annulus (short coaxial tube)

L/R_2	0·25	R_1/R_2 0·50	0·75	0·90
0·25	—	—	—	0·488
0·50	—	—	0·541	0·351
0·75	—	—	—	6·279
1·00	0·605	0·524	0·388	0·230
1·25	—	—	—	0·197
1·50	—	—	0·305	0·173
2·00	0·449	0·360	0·252	0·144
2·50	—	—	0·217	0·123
3·00	0·362	0·289	0·193	0·102
3·50	—	—	—	0·094
4·00	0·304	0·238	0·145	0·083

TABLE 1.20 (*continued*)

L/R_2	R_1/R_2 0·25	0·50	0·75	0·90
5·00	0·260	0·203	0·128	–
6·00	0·233	0·177	0·115	–
7·00	0·205	0·162	–	–
8·00	0·185	0·146	0·086	–
9·00	–	0·136	–	
10·00	0·154	0·124	–	
12·00	0·134			
14·00	0·118			
16·00	0·101			

(*c*) *Molecular flow in short rectangular tubes*

Transmission probabilities from Monte Carlo calculation of Levenson et al. (1963).

TABLE 1.21 *W* for short rectangular tubes

L/b	a/b 10	5	1
1	0·664	0·650	0·541
2	0·519	0·500	0·383
3	0·430	0·407	0·300
4	0·368	0·347	0·245
5	0·326	0·300	0·208
6	0·295	0·267	0·182
7	0·272	0·240	0·166
8	0·250	0·221	0·146
9	0·226	0·201	0·135
10	0·212	0·187	0·123

(*d*) *Molecular flow through short slits* (*or between large flat plates*)

L = length, d = spacing, b = width, $b \gg d, b \gg L$.

Values of W were first calculated by Clausing and are given in Table 1.22. Berman (1965) has given equations for the direct calculation of W for any value of L and d and to a greater accuracy than the tabulated values.

Define

$$W = \omega_1 - \omega_2, \qquad a = L/d,$$

$$\omega_1 = \tfrac{1}{2}[1 + (1 + a)^{1/2} - a],$$

$$\omega_2 = \frac{\tfrac{3}{2}\{a - \ln(a + (a^2 + 1)^{1/2})\}^2}{a^3 + 3a^2 + 4 - (a^2 + 4)(1 + a^2)^{1/2}}.$$

For small a,

$$W = 1 - (a/2) + (a^2/4) - (a^3/24) - (a^4/16) + \ldots,$$

giving an error of about 0·2% at $a = 0{\cdot}5$, less than 10^{-4}% at $a = 0{\cdot}1$.

For large a (short thin slits),

with $x = 1/a = d/L$

$$W \sim x\left\{\frac{(96 - 24x^3 + 9x^5)\ln(2/x) - 48x[\ln(2/x)]^2 - 48 + 28x + 20x^2 - 6x^3 + 7x^4}{96 - 144x + 128x^2 - 60x^3 + 14x^5}\right\}$$

with an error of about 0·7% at $L/d = 1{\cdot}5$ and less than 10^{-4}% at $L/d = 5$.

For long and very closely spaced slits (when $x \to 0$, $L/d \to \infty$)

$$W \sim x(\ln 2/x - 1/2) = (d/L)[\ln(2L/d) - 1/2]$$

$$= (d/L)[\ln(L/d) + 0{\cdot}193] \to (d/L)\ln(L/d).$$

For $L/d \geqslant 4$, the error in the asymptotic limit $(d/L)[\ln(L/d) + 0{\cdot}193]$ is less than 2%.

TABLE 1.22 *W* for short thin slits $b \gg d$, $b \gg L$ (following Clausing)

L/d	*W*	*L/d*	*W*
0	1·0	1·0	0·685
0·1	0·952	1·5	0·602
0·2	0·910	2·0	0·542
0·3	0·871	3·0	0·457
0·4	0·836	4·0	0·400
0·5	0·805	5·0	0·358
0·7	0·750	8·0	0·279
0·8	0·727	10·0	0·246
		∞	$(d/L)\ln(L/d)$

1.4.6 Intensity distribution of gas issuing from tubes under molecular flow conditions

The intensity distribution of gas flowing through a *thin* orifice situated between two large chambers is completely spherical as shown in Fig. 1.3(a). It is of the same form as that for gas leaving a surface in the plane of the orifice under diffuse reflection with a cosine distribution. The flow pattern is modified in the presence of a short cylindrical tube between the two chambers as indicated in Fig. 1.3(a) (following Clausing). The effect of different lengths of cylindrical tubes on the flow pattern is shown in Fig. 1.3(b) (as calculated by Dayton, 1957) which indicates the increasing beaming effect. This also shows the intensity distribution of the back-scattered beam giving rise to the two lobes of increasing size with the thinner or longer tube. In these calculations gas is considered to enter the tube randomly from a large chamber with an intensity distribution described by a sphere tangential to the entrance aperture. Depending on the angular distribution of the impinging molecules, there are three possible interactions with the tube:

(a) molecules are transmitted without wall collisions
(b) molecules collide with the tube walls one or more times and are ultimately reflected in the forward direction to contribute to the transmitted flux and
(c) molecules collide with the wall one or more times and are ultimately reflected back into the chamber to contribute to the back-scattered flux.

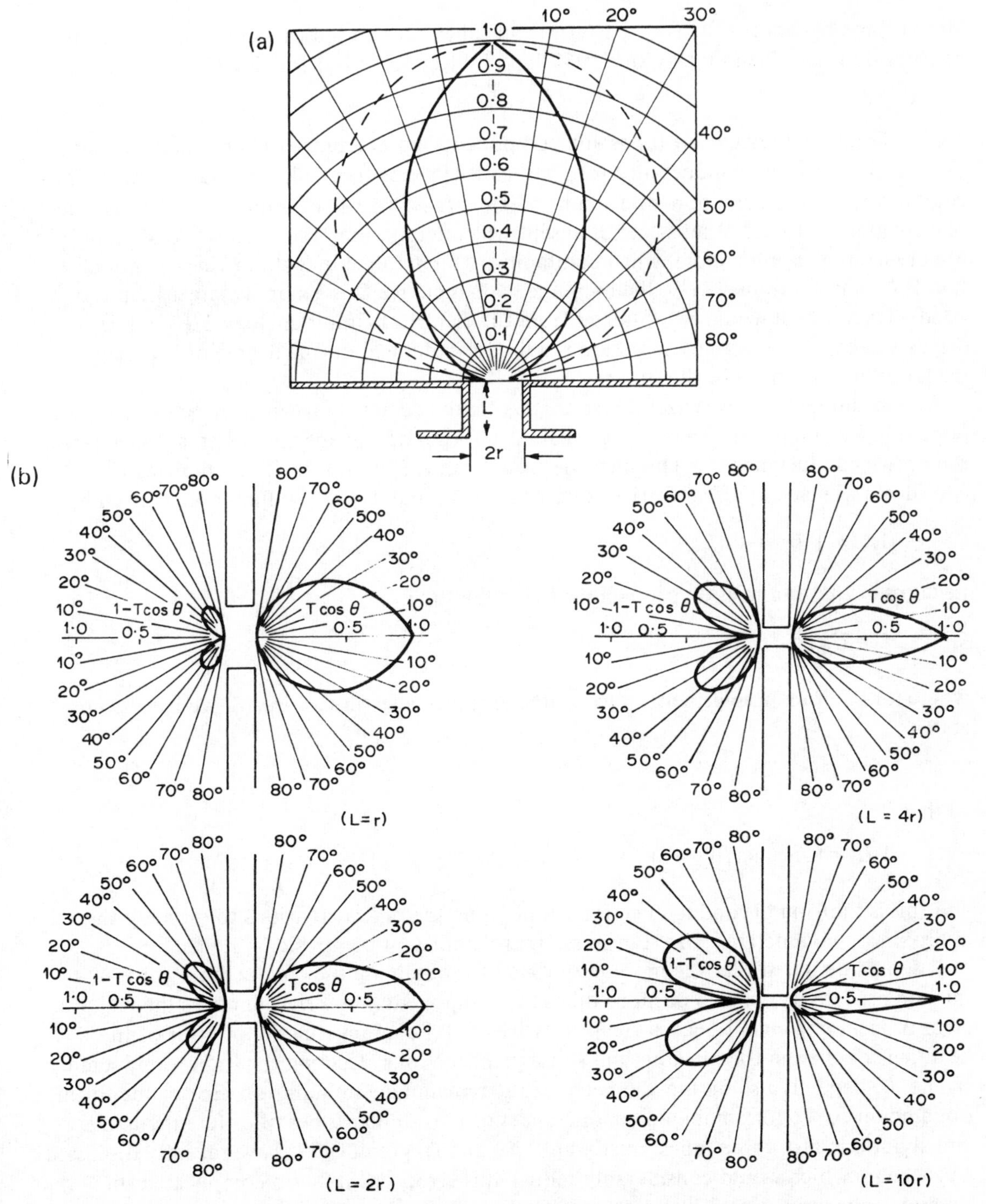

Fig. 1.3 (a) Flow pattern of gas through thin aperture and short tube (following Clausing). (b) Flow pattern of gas issuing through short tubes of various ratios of length diameter.

More complete data for the flow pattern from tubes with an indication of these different contributions are given by Dayton (1957) and by Reynolds and Richley (1964).

1.4.7 Combined systems of tubes and components, conductance of composite systems

To obtain the correct conductance for composite systems from a knowledge of the intrinsic conductance of each component account must be taken of the beaming effect of gas from one component to the other which is absent when determining the intrinsic conductance of each separately. For example, when combining two components with intrinsic conductances U_1 and U_2 respectively, the combined conductance U is generally larger than the conductance which would be obtained from the electrical circuit analogy $1/U_1 + 1/U_2$ discussed above. The actual correction required will depend on the degree of beaming from one component into the other.

A procedure which gives useful results, takes into account that when gas enters a component in a chaotic state, there is a probability of finding the aperture of area A given by the orifice conductance U_0. The intrinsic conductance U of a tube for example can be considered as made up of an orifice conductance U_A and 'tube' conductance U_t given by

$$1/U = 1/U_A + 1/U_t.$$

In terms of the transmission probability W for the tube:

$$U = U_A W,$$

where for random impingement of gas at the entrance aperture of area A, we have from above

$$U_A = A(kT/2\pi m)^{1/2},$$

so that

$$1/U_t = (1/U_A)(1/W - 1),$$

and this is the form of conductance for a tube to be used in calculations to estimate the approximate conductance and transmission probability of composite systems.

It should be noted that the method outlined for obtaining the conductance (or transmission probability) of two or more tubes (or components) in series, removes the major error of including an entrance correction twice, it still contains the minor error of the beaming effect from the previous and subsequent sections affecting the 'tube' conductance U_t; i.e. U_t depends also on the geometry of the remaining configuration. Errors due to this can amount to 4% for combined conductances of two similar tubes of different lengths and about 6·5% for three tubes. Fustoss (1970) and Dayton (1972) have recently discussed a procedure which would considerably reduce this error, but it is too complicated for general application, especially when only fairly rough estimates of the conductance to predict the performance of pumping systems are required.

Some examples of conductances and transmission probabilities for composite systems

(*a*) *Two tubes of same cross-section in series* (see Fig. 1.4(a))

U_1, U_2 = conductances of tubes, A = cross-sectional area,
W_1, W_2 = transmission probabilities of each tube,
U_A = conductance of entrance aperture.

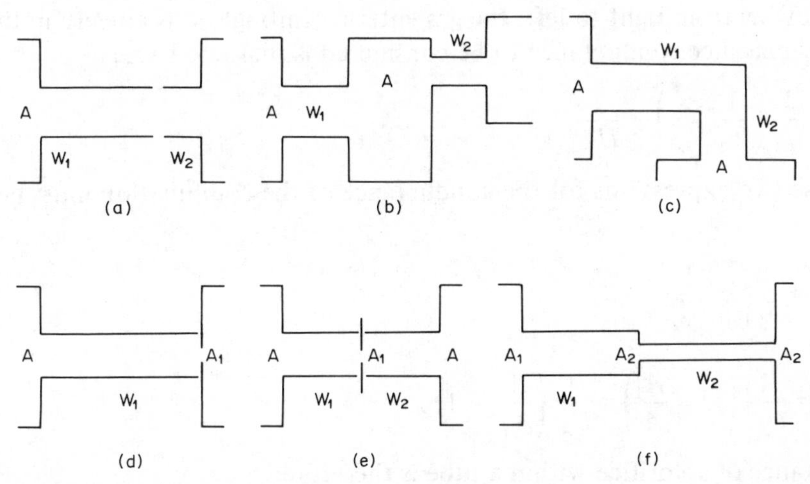

Fig. 1.4 Composite systems: (a) two tubes, same cross-section, in series; (b) two tubes, large chamber between them; (c) elbow two tubes; (d) tube with thin orifice at one end; (e) two tubes with orifice plate between them; (f) two tubes in series with different cross-sectional area.

Since there is only one entrance aperture in the combined tube

$$\frac{1}{U} = \frac{1}{U_A}\left(\frac{1}{W_1} - 1\right) + \frac{1}{U_A} + \frac{1}{U_A}\left(\frac{1}{W_2} - 1\right),$$

therefore

$$\frac{1}{W} = \frac{1}{W_1} + \frac{1}{W_2} - 1.$$

(*b*) *Two tubes of same cross-section with large chamber between them*
The tubes are assumed to be in series, but not in line so that the beaming effect of gas from the first tube is lost when gas enters the second (see Fig. 1.4(b)). For each tube the intrinsic conductance applies:

$$\frac{1}{U} = \frac{1}{U_1} + \frac{1}{U_2} \quad \text{and} \quad \frac{1}{W} = \frac{1}{W_1} + \frac{1}{W_2}.$$

A similar argument applies to an *ordinary elbow of two tubes with same cross-section* (Fig. 1.4(c)).

(*c*) *Tube with thin orifice plate at one end* (see Fig. 1.4(d))
For gas flowing from left to right

$$\frac{1}{U} = \frac{1}{U_A}\left(\frac{1}{W_1} - 1\right) + \frac{1}{U_A} + \frac{1}{U'_{A_1}}.$$

Where U'_{A_1} is the conductance of the thin orifice of area A_1 situated in a tube of cross-sectional area A.

For gas flowing from right to left, the gas entering through A, is already in the tube, so that the only entrance conductance to be considered is that due to A_1,

$$\frac{1}{U} = \frac{1}{U_A}\left(\frac{1}{W_1} - 1\right) + \frac{1}{U_{A_1}}.$$

Since these two expressions for the conductance of the combination must be the same, we have

$$\frac{1}{U_{A_1}} = \frac{1}{U_A} + \frac{1}{U'_{A_1}},$$

or

$$\frac{1}{U'_{A_1}} = \frac{1}{U_{A_1}}\left(1 - \frac{A_1}{A}\right) = \frac{1}{U_A}\left(\frac{A}{A_1} - 1\right).$$

The conductance of an orifice within a tube is therefore

$$U'_{A_1} = \frac{U_A U_{A_1}}{U_A - U_{A_1}} = U_{A_1}\frac{A}{A - A_1} = U_A\frac{A_1}{A - A_1}.$$

The combined transmission probability is given by

$$\frac{1}{W} = \frac{1}{W_1} + \frac{A}{A_1} - 1.$$

(*d*) *Two tubes in series, with thin orifice plate area A, between them* (see Fig. 1.4(e))

A = cross-sectional area of tubes

$$\frac{1}{U} = \frac{1}{U_A}\left(\frac{1}{W_1} - 1\right) + \frac{1}{U_A} + \frac{1}{U'_{A_1}} + \frac{1}{U_A}\left(\frac{1}{W_2} - 1\right),$$

using the previous result for U'_{A_1}

$$\frac{1}{U} = \frac{1}{U_A W_1} + \frac{1}{U_{A_1}}\left(1 - \frac{A_1}{A}\right) + \frac{1}{U_A}\left(\frac{1}{W_2} - 1\right),$$

or

$$\frac{1}{W} = \frac{1}{W_1} + \frac{1}{W_2} + \frac{A}{A_1} - 2,$$

clearly when $A = A_1$, this reduces to example (a) above; when $W_2 = 1$, it reduces to example (c).

(*e*) *Combination of two tubes with different cross-sectional areas*

$$A_1 > A_2$$

for flow from 1 to 2 (referring to Fig. 1.4(f))

$$\frac{1}{U_{12}} = \frac{1}{U_{A_1}W_{12}} = \frac{1}{U_{A_1}}\left(\frac{1}{W_1} - 1\right) + \frac{1}{U_{A_1}} + \frac{1}{U_{A_2}}\left(\frac{1}{W_2} - 1\right) + \frac{1}{U'_{A_2}},$$

where for orifice A_2 in A_1 from above

$$\frac{1}{U'_{A_2}} = \frac{1}{U_{A_2}}\left(1 - \frac{A_2}{A_1}\right),$$

$$\frac{1}{U_{12}} = \frac{1}{U_{A_1}W_{12}} = \frac{1}{U_{A_1}W_1} + \frac{1}{U_{A_2}W_2} - \frac{1}{U_{A_2}} \cdot \frac{A_2}{A_1},$$

hence,

$$\frac{1}{W_{12}} = \frac{1}{W_1} + \frac{A_1}{A_2}\frac{1}{W_2} - 1,$$

for flow from 2 to 1 (referring to Fig. 1.4(f))

$$\frac{1}{U_{21}} = \frac{1}{U_{A_2}W_{21}} = \frac{1}{U_{A_2}}\left(\frac{1}{W_2} - 1\right) + \frac{1}{U_{A_2}} + \frac{1}{U_{A_1}}\left(\frac{1}{W_1} - 1\right),$$

$$= \frac{1}{U_{A_2}W_2} + \frac{1}{U_{A_1}W_1} - \frac{1}{U_{A_1}},$$

hence

$$\frac{1}{W_{21}} = \frac{1}{W_2} + \frac{A_2}{A_1} \cdot \frac{1}{W_1} - \frac{A_2}{A_1}.$$

We also note that $U_{12} = U_{A_1}W_{12} = U_{A_2}W_{21} = U_{21}$, and $A_1W_{12} = A_2W_{21}$ as it should be.

(*f*) *Combination of pumps with components*

When considering the net speed of a pump connected to a component, the same considerations apply as for the determination of the combined conductance of two components. Generally the intrinsic speed of a pump is known, i.e. the speed determined when the pump is connected to a large chamber. If such a pump is connected to a tube which produces some gas beaming, account must be taken of the entrance aperture of the pump and the tube respectively. If on the other hand the component connected to the pump is a baffle system which randomizes the gas flow it is more accurate to take the intrinsic speed of the pump, i.e. without an aperture effect. Some simple examples illustrate calculations for combined systems.

If the 'intrinsic' speed of a pump is S, then $S = \sigma U_A$, where σ is a 'capture coefficient' for the pump, so that for a 'perfect' pump $\sigma = 1$, whose speed becomes equal to the aperture conductance U_A. Analogous to the case of composite pipe systems, it becomes necessary to consider the 'intrinsic' pump speed as made up of an aperture effect and a 'basic' speed S_B, where

$$\frac{1}{S} = \frac{1}{U_A} + \frac{1}{S_B},$$

and hence

$$S_B = U_A \sigma/(1 - \sigma).$$

For a 'perfect' pump (equivalent to merely an aperture connected to an infinitely large evacuated space) it is clear that $S_B = \infty$.

For example consider the net speed S_n of a pipe of diameter $2R$ connected to a pump of the same diameter. The only aperture to be considered in this case, is that at the inlet to the pipe and hence

$$\frac{1}{S_n} = \frac{1}{U} + \frac{1}{S_B}, \quad \text{also} \quad \frac{1}{U} = \frac{1}{U_A} + \frac{1}{U_T},$$

and in terms of the 'intrinsic' speed S

$$\frac{1}{S_n} = \frac{1}{S} + \frac{1}{U_T},$$

where U_T is given by $U_A W/(1 - W)$ as above.

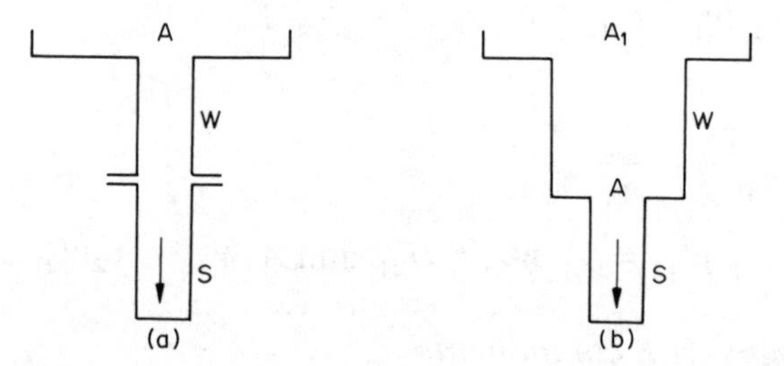

Fig. 1.5 Combination of pumps and components: (a) pump connected to tube of same diameter; (b) pump connected to tube of different diameter.

The same procedure would be used if the net speed was required of composite systems involving additional pipes, baffles etc. Thus if the pipe to be connected, had a larger radius R_1, so that the aperture area is $A_1 = \pi R_1^2$ then the net speed is given by

$$\frac{1}{S_n} = \frac{1}{U_1} + \frac{1}{U_A'} + \frac{1}{S_B}, \quad \text{where} \quad U_1 = U_{A_1} W_1 \quad \text{and} \quad \frac{1}{U_A'} = \frac{1}{U_A}\left(1 - \frac{A}{A_1}\right)$$

which becomes

$$\frac{1}{S_n} = \frac{1}{S} + \frac{1}{U_1} - \frac{1}{U_A} \cdot \frac{A}{A_1}.$$

1.4.8 Transient conditions

Pumpdown characteristics

(a) Consider the simplified conditions of a pump connected to a chamber of volume V.
  (i) constant pumping speed S
  (ii) no additional gas load to that in volume V, at any time $Q = pS$

$$Q = -\frac{\mathrm{d}}{\mathrm{d}t}(pV) = -V\frac{\mathrm{d}p}{\mathrm{d}t},$$

$$\therefore pS = -V\frac{\mathrm{d}p}{\mathrm{d}t}, \quad \text{or} \; -\frac{S}{V} = \frac{1}{p}\frac{\mathrm{d}p}{\mathrm{d}t},$$

which can be integrated to give

$$\ln p = -St/V + K$$

for initial condition $t = 0$, $p = p_0$, we have $\ln(p/p_0) = -St/V$, or

$$p = p_0 \exp(-St/V) \tag{1}$$

$$t = \frac{V}{S}\ln\frac{p_0}{p} = 2{\cdot}30\,\frac{V}{S}\log_{10}\frac{p_0}{p}. \tag{2}$$

It is seen that the time taken to pump down to 1/10th of initial pressure ($p_0/p = 10$) is given by

$$t_{(0{\cdot}1)} = 2{\cdot}3\; V/S, \tag{3}$$

i.e. the ratio volume/pump speed (V/S) is a time constant for the vacuum system.

(b) As above, but with additional constant gas load Q_0 – e.g. due to outgassing. The pressure is limited to an ultimate p_u given by

$$Q_0 = Sp_u,$$

$$Q = Sp - Q_0,$$

$$= -V\,\mathrm{d}p/\mathrm{d}t,$$

the solution is

$$p = (p_0 - p_u)\exp(-St/V) + p_u, \tag{4}$$

or

$$t = \frac{V}{S}\ln\frac{p_0 - p_u}{p - p_u}. \tag{5}$$

Note:

$$Q = Sp\left(1 - \frac{Q_0}{Sp}\right) = Sp\left(1 - \frac{p_u}{p}\right),$$

or defining $S_1 = Q/p$, $S_1 = S(1 - p_u/p)$.

1.4.9 Pump specifications and tests

Pumping speed (volume rate of flow)

The speed (volume rate of flow) of a pump for a given gas is determined from the ratio of throughput of the gas divided by equilibrium pressure measured at a specified place on a test dome.

Units: $1 . s^{-1}$ (or $dm^3\ s^{-1}$)

Note: $1\ 1 . s^{-1} = 10^{-3}\ m^3\ s^{-1}$ (for other conversions see pp. 50– 51)

During the measurement, the pressure p at the gauge is kept constant, and is determined for each value of throughput Q, so that the speed S can be calculated from $S = Q/p$. The procedures and test dome arrangement to be used are covered by international standards and reviewed below.

(a) Vapour pumps

The types of pumps considered are all types of vapour pumps using pump fluids of relatively low vapour pressures (say <100 Pa) at room temperature (20°C), i.e. they exclude steam jets, but include vapour diffusion pumps, diffusion ejector pumps and vapour booster pumps.

Test dome (for pumping speed measurement). This is of the same internal diameter D as the mouth of the pump and of the form shown in Fig. 1.6.

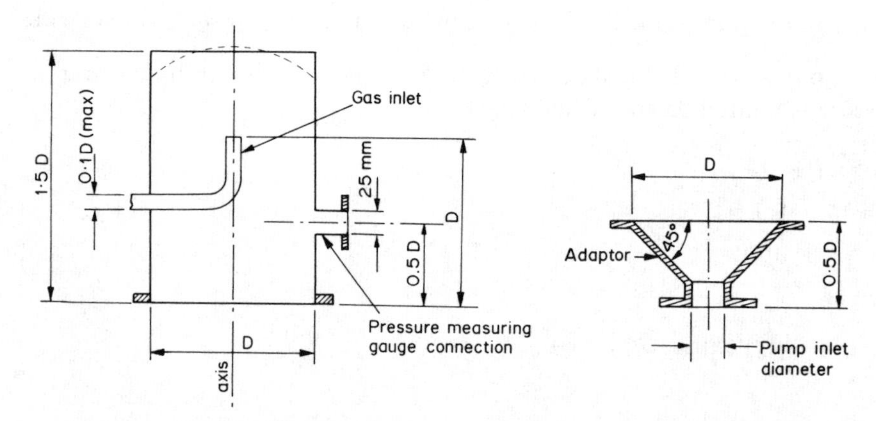

Fig. 1.6 Test dome for measuring pumping speed (volume rate of flow) in vapour pumps and positive displacement pumps.

The location and orientation of the gauge tube and gas inlet on such a test dome is critical due to the dynamic flow conditions under which the test needs to be carried out. The position of the gauge indicated in the diagram is the one chosen to give speed values in the molecular flow range which are a good approximation to the intrinsic speed of the pump (i.e. as when it is connected to a large vessel). The test gas is beamed towards the roof of the test dome from where it is reflected diffusely to reduce the beaming effect of the gas inlet tube. It has now been suggested that a perforated plate to spread the gas evenly over the test dome would be preferable (Dayton, 1968), but the arrangements in Fig. 1.6 is the one agreed internationally (ISO R.1608, 1st Ed. 1970). It should be noted when comparing catalogue data for pumps that test procedures based on this ISO recommendation may not yet have been universally adopted by all manufacturers or in some cases only recently. Some arrangements previously used, placed the gauge tube as near as possible to the mouth of the pump, resulting at a given gas flow Q in a lower pressure reading p and therefore giving rise to higher apparent pump speeds $S = Q/p$. It has been shown that increased speed values of 20% to 50% were possible in practice depending on the types of pump; in fact for a 100% efficient 'ideal' pump the factor S/S_0 (ratio of measured to orifice speed) was shown to be as high as 2·7 (Steckelmacher, 1966).
Flowrate and pressure measurements. The flow rates Q are usually determined by a suitable liquid displacement flowmeter at the atmospheric pressure side of an adjustable leak-valve to admit the calibrating gas (dry air, unless otherwise specified) Ambient temperature needs to be held fairly constant during the period of a measurement (usually ±1°C). The pressure is measured by a suitable vacuum gauge – typically a McLeod gauge or micro-manometer at higher pressure levels, and a calibrated ionization gauge or residual gas analyser at lower pressures.
Ultimate pressure. This is determined with no gas input. The ultimate pressure is an equilibrium condition established between the outgassing from the surfaces above the pump mouth (e.g. test dome) and available pumping speed and hence not a very meaningful characteristic – except to indicate the condition of a test system. The speed should be determined at pressures some factor (e.g. 10) higher than the ultimate pressure.
Maximum tolerable backing pressure. This is determined by admitting gas to the output side of the pump and raising the backing pressure until there is a noticeable deterioration in the inlet pressure. An ISO recommendation is being prepared.

(b) Sputter ion pumps, pumping speed and performance tests
To cover the lower throughput and lower pressure measurements required for sputter ion pump tests a modified test dome as shown in Fig. 1.7 is used incorporating a small orifice of diameter d having a known molecular flow conductance U. In this case the throughput

$$Q = U(p_1 - p_2).$$

For molecular flow at the orifice to prevail, p_1 should be less than $p_{1\max} = 0{\cdot}6/d$ (Pa) (d in mm). The pumping speed $S = Q/p_2 = U(p_1/p_2 - 1)$ so that instead of measuring Q

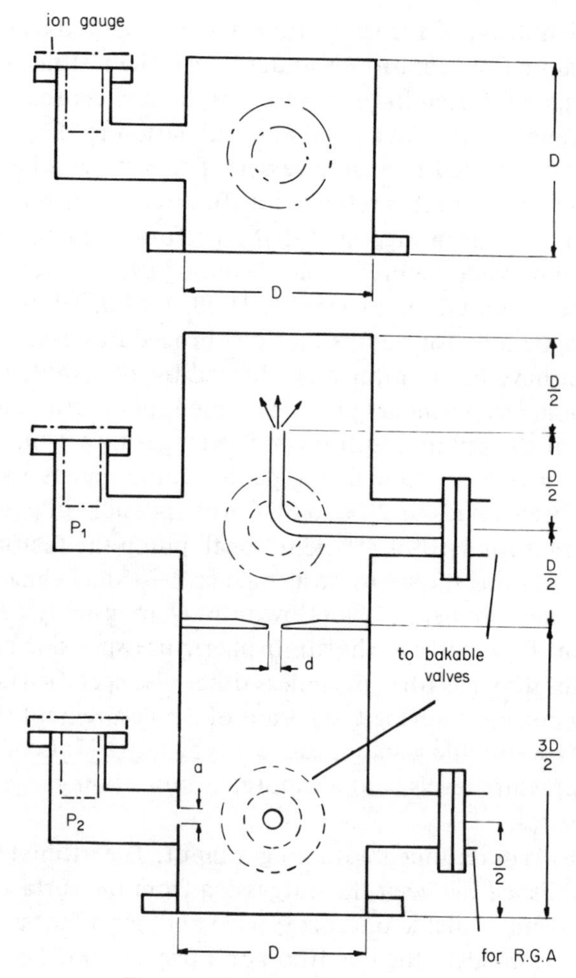

Fig. 1.7 Test dome for sputter ion pumps.

and p_2, it is sufficient to know U and measure the ratio p_1/p_2. It should be noted that this ratio is independent of gauge calibration factors, provided that this factor remains constant over the range of measurement (and that the gas composition remains unchanged).

The location and orientation of gauge tubes on the test dome are governed by similar considerations to those noted for the vapour pump test dome.

Test procedure. An ISO recommendation is being prepared (see also Steckelmacher, 1968). To obtain consistent results particularly at low pressures, it is desirable first to condition the pump. A long period low pressure bakeout (e.g. 4 h at 300°C) with roughing pump in operation, followed by sealed off pump bakeout (for 10 h) with voltages applied should first produce a low ultimate pressure. Under this condition sputter ion pumps tend to show an enhanced pumping speed, denoted as 'regenerated' pumping speed. These may be determined up to pressures of about 5 x 10^{-9} Pa.

After operating with a sufficient quantity of gas (e.g. $S \times 4$(Pa. 1) for a pump of nominally S l/s speed), the 'regenerated' pumping speed deteriorates to the 'saturated' speed – the highest value speed obtained in this state should be quoted as the 'nominal' speed of the pump.

Reversibility of the 'regeneration' procedure following conditioning may be checked by repeating the bakeout schedule and checking that the low ultimate pressure is again obtained.

(*c*) *Turbomolecular pump*

The test dome of Fig. 1.7 recommended for sputter ion pumps may be readily adapted – but bakeout temperature particularly near the pump would be more limited. An additional important parameter to be determined is the compression ratio for various gases – e.g. helium or hydrogen compared with dry air.

(*d*) *Positive displacement pumps*

These include all kinds of (oil sealed usually) mechanical pumps which discharge gas against atmospheric pressure and achieve an inlet pressure of less than 100 Pa in one stage.

Test dome for pumping speed measurement. The test dome and adaptor recommended is shown in Fig. 1.6 to conform to the ISO R.1607 (1st Ed. 1970), the volume of the dome should be at least five times the volumes swept by the pump during one compression cycle. A recommended adaptor is also shown in the figure. Dimensions of test domes to suit different sized pumps are listed in Table 1.23.

TABLE 1.23

Swept volume litres (l)	Test dome	
	Volume (l)	Diameter D (mm)
0 to 0·26	1·3	100
0·26 to 1·1	5·4	160
1·1 to 4·2	21	250
4·2 to 17	84	400
17 to 65	325	630
65 to 260	1300	1000

Pumping of water vapour (for gas ballast pumps). The quantity of water vapour which can be pumped by a gas ballast pump depends on the gas ballast volume rate of flow B in relation to the volume rate of flow (or speed) S of the pump. Often B/S is expressed as a percentage, and typically B/S = 10–15% for gas ballast pumps. The maximum partial pressure of water vapour P_{WL} which may be pumped without causing condensation in the pump is determined additionally by:

(a) the saturated vapour pressure of water P_S corresponding to the temperature of the coolest part of the pump

(b) the partial pressure P_0 of water vapour in the air intake at atmospheric pressure. Typically $P_0 = 1{\cdot}5 \times 10^5$ Pa (or 11·4 Torr) for air at 20°C, 65% relative humidity
(c) the partial pressure P_L of air being pumped
(d) atmospheric pressure P_A

Then, as shown by Thees (1957),

$$P_{WL} \leqslant (B/S)\,\frac{P_A(P_s - P_0) + P_L P_s}{P_A - P_s}. \tag{1}$$

This quantity has its lowest value when $P_L = 0$, i.e. when pumping pure water vapour. One may define the partial pressure of water vapour under this condition as the *water vapour tolerance* P_W, where from equation (1), with $P_L = 0$

$$P_W = (B/S)P_A(P_s - P_0)/(P_A - P_s). \tag{2}$$

If P_W is known, then for a fixed gas ballast flow B, the maximum partial pressure of water vapour P_{WL} which may be pumped in the presence of the partial pressure P_L of air, is obtained from (1) as:

$$P_{WL} \leqslant P_W + \frac{P_L P_s}{P_A - P_s}. \tag{3}$$

In the absence of gas ballast flow, i.e. when $B = 0$, the water vapour tolerance $P_W = 0$. In this case, it is still possible to pump some water vapour in the presence of a partial pressure of air P_L. When $B = 0$, the maximum partial pressure P_{WL} of water vapour is given by:

$$P_{WL} \leqslant \frac{P_L \,.\, P_s}{P_A - P_s}. \tag{4}$$

Water vapour tolerance. This is the highest inlet pressure at which the pump can continuously exhaust water vapour (at normal ambient conditions 20°C, 10^5 Pa). This is usually specified by the manufacturer (in Pa) and may be checked by an agreed specification (ISO recommendation under preparation).
Water vapour pumping capacity. This is the maximum mass of water per unit time which can be continuously exhausted by the pump (in form of water vapour) under normal ambient conditions and with gas ballast in operation. It is specified by the manufacturer (in mg s^{-1} or g h^{-1}). It may be estimated to sufficient accuracy from the water vapour tolerance P_W using the relationship

$$C_W = 2{\cdot}2 \times 10^5\, SP_W/P_A T_s \text{ g h}^{-1},$$

where S = speed of pump at inlet (l/s), P_A = atmospheric pressure and T_s = the absolute temperature of the saturated water vapour in the test dome (K).

REFERENCES

DUSHMAN, S. (1949) 'Scientific foundations of vacuum techniques', 2nd Ed. (John Wiley & Sons Inc., New York).
GUTHRIE, A., and WAKERLING, R. K. (1949) 'Vacuum equipment and techniques' (McGraw Hill, New York).
HANDBOOK OF MATHEMATICAL FUNCTIONS (1971). N.B.S. Applied Mathematics Series, 55 (US Govt. Printing Office, Washington DC), 9th Printing.
JAHNKE, E., and EMDE, F. (1945) 'Tables of functions' (Dover Publications, New York).
KENNARD, E. H. (1938) 'Kinetic theory of gases' (McGraw Hill Book Group Inc., New York).
KNUDSEN, M. (1934) 'Kinetic theory of gases' (Methuen, London).
LOEB, L. B. (1934) 'Kinetic theory of gases', Chapter VII (McGraw Hill Book Co. Inc., New York).
PIRANI, M., and YARWOOD, J. (1961) 'Principles of vacuum engineering' (Chapman & Hall, London).
WUTZ, M. (1965) 'Theorie und Praxis der Vakuumtechnik', (F. Vieweg & Sohn, Braunschweig).
BERMAN, A. S. (1965) Free molecule transmission probabilities, *J. Appl. Phys.*, **36** (10), 3356.
BERMAN, A. S. (1969) Free molecule flow in an annulus, *J. Appl. Phys.*, **40**, 4991–4992.
CLAUSING, P. (1926) On the stationary streaming of highly rarefied gases through round cylindrical tubes of arbitrary lengths (in Dutch), *Versl. Afd. Nat. Kon. Akad. Wet. Amst.*, **35** (2), 1023–1035.
CLAUSING, P. (1932) On the streaming of highly rarefied gases through tubes of arbitrary lengths (in German), *Ann. d. Phýs.*, **12** (8), 961-989.
DAVIS, D. H. (1960) A Monte Carlo calculation of molecular flow rates through a cylindrical elbow and pipes of other shapes, *J. Appl. Phys.*, **31** (7), 1169–1176.
DAYTON, B. B. (1957) Gas flow patterns at entrance and exit of cylindrical tubes, *Vac. Symp. Trans.* (1956), 5-11 (Pergamon Press, London).
DAYTON, B. B. (1968) Problems in vacuum physics influencing the development of standard measuring techniques, *Proc. 4th Int. Vac. Congr.*, Part 1, 57–66.
DAYTON, B. B. (1972) Gas flow in vacuum systems, *J. Vac. Sci. Tech.*, **9**, 243–245.
DELAFOSSE, J., and MONGODIN, G. (1961) Les calculs de la technique du vide, *Le Vide*, **16** (92), 3–108.
DE MARCUS, W. C. (1956) The problem of Knudsen flow. Pts I–VI, U.S. A.E.C. Report K 1302, AD 12457.
DONG, W. (1956) U.C. Rad. Lab. Report 3353.
DONG, W., and BROMLEY, L. A. (1961) Vacuum flow of gases through channels with circular, annular and rectangular cross-section, *1961 Trans. 8th Vac. Symp. & 2nd Int. Congr.*, **2**, 1116–1132.
FÜSTOSS, L. (1970) Calculation of transmission probability in molecular flow through straight tubes, *Vacuum*, **20** (7), 279–283.
FÜSTOSS, L., and TOTH, G. (1972) The problem of the compounding of transmission probabilities for composite systems, *J. Vac. Sci. Tech.*, **9** (4), 1214–1217.
LEVENSON, L. L., MILLERON, N., and DAVIS, D. H. (1963) Molecular flow conductance, *Le Vide.*, **103**, 42–54.
MOORE, B. C. (1972) Gas flux patterns in cylindrical flow systems, *J. Vac. Sci. Tech.*, **9**, 1090–1099.
REYNOLDS, T. W., and RICHLEY, E. A. (1964) Flux patterns resulting from free molecule flow through converging and diverging slots, NASA TN.D.1864.
SCHUMACHER, B. W. (1961) Dynamic pressure stages for high pressure/high vacuum systems, *Trans. 8th Vac. Symp. & 2nd Int. Congr.*, **2**, 1192–1200.
STECKELMACHER, W. (1966) A review of the molecular flow conductance for systems of tubes, and components and the measurement of pumping speed, *Vacuum*, **16** (11), 561–584.
STECKELMACHER, W. (1968) Performance specification of high vacuum pumps, *Proc. 4th Int. Vac. Congr.*, Part 1, 67–72.
THEES, R. (1957) Vakuumpumpen und ihr Einsatz zum Absaugen von Dämpfen, *Vakuum Technik.*, **6** (Heft 7), 160–170.

1.4.10 Conversion tables

(a) *Units of pressure*

1 Torr = 1 mmHg closer than rounded values in table.

	pascal Pa	(mmHg) Torr	Standard atmosphere atm	dyne per square cm dyn/cm^2	millibar mbar	(micron) micrometer of mercury μmHg
1 Pa = $1(N/m^2)$	1	$7{\cdot}50062 \times 10^{-3}$	$9{\cdot}86923 \times 10^{-6}$	10	10^{-2}	7·50062
1 Torr (mmHg)	1·33·322	1	$1{\cdot}31579 \times 10^{-3}$	1333·22	1·33322	10^3
1 atm	101 325	760	1	1 013 250	1013·25	760 000
1 dyn/cm^2	10^{-1}	$7{\cdot}50062 \times 10^{-4}$	$9{\cdot}86923 \times 10^{-7}$	1	10^{-3}	$7{\cdot}50062 \times 10^{-1}$
1 mbar	10^2	0·750062	$9{\cdot}86923 \times 10^{-4}$	10^3	1	750·062
1 μmHg (micron)	0·13322	10^{-3}	$1{\cdot}31579 \times 10^{-6}$	1·33322	$1{\cdot}33322 \times 10^{-3}$	1

Note:
1 Torr = 1 mmHg closer than rounded values in table.

(b) *Volume*

	m^3	(dm^3) litre l	ft^3	UK gal	US gal	(ml) cm^3
1 cubic metre = m^3	1	1000	35·315	219·97	264·18	10^6
1 litre = 1 (dm^3)	10^{-3}	1	$3{\cdot}5315 \times 10^{-2}$	0·21997	0·26418	10^3
1 cubic ft = ft^3	$2{\cdot}8317 \times 10^{-2}$	28·317	1	6·23	7·48	$2{\cdot}8317 \times 10^4$
1 UK gal	$4{\cdot}546 \times 10^{-3}$	4·546	0·161	1	1·201	$4{\cdot}546 \times 10^3$
1 US gal	$3{\cdot}785 \times 10^{-3}$	3·785	0·134	0·833	1	$3{\cdot}785 \times 10^3$
1 cm^3 = 1(ml)	10^{-6}	10^{-3}	$3{\cdot}5315 \times 10^{-5}$	$2{\cdot}1997 \times 10^{-4}$	$2{\cdot}6418 \times 10^{-4}$	1

(c) *Volume rate of flow (speed of pumping)*

	m^3/s	l/s	l/min	m^3/h	ft^3/min	ft^3/h
1 $m^3\ s^{-1}$	1	1000	60 000	3600	2118·9	$1{\cdot}271 \times 10^5$
1 litre per second = 1 l s^{-1}	10^{-3}	1	60	3·60	2·119	127·1
1 litre/minute = 1 l/min	$1{\cdot}6667 \times 10^{-5}$	$1{\cdot}6667 \times 10^{-2}$	1	6×10^{-2}	$3{\cdot}53 \times 10^{-2}$	2·119
1 m^3/hour = 1 $m^3\ h^{-1}$	$2{\cdot}78 \times 10^{-4}$	0·278	16·67	1	0·589	35·3
1 ft^3/min	$4{\cdot}72 \times 10^{-4}$	0·472	28·32	1·70	1	60
1 ft^3/h	$7{\cdot}87 \times 10^{-6}$	$7{\cdot}87 \times 10^{-3}$	0·472	$2{\cdot}83 \times 10^{-2}$	$1{\cdot}6667 \times 10^{-2}$	1

(d) *Quantity of gas (amount of substance) and equivalents*

	mol	Number of molecules	Pa m^3 (at 20°C)	Torr l (at 20°C)
1 mole (1 mol)	1	$6{\cdot}0222 \times 10^{23}$	$2{\cdot}44 \times 10^3$	$1{\cdot}83 \times 10^4$
1 molecule	$1{\cdot}66035 \times 10^{-24}$	1	$4{\cdot}04 \times 10^{-21}$	$3{\cdot}04 \times 10^{-20}$
1 Pa m^3 (at 20°C)	$4{\cdot}1 \times 10^{-4}$	$2{\cdot}474 \times 10^{20}$	1	7·5
1 Torr-l (at 20°C)	$5{\cdot}47 \times 10^{-5}$	$3{\cdot}29 \times 10^{19}$	0·1444	1

Note:
Mass per mole (atoms or molecules) = M(g) = $M \times 10^{-3}$ (kg), where M = mol. wt.
Mass per molecule = M/N_A(g), where N_A = Avogadro's number = Number molecules per mole.

(e) Time

	seconds s
1 minute	60
1 hour	3600
1 day	$8{\cdot}640 \times 10^4$
1 week	$6{\cdot}048 \times 10^5$
1 month	$2{\cdot}417 \times 10^6$
1 year	$3{\cdot}154 \times 10^7$

(f) Throughput (pV – units/time)

	Pa m^3 s^{-1}	Pa l s^{-1}	Torr l s^{-1}	atm cm^3 s^{-1}	atm cm^3 min^{-1}	atm ft^3 h^{-1}
1 Pa m^3 s^{-1}	1	1000	7·50062	9·86923	592·154	1·26
1 Pa l s^{-1}	10^{-3}	1	$7{\cdot}5 \times 10^{-3}$	$9{\cdot}86923 \times 10^{-3}$	$5{\cdot}92154 \times 10^{-1}$	$1{\cdot}26 \times 10^{-3}$
1 Torr l s^{-1} (1000 l lusec)	0·133322	133·322	1	1·316	78·97	0·167
1 atm cm^3 s^{-1}	0·101325	101·325	0·76	1	60	0·127
1 atm cm^3 min^{-1}	$1{\cdot}68875 \times 10^{-3}$	1·68875	$1{\cdot}27 \times 10^{-2}$	$1{\cdot}667 \times 10^{-2}$	1	$2{\cdot}12 \times 10^{-3}$
1 atm ft^3 h^{-1}	0·794	794	5·97	7·87	472	1

TABLE 1.24(a) General physical constants ||

Constant	Symbol	Value	Uncertainty*	Units: Systeme Internat. (SI)		Units: Centimeter-gram-second (CGS)	
Speed of light in vacuum	c	2·9979250	±10	$\times 10^8$	m/s	$\times 10^{10}$	cm/s
Elementary charge	e	1·6021917	70	10^{-19}	C	10^{-20}	cm$^{1/2}$ g$^{1/2}$ †
		4·803250	21			10^{-10}	cm$^{3/2}$ g$^{1/2}$ s^{-1} ‡
Avogadro constant	N_A	6·022169	40 §	10^{23}	mol^{-1}	10^{23}	mol^{-1}
Atomic mass unit	u	1·660531	11 §	10^{-27}	kg	10^{-24}	g
Electron rest mass	m_e	9·109558	54	10^{-31}	kg	10^{-28}	g
Proton rest mass	m_p	1·672614	11 §	10^{-27}	kg	10^{-24}	g
Faraday constant	F	9·648670	54 §	10^4	C/mol	10^3	cm$^{1/2}$ g$^{1/2}$ mol^{-1} †
Planck constant	h	6·626196	50	10^{-34}	J s	10^{-27}	erg s
Fine structure constant	α	7·297351	11	10^{-3}		10^{-3}	
Charge to mass ratio for electron	e/m_e	1·7588028	54	10^{11}	C/kg	10^7	cm$^{1/2}$/g$^{1/2}$ †
		5·272759	16			10^{17}	cm$^{3/2}$ g$^{-1/2}$ s^{-1} ‡
Rydberg constant	R_∞	1·09737312	11	10^7	m^{-1}	10^5	cm^{-1}
Gyromagnetic ratio of proton	γ_p	2·6751965	82	10^8	rad s^{-1} T^{-1}	10^4	rad s^{-1} G^{-1} †
(uncorrected for diamag., H_2O)	γ'_p	2·6751270	82	10^8	rad s^{-1} T^{-1}	10^4	rad s^{-1} G^{-1} †
Bohr magneton	μ_B	9·274096	65	10^{-24}	J/T	10^{-21}	erg/G†
Gas constant	R	8·31434	35	10^0	J K^{-1} mol^{-1}	10^7	erg K^{-1} mol^{-1}
Boltzmann constant	k	1·380622	59	10^{-23}	J/K	10^{-16}	erg/K
First radiation constant ($8\pi hc$)	c_1	4·992579	38	10^{-24}	J m	10^{-15}	erg cm
Second radiation constant	c_2	1·438833	61	10^{-2}	m K	10^0	cm K
Stefan-Boltzmann constant	σ	5·66961	96	10^{-8}	W m^{-2} K^{-4}	10^{-5}	erg cm^{-2} s^{-1} K^{-4}
Gravitational constant	G	6·673 2	31	10^{-11}	N m^2/kg^2	10^{-8}	dyn cm^2/g^2

* Based on 1 std. dev.; applies to last digits in preceding column.
† Electromag. system.
‡ Electrostatic system.
§ These values may be in conflict with data available since the Taylor, Parker, Langenberg review (*Rev. Mod. Phys.*, **41**, 375 (1969)). Pending a complete new readjustment of the constants, it would be prudent to multiply the above uncertainties by 3.
|| Based on data in 'Handbook of Mathematical Functions' (1971), chapter 2.

TABLE 1.24(b) Defined values and conversion factors†

Atomic mass unit (u)	1/12 the mass of an atom of the ^{12}C nuclide
Standard acceleration of free fall	9·80665 m/s^2, 980·665 cm/s^2
Standard atmosphere	101325 N/m^2, 1013250 dyn/cm^2
Thermochemical calorie	4·184 J, 4·184 x 10^7 ergs
Int. Steam Table calorie	4·1868 J, 4·1868 x 10^7 ergs
Litre	0·001 cubic metre
Mole (mol)	amount of substance comprising as many elementary units as there are atoms in 0·012 kg of ^{12}C
Inch	0·0254 m, 2·54 cm
Pound, avdp.	0·45359237 kg, 453·59237 g

† Based on data in 'Handbook of Mathematical Functions' (1971), chapter 2.

TABLE 1.25 Sensitivity of hot-cathode ionization gauges to different gases and vapours

The data refers to the relative sensitivities to that for nitrogen S/S_{N_2}, where S is defined in the usual way in units of reciprocal pressure. There has been much conflicting data in the literature obtained by direct calibration or by estimation from physical properties of the gases, such as ionization cross-section. A useful summary was given by Summers (1969)[1]. Other sources are indicated in the table. Ionization cross-sections were used when no direct calibration data was available and is shown where there are discrepancies. The proposal made by Found and Dushman (1924)[14] and again by Flaim and Ownby (1971)[15] to relate sensitivity for different gases to the number of electrons per molecule, although useful for some gases, was shown by Summers (1969)[1] not to give the best correlation with measured sensitivities, whereas comparison with ionization cross-section data appears to be more reliable. Estimated (or derived) values are shown with e.

Substance		Formula	S/S_{N_2}	Source of Data	
Acetaldehyde		C_2H_4O	2·6 e	1, 2	
Acetone		$(CH_3)_2CO$	4·0 e	1, 2	
Acetylene		C_2H_3	2·0 e	1, 3	
Air		—	1·0	1	
Ammonia		NH_3	1·2	1, 3, 16	
Argon		Ar	1·3	1	
			1·4	11	
Benzene		C_6H_6	5·7 e	1, 3	
			3·5	4	
Bromine		Br	3·8 e	1	
Butane	n-	n-C_4H_{10}	4·9	1, 9	
	iso-	iso-C_4H_{10}	4·6	1, 9	
Cadmium		Cd	2·3	1	
			3·4 e	5	
Carbon dioxide		CO_2	1·4	1, 11	
Carbon disulphide		CS_2	4·8 e	1, 2	
Carbon monoxide		CO	1·05	1, 11	
Carbon tetrachloride		CCl_4	6·0 e	1	
Caesium		Cs	4·3	1, 6	
			2·0	1, 7	
Calor gas		mainly propane	4·6	13	Conventional gauge
		and butane mixture	3·7	13	B.A. gauge
Coal gas		mainly H_2 (47%) CH_4 (36%), CO (8%)	1.25	13	
Chlorine		Cl_2	0·68	1	
Deuterium		D_2	0·35	1	

TABLE 1.25 (*continued*)

Substance	Formula	S/S_{N_2}	Source of Data
Dibutylpthalate	pump fluid	9·15	12
Dichloro-difloromethane	CCl_2F_2	2·7	1, 9
Ethane	C_2H_6	2·6	1, 9
		2·5	4
Ethylene	C_2H_4	2·0	4
		2·2 e	1, 3
Helium	He	0·18	1, 11
		0·13 e	1, 3
Hydrogen	H_2	0·46	1
		0·41	11
		0·41 e	1, 3
Hydrogen bromide	HBr	2·0	1
Hydrogen chloride	HCl	1·5 e	1, 3
		1·6	16
Hydrogen fluoride	HF	1·4	1
Hydrogen sulphide	H_2S	2·2 e	1, 3
Iodine	I	5·4	1
Krypton	Kr	1·9	1, 11
		1·7 e	1, 3
Lithium	Li	1·9 e	1
Mercury	Hg	3·6	1
Methane	CH_4	1·4	1, 9, 4
Narcoil 40	pump fluid	13·1	12
Neon	Ne	0·30	1
		0·32	11
Nitric oxide	NO	1·3	1
Octoil	pump fluid	13	12
Oxygen	O_2	1·0	1
		0·78	11
Propane	C_3H_8	4·2	1, 9
		3·7 e	1, 3
Sulphur hexafluoride	SF_6	2·3	10
		2·2	11
		2·5	16
Toluene	$C_6H_5CH_3$	6·4	4
		6·8 e	1
Water	H_2O	1·1	1
Xenon	Xe	2·9	1
		2·8	11
		2·4 e	3

References

(1) Summers, R. L. (1969) 'Empirical observations on the sensitivity of hot cathode ionisation type vacuum gauges.' NASA Report TND. 5285 (June).

(2) Otvos, J. W., and Stevenson, D. P. (1956) *J. Am. Chem. Soc.*, **78**, 546–551.

(3) Lampe, F. W., Franklin, J. L., and Field, F. H. (1957) *J. Am. Chem. Soc.*, **79**, 6129–6132.

(4) Young, J. R. (1973) *J. Vac. Sci. Tech.*, **10**, 212–214.

(5) Pottie, R. F. (1966) *J. Chem. Phys.*, **44**, 916–922.

(6) Witting, H. L. (1964) 'The absolute value of the direct electron cross-section of Caesium.' Thermionic Conversion Conference Report. Ed. R. E. Stickney, I.E.E.E., pp. 214–128.

(7) Benninghoven, A. (1963) *Z. Angew. Phys.*, **15**, 326–327.

(8) Riddiford, L. (1951) *J. Sci. Instrum.*, **28**, 375–379.

(9) Moesta, H., and Renn, R. (1957) *Vakuum Technik.*, **6**, 35–36.

(10) Schulz, G. J. (1957) *J. Appl. Phys.* **28**, 1149–1152.

(11) Holanda, R. (1972) 'Sensitivity of hot cathode ioniastion vacuum gauges in several gases.' NASA Report TN.D.6815 (July)
(12) Reich, G. (1957) *Z. Angew. Phys.*, **9**, 23–29.
(13) Elsworth, L. (1960) *Vacuum.*, 10, 266–267.
(14) Found, C. G., and Dushman, S. (1924) *Phys. Rev.*, **23**, 734–743.
(15) Flain, T. A., and Ownby, P. D. (1971) *J. Vac. Sci. Tech.*, **8**, 661–662.
(16) Bennewitz, H. G., and Dohmann, H. D. (1965) *Vakuum Technik.*, **14**, 8–12.

TABLE 1.26 Physical properties of some common gases and vapours

Gas	$M \times 10^3$ kg mole^{-1}	$\lambda . P$ (At 20°C) m. Pa	$\eta \times 10^5$ (At 20°C) N s m^{-2}	δ (from viscosity) $\times 10^{10}$ (At 20°C) m	ρ (At 0°C and 1 atm) kg m^{-3}
He	4·003	1·00	1·96	2·18	0·1785
Ne	20·18	0·705	3·10	2·60	0·8999
A	39·94	0·354	2·22	3·67	1·7839
Kr	83·7	0·272	2·46	4·15	3·74
Xe	181·3	0·196	2·26	4·91	5·89
H_2	2·016	0·66	0·88	2·75	0·0899
N_2	28·02	0·337	1·75	3·70	1·2505
O_2	32·00	0·362	2·03	3·64	1·42895
H_2O	18·02	0·222	8·80	4·68	—
CO	28·01	0·336	1·77	4·65	1·250
CO_2	44·01	0·222	1·47	4·19	1·9768
Hg	200·6	0·165	2·28	5·11	9·021
Air	28·98 (from density) ρ	0·342	1·81	3·75	1·2928

1.5 Pump fluids, sealing compounds and greases

1.5.1 Impurities and degradation

Fluids*, greases and waxes etc., can emit vapour molecules and fragments from their parent substances together with entrapped gases and these characteristics determine working performance when such substances are used in vacuum systems. One can estimate the deleterious effect of a substance on a vacuum atmosphere from its saturated vapour pressure at the given operating temperature but for a compound this is *only an initial guide* to its possible utility. Polymeric substances of high molecular weight, e.g. organic materials >1000 a.m.u. may have intrinsic vapour pressures in the uhv range, but as with low molecular weight materials these substances invariably contain volatile impurities even if the basic substance by the mode of manufacture tends to a constant molecular weight. Further production of volatiles may result from thermal and catalytic processes as when chemical compounds are used in vapour stream pumps. Thus petroleum distillates begin to show dissociation at ~200°C if not lower. Degradation also occurs when compounds are used as fluid lubricants in shaft bearings and mechanical pumps, e.g. hydrogen and low molecular weight fragments are released from mineral oils employed in rotary mechanical pumps when the coefficient of sliding friction is above ~0·1. High molecular weight fluids (for organic substances ~500 a.m.u.) are used as evaporants in vapour stream pumps and lubricants in mechanical pumps.

* The vapour pressure characteristics of mercury which is used as a fluid in vapour stream pumps are given in Table 1.10.

Greases and waxes with a high average molecular weight can be used for sealing purposes and residual products from distillation are available for which vapour pressures in the uhv-range are claimed. However, all these substances can trap gas which is subsequently desorbed and heating in use can promote thermal degradation as with pump oils. Solid polymers may contain volatile molecular components remaining after chemical reaction and a hydrocarbon such as polyethylene with an average molecular mass of >20 000 a.m.u. can freely emit methane and related molecular species at normal temperature and further dissociate liberating hydrogen and $[CH_x]_n$-fragments when heated.

To summarize, the emission characteristics of a given fluid used in vacuum will depend on its

(a) *initial composition* – molecular weight distribution and impurity and absorbed gas content;
(b) *degradation in a specific use* – by thermal, catalytic, rubbing friction and, if exposed to air, oxidation effects.

Thus if employed in a diffusion pump the existence of fractionating and purifying stages can further modify the effects of the foregoing on the composition of the vapour and gases released.

1.5.2 Pressure measurement problems

Having established that *a pure substance* with a specified vapour pressure at a given temperature may be dissociated, in the condensed or vapour phase, thereby enhancing the pressure in a vacuum envelope we shall now consider the problem of measurement under such conditions. Assuming that a sensing head: (i) is truly immersed in the atmosphere created by the working substances* and (ii) by sorption effects negligibly raises or lowers the local equilibrium gas pressure – then for knowledge of the vacuum atmosphere produced we should determine the *molecular mass* and *equilibrium partial pressure* of each emitted component. As the equilibrium pressure of a component depends on its molecular emission rate and its velocity of removal, e.g. by trapping, complete knowledge of a system requires measurement of the emission rate. Under high vacuum conditions free molecules are usually detected by ionization and collection and one has the added problem that electron impact and hot cathode emitters can cause further molecular fragmentation. *This raises the pressure measured and increases the complexity of mass spectra in partial pressure analysers.* Emission rates can be found from adsorption or condensation trapping, e.g. using mechanical or h.f. crystal microbalances. The observed result can depend on the conductance for molecular flow of the path between source and sensing head and when pumping at low pressure on gas diffusion and counter gas flow.

* Blears (1944, 1947) has shown how sorption in an ionization gauge head can lower the observed ultimate pressure when gas and vapour flow are limited by gauge tubulation. His immersed 'high speed' gauge is now generally known as a nude gauge. However, gauge contamination can be rapid if exposure is in hydrocarbon or silicone vapours at $\geqslant 10^{-7}$ Torr.

1.5.3 Pump and trap effects

The vacuum performance of a fluid will be influenced by the design of pump used, e.g. fractionating, its operating temperature and the constructional materials employed. Vapour issuing from pump jets may, because of bad jet design or gas collision, backstream with that evaporating from the condensed fluid. Mercury, the first fluid used in vapour stream pumps and which remains in common use, obviously cannot be degraded but its backstreaming can be enhanced by a direct contribution from a vapour jet.

As vapour stream (and lubricated mechanical pumps) are often fitted with devices to trap backstreaming substances by condensation or adsorption an overall assessment of the behaviour of a pump fluid requires measurement of the backstreaming rate and/or equilibrium pressure before and after trapping. Inclusion of a 'trap' in a pump line will reduce the pipe conductance between vessel and pump. This restrictive effect will lower the pump speed and backstreaming rate at the vessel and *to determine trapping efficiency one must compare the reduction in backstreaming rate with loss in pump speed.*

A cold trap may be ineffective if emitted molecular fragments are too volatile for condensation with the parent molecules at the temperature used, e.g. CH, CH_2-groups etc. released from hydrocarbons require use of a LN_2-cooled trap. If an outgassed trap is employed the attainment of equilibrium may be delayed as the cleaned surface initially strongly adsorbs backstreaming components. On the other hand condensation may not immediately occur at low saturated vapour pressures should the trap surface lack adsorption affinity for the incident molecules and nucleating aggregates must be formed for condensation to commence.

When the temperature of a trap is sufficiently low for the v.p. of a fluid to be negligible and an outgassed vessel is free of fluid contamination, one can assume that the measured pressure is that of degraded molecules and desorbed gases from the pump fluid. Raising the trap temperature should increase the vapour content of the vessel atmosphere until it becomes the dominant component. This was observed by Reich (1957) who found that heating the trap above room temperature gave a measured pressure which followed a saturated vapour pressure relation.

Rarely is information on vapour and trap characteristics available *in toto.* Although LN_2-traps usually operate at a temperature effective for condensing (and trapping) backstreaming products it is dangerous to assume that *sorbent* materials will be equally active for all such substances. As a device can show an apparent 'trapping' effect although it is purely restrictive reducing the pumping speed in the same ratio as the backstreaming rate it is essential to make the measurements described above.

1.5.4 Vapour pressure – measurement and use

It is usual when compiling data on vacuum fluids (and greases etc.) to give their saturated vapour pressure over an operating temperature range, but, as shown above, the vacuum performance depends on the fluid *as supplied* and the pump, trap and vacuum system chosen. Also the *observed* performance depends on gauge interaction with the vacuum atmosphere. Thus as stated in the introduction v.p. data is only an initial guide to vacuum performance. The writer has therefore compiled additional information showing the behaviour of fluids when used in common types of vacuum pump. Although this lacks the *absolute* nature of thermodynamic data it will more usefully aid estimation of equilibrium pressure and backstreaming rate to be expected under practical pumping circumstances. The compilation also contains information on the chemical nature of the chief fluids, greases and sealing compounds in use.

Free evaporation. Assuming that the specified v.p. of a fluid is for a pure compound its rate of free evaporation from a surface can be found from the Langmuir equation

$$E = 5{\cdot}85 \times 10^{-1}\, a_e A P \sqrt{M/T}\ (\mathrm{kg\ s^{-1}}) \tag{1}$$

where P is the v.p. (Torr),* T the absolute temperature (°K), M the molecular mass, A the evaporant area (m^2), and a_e the accommodation coefficient for evaporation.

The evaporation rate E under conditions of free molecular flow (i.e. negligible collisions between vapour and vapour, or vapour and gas molecules) can be found from the relation for the condensation rate (kg s^{-1}) on a cooled receiver

$$C = a_c E A_c \cos\theta \cos\phi/(\pi l^2), \tag{2}$$

where A_c is the condensate area, a_c the condensation coefficient, l the distance between condenser and evaporator and θ and ϕ the vapour emittance and incidence angles at their surfaces respectively. From these relations $a_c a_e P$ can be found if M is known.

If the operative vapour pressure is less than 10^{-4} Torr the vapour flow should tend to be free molecular and if the condenser is at LN_2-temperature a_c can generally be assumed to equal one. The dimensions of the evaporation area are assumed small compared with that of l.

Using a Knudsen cell with an effusion aperture, whose area is small compared with that of the evaporating liquid, one can calculate P from the derived evaporation rate; the evaporant area A is that of the cell aperture and $a_e = 1$. The mean free path of the molecules should be greater than the cell aperture but smaller than the cell inner diameter to ensure random arrival at the aperture.

One can use evaporators for which the evaporant rate E will be determined by the evaporant area and therefore the product $a_e P$. From measurements using both types of vapour source a_e can be found. The evaporation coefficient will be less than unity if low volatility impurities form at the evaporant surface. Measuring either evaporant weight change with time or condensation rate permits the detection of volatile components as these are preferentially emitted initially, whereas measurement of saturated v.p. at a given temperature can only show a departure from a standard value indicating composition changes.

* As data on pump fluids in the literature gives pressures in torr, this is the unit which has been adopted in this section up to 1.5.17. For conversion from p (SI units, i.e. Pa) to P (Torr), and vice-versa, use conversion tables 1.4.10 (see page 50). Equation (1) can be derived directly (in SI units) from the impingement rate N defined in section 1.4.1. and the relations in 1.4.3. With p (in Pa), the equivalent expression to equation (1) is $E = a_e N m A$, so that

$$E = 3{\cdot}64 \times 10^{-2}\, a_e A p \sqrt{M/T}/R_0 = 4{\cdot}38 \times 10^{-3}\, a_e A p \sqrt{M/T}\quad (\mathrm{kg/s}) \tag{1a}$$

Finally, gas/liquid chromatography is a simple but effective technique for showing the existence of volatile impurities in – and batch variations between – pump fluids.

Determining v.p. A simple method for finding the vapour pressure of a pump fluid is to evaporate the fluid under controlled conditions *in vacuo* using a h.f. crystal microbalance cooled by liquid nitrogen to serve as a vapour condenser. If one employs different types of evaporators information can also be obtained on whether the fluid contains more than one component; for further information see Holland, Laurenson and Deville (1965) and Deville, Holland and Laurenson (1965). However, it must be noted that many of the curves for v.p. as a function of temperature given below have been derived from extrapolated plots with measured values obtained in a pressure range permitting use of manometers. If the fluid is a mixture of complex molecules their binding energies in the fluid could be concentration dependent and the vapour pressure may not follow the Clausius–Clapeyron equation. This relation for a pure compound with molecular evaporation can be written

$$P = \text{constant}\ e^{-L/RT}, \qquad (3)$$

and as generally expressed

$$\log_{10} P = A - \frac{B}{T}, \qquad (4)$$

where A and B are constants and $L = BR/\log_e e$

$$R/\log_e e = 4{\cdot}57 \text{ cal deg}^{-1} \text{ mole}^{-1} \text{ or } 19{\cdot}1 \text{ J deg}^{-1} \text{ mole}^{-1}.$$

Dushman (1949) has reviewed reported data on the v.p. of pump oils and discussed the validity of equation (4). He gives values of A and B for pump fluids based on phthalates, distilled mineral oils, straight chain hydrocarbons, chlorinated hydrocarbons and linear silicones. Generally L for pump fluids is about 25 kcal/mole. Vapour pressures as a function of temperature for many of the pump oils in common use are shown in Figs 1.8(a), (b) and (c).

Smith (1959) has shown how the boiler and jet assembly in vapour stream pumps can have greatly differing temperatures, e.g. in a 150 mm pump the boiler temperature was 20°C higher than that of the top cap. Also, changes in the boiler temperature with power input showed that free evaporation did not occur at a rate obtained from the Langmuir equation using the corresponding saturated vapour pressures. Smith considered that the oil temperature rose but vaporization was inhibited by the boiler design. However, one would not expect free evaporation but diffusion and resistance to flow when the vapour pressure is in the fraction of a Torr region and vapour must pass through the jet stack to escape from the boiler.

1.5.5 Pump fluids – chemical types

Classification. Oils and greases developed for vacuum use are often classified according to their specific employment, e.g. *diffusion pump fluids, rotary pump lubricants etc.* Whilst

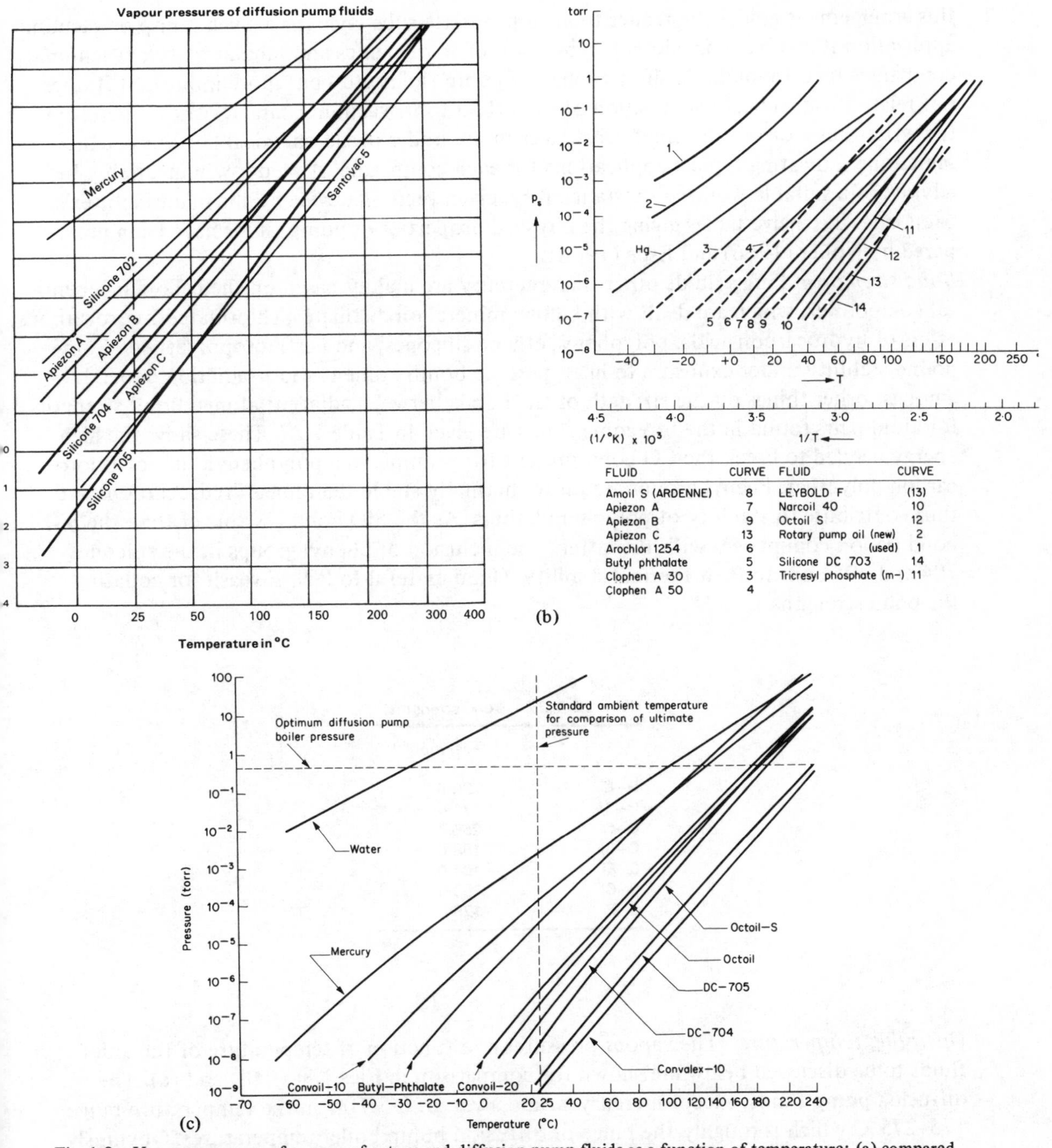

FLUID	CURVE	FLUID	CURVE
Amoil S (ARDENNE)	8	LEYBOLD F	(13)
Apiezon A	7	Narcoil 40	10
Apiezon B	9	Octoil S	12
Apiezon C	13	Rotary pump oil (new)	2
Arochlor 1254	6	(used)	1
Butyl phthalate	5	Silicone DC 703	14
Clophen A 30	3	Tricresyl phosphate (m−)	11
Clophen A 50	4		

Fig. 1.8 Vapour pressures of a range of rotary and diffusion pump fluids as a function of temperature: (a) compared with mercury. Apiezons are petroleum distillates and Santovac 5 is a 5 ring polyphenylether (from data compiled by Edwards Vacuum Components); (b) from data compiled by Espe (1968); (c) from data compiled by Bendix Scientific Instruments and Equipment.

this arrangement guides the reader to an appropriate substance for a given pump or vacuum application it involves considerable repetition of trade names and tabular matter (vacuum companies tend to market a similar range of pump fluids etc.) but more important it does not relate the chemical composition of each fluid to its vacuum characteristics. Therefore the writer has grouped the chief substances in use under broad chemical headings at the same time indicating typical applications for each compound. A reader will invariably be advised of a suitable working substance for a given vacuum device by the manufacturer's catalogue. Extensive tables giving the physical properties of pump fluids have been prepared by Noller (1966) and Espe (1968).

Bond strengths. Pump fluids other than mercury are mainly based on the following chemical compounds, which are dealt with below: mineral oil distillates; chlorinated hydrocarbons; esters of hydrocarbon acids; polyphenyl ethers; silicones; and perfluoropolyethers. Compound stability under exposure to heat, particle bombardment and irradiation depends amongst other things on the strength of the bonds between adjacent atoms. Bond strengths for atom pairs found in the foregoing fluids are given in Table 1.27. These show the high energy needed to break the CO bond present for example in a polyphenylether or fluorocarbon polyether. Phenyl groups are more thermally stable than linear hydrocarbons and thus contribute to stability of polyphenylethers. As the SiO bond is stronger than the SiC bond fission commences with the latter. The inclusion of phenyl groups in the silicones 704 and 705 adds to their thermal stability. (I am grateful to P.N. Fiveash for collating the bond strengths.)

TABLE 1.27 Bond strengths

	kcal mole^{-1}
H–C	80·9
H–Si	71·4
O–C	256·7
O–Si	188·0
C–Si	104·0
C–Cl	93·0
C–F	128·0

Operating temperature. The vapour pressures as a function of temperature of the chief fluids to be discussed below are shown for comparison in Figs 1.8(a), (b) and (c). The diffusion pump fluids, except mercury, attain a v.p. of 0·5 Torr in the temperature range 175–275°C which is roughly the range of diffusion pump boiler temperatures. Obviously no problems are experienced in providing high boiler pressures with mercury as it cannot be dissociated, and pumps with backing pressure of tens of Torr can be designed. With organic fluid the boiler pressure (and related backing pressure) is limited to a few Torr.

TABLE 1.28 Physical properties of Apiezon diffusion pump oils. (Reproduced from data compiled by Shell Chemical)
a Assumes that a suitably designed pump is used with the pump cooling water at 20°C.
b Calculated from the density values.
Note: Latent heats of fusion of Apiezon oils cannot be measured as they have glassy, disordered structures and exhibit no fusion peaks between −80°C and −60°C.

	Oil A	Oil B	Oil BW	Oil C	Oil G
Ultimate pressure obtainable, Torr (a)	5×10^{-5}	10^{-6}	10^{-6}	10^{-7}	10^{-5}
Average boiling point, °C, at 1 Torr pressure	190	220	225	255	210
Specific gravity at: 20°C/15·5°C	0·865	0·873	0·882	0·876	0·868
30°C/15·5°C	0·859	0·869	0·876	0·869	0·862
Density, g/ml at: 10°C	0·871	0·878	0·888	0·881	0·874
20°C	0·865	0·872	0·882	0·875	0·868
30°C	0·859	0·866	0·875	0·869	0·861
40°C	0·852	0·859	0·869	0·863	0·855
Flash point, Pensky-Martens, °F, closed	410	470	455	475	430
open	410	470	480	510	440
fire	450	505	525	560	485
Viscosity, kinematic, cS at: 20°C	59	142	361	283	97
40°C	23·4	49·3	98·1	90	35
100°C	4·5	7·0	10·1	10·6	6·1
Viscosity, Engler (°E) at: 20°C	7·8	18·8	16·5	37·3	12·8
50°C	2·44	4·21	7·95	7·3	3·35
100°C	1·35	1·56	1·85	1·90	1·49
Viscosity, dynamic (cP) at: 40°C	19·9	42·4	85·2	77·2	29·9
Pour point, ASTM, °F	20	15	−10	5	15
Coefficient of expansion per °C over 20°C–30°C	0·00070	0·00073	0·00070	0·00069	0·00070
Coefficient of expansion per °C over 10°C–40°C (b)	0·00083	0·00080	0·00083	0·00080	0·00087
Average molecular weight	354	420	462	479	397
Refractive index at 20°C	1·4780	1·4815	1·4825	1·4830	1·4795
Thermal conductivity Btu in/ft^2 h °F	0·91	0·91	0·89	0·96	0·94
w/m °C	0·132	0·132	0·129	0·139	0·134
Specific heat at 25°C: cal/g	0·46	0·49	0·46	0·46	0·48
joule/g	1·9	2·0	1·9	1·9	2·0

1.5.6 Mineral oil distillates

Diffusion pump fluids

Apiezon. Burch (1928, 1929) was the first to replace mercury in a diffusion pump by an organic fluid using a high boiling point petroleum distillate and to prepare a low vapour pressure petroleum grease suitable as a vacuum sealing compound. A useful account of molecular distillation as used in the preparation of these substances (and those discussed below) is given in a work by Burrows (1960) together with a description of the early work by Burch in the Research Laboratories of the then Metropolitan Vickers Electrical Co. Ltd. The mineral oil distillates, oils and greases, were marketed under the trade name 'Apiezon' (derived from Greek meaning 'without pressure') and are now the products of Shell Chemicals. The vapour pressures of a range of oils as a function of temperature are shown in Fig. 1.9 (these curves are based on Burrows 1946 and acknowledgments are made to GEC Power Engineering Ltd for their reproduction). Physical properties of the diffusion pump fluids are given in Table 1.28.

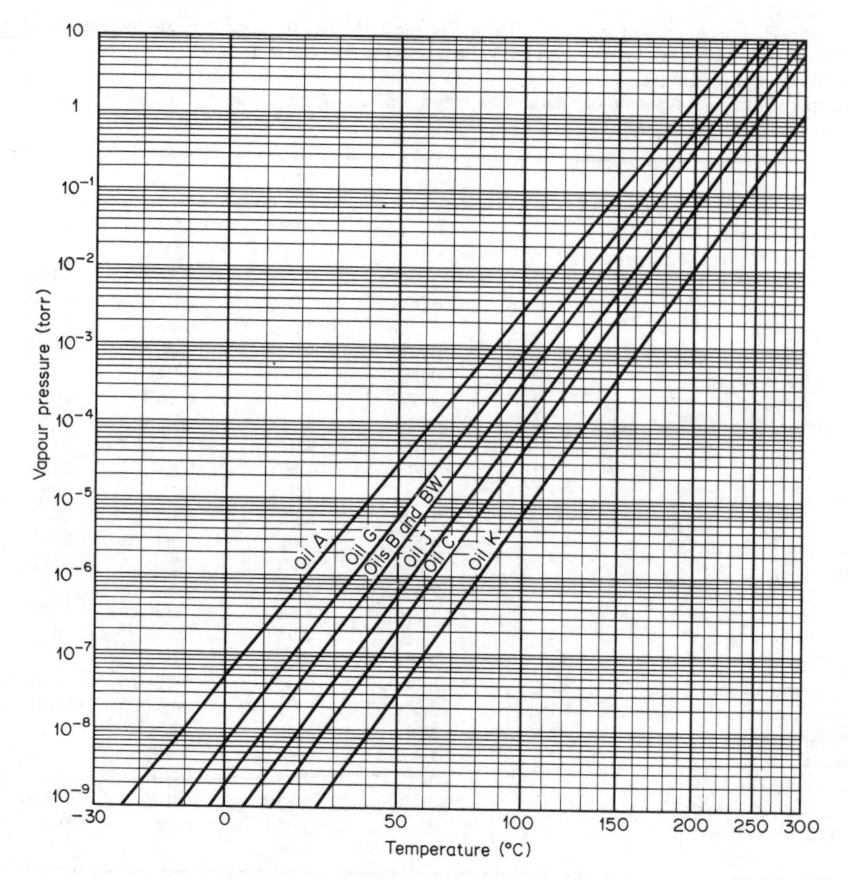

Fig. 1.9 Vapour pressure curves of Apiezon oils.

Apiezon C is a low v.p. diffusion pump oil, and oil G is intended for use in large vapour stream pumps with an ultimate baffled pressure of 10^{-5} Torr (booster pump fluids are dealt with below). Oil BW is a high purity fluid suitable for use in mass spectrometers to avoid trace impurities in the spectra.

Diffelen. Many vacuum companies market Apiezon and also their own petroleum distillates. Thus Leybold–Heraeus report ultimate pressures for their mineral oil fluid known as 'Diffelen' as given in Table 1.29.

Convoil is the trade name of petroleum distillates obtainable from Bendix Scientific Instruments and Equipment; v.p. values are in Table 1.29.

Mono-eicosyl naphthalene produced by Lion Fat & Oil Co. of Tokyo is marketed under the trade name 'Lion S' for use in diffusion pumps. Its properties are:

mol. wt. 407
specific gravity 0·93
viscosity cS 38°C 41·12

H. Oikawa (JOEL) and O. Mikami (Lion Fat & Oil) have reported its vacuum properties in a paper entitled 'Vacuum characteristics of a new diffusion pump fluid'. It appears to have v.p. characteristics similar to Apiezon C.

TABLE 1.29 Diffelen and Convoil fluid properties

	Mol. wt.	Density 20°C g cm^{-3}	v.p. Torr at 25°C
Diffelen – light	335	0·884	1·8 . 10^{-8}
Diffelen – normal	435	0·88	8·5 . 10^{-9}
Diffelen – ultra	450	0·876	2·6 . 10^{-9}
Convoil – 10	250	ρ 25/25°C 0·91	2 . 10^{-3} ult. P G-4 pump
Convoil – 20	400	ρ 25/25°C 0·86	8 . 10^{-6} ult. P G-4 pump

Vapour booster pump fluids

An inexpensive fluid is usually needed for the large charge required by a booster pump. As the fluid must attain several Torr vapour pressure at normal boiler temperature the substances used are distillate cuts of higher vapour pressure than those employed in diffusion pumps.

Apiezon AP 200 series. These are hydrocarbon based fluids treated with anti-oxidants that can be used in vapour stream pumps with backing pressures up to 1 Torr and 500–1000 atmospheric exposures whilst hot. AP201 is an inhibited oil for use in booster pumps. Unlike chlorinated compounds discussed below it does not attack Nitrile materials and is non-irritant.

1.5.7 Chlorinated hydrocarbons

Booster pump fluids based on the pentachlorodiphenyls such as Arochlor made by Monsanto, Narcoil 10 and Convaclor 8 and 12, have good thermal and chemical resistance and are used to obtain ultimate pressures of $\sim 10^{-4}$ Torr. Unfortunately emitted vapours are irritants and harmful to health. Arochlor 1254 has a molecular weight of 339 and a v.p. of 2×10^{-5} Torr at 25°C. *Clophen* is the trade name of a diphenyl chloride based fluid marketed by Leybold Heraeus and fluid A50 has a v.p. of 10 Torr at 180° C.

1.5.8 Hydrocarbon waxes and greases

Apiezon greases. Their vapour pressures are given in Fig. 1.10. The greases are made from mineral oils containing a proportion of branched and unsaturated hydrocarbons which produce high-molecular weight products. Type numbers and properties are as follows:

AP 100 Anti-seize grease. A blend of low v.p. petroleum and PTFE ($<10^{-10}$ Torr at 20° C).

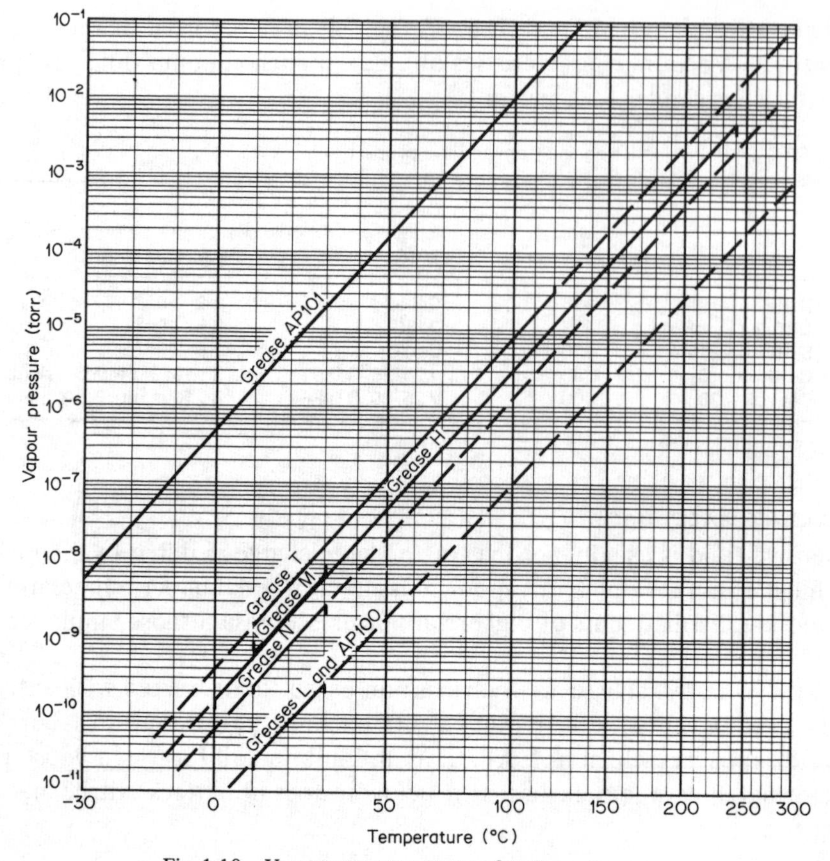

Fig. 1.10 Vapour pressure curves of Apiezon greases.

AP 101 Only product not containing a distilled grease. It is based on heavy duty lubricating grease gelled with lithium stearate, a silicious earth and PTFE. (<10^{-5} Torr at 20°C usable from −40 to 180°C.)

L Hydrocarbon lubricant grease without additive, softening at 30°C.

M Hydrocarbon lubricant grease containing more wax than L and with higher v.p.

N Similar to L but with added high molecular weight hydrocarbon polymer giving rubber consistency.

T Similar to N but can be used over wide temperature range. Contains a gelling agent dispersable in solvents for cleaning.

H Similar to T but can be used over a wider temperature range (−10 to 24°C). Does not melt at high temperature but *stiffens above 40°C.*

Lithelen grease (Leybold Heraeus). Vacuum treated grease to remove volatiles with added lithium compounds permitting use from 0 to 150°C.

Waxes and sealing compounds. The vapour pressures of Apiezon products are given in Fig. 1.11.

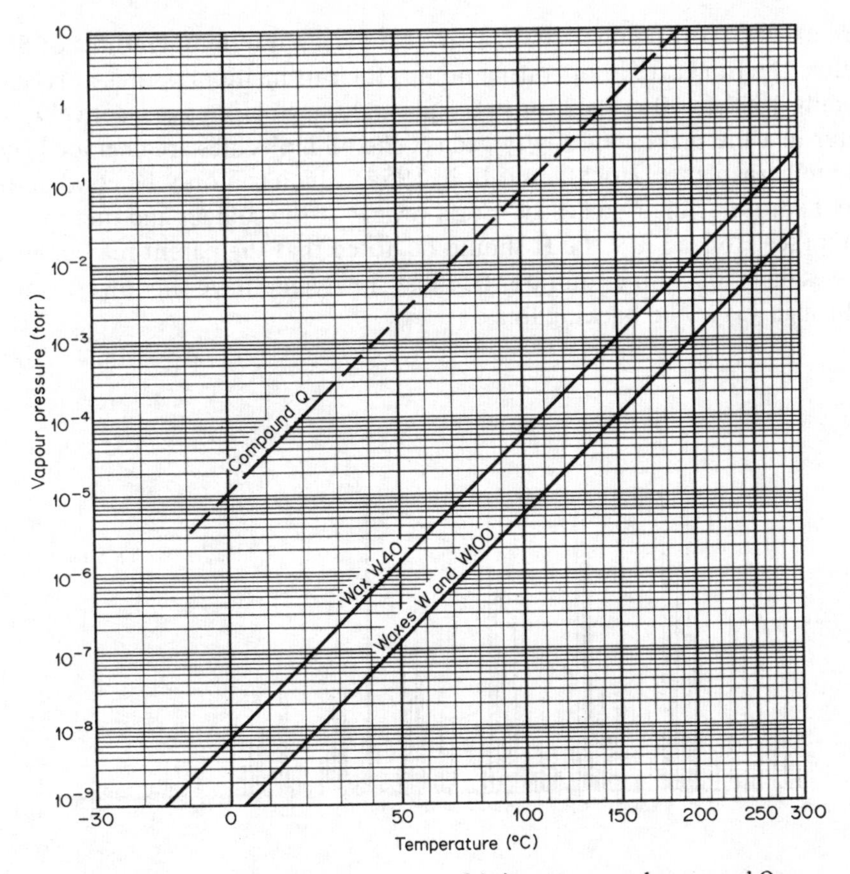

Fig. 1.11 Vapour pressure curves of Apiezon waxes and compound Q.

TABLE 1.30

Wax	Softening temp. °C	Av. mol. wt.
W	85	1214
W100	55	1160
W40	45	1140

Compound Q (Apiezon). Plasticine – like substance can be kneaded into cord for sealing between flanges. Contains petroleum distillate and graphite.
Picein. This is a classical vacuum wax – sticks well to metal and glass, is resistant to dilute acid and alkalies, is thermoplastic and made from bituminous substances.

1.5.9 Mineral oil lubricants

Rotary pump fluids. Highly refined oils with average molecular weights chosen to give suitable viscosities are generally used in rotary pumps. Such fluid can have a v.p. at normal temperature of 10^{-7} Torr rising to 10^{-4} Torr at typical pump temperatures of 50 to 60°C. However when the fluids are in operating pumps their ultimate pressures at room temperature are usually 10^{-3} to 10^{-2} Torr for two and single stage pumps respectively; the pressure readings refer to a thermal conductivity gauge. The oil molecules are degraded by frictional effects and emit fragments (Hockly and Bull, 1954) which can only be condensed at LN_2-temperature because of their volatility. For a review of the subject and methods of reducing backstreaming see Holland (1971). It should be noted that the parent molecules with average masses of ~500 amu evaporate the most slowly and have the lowest rate of back-diffusion through gas in the backing line.

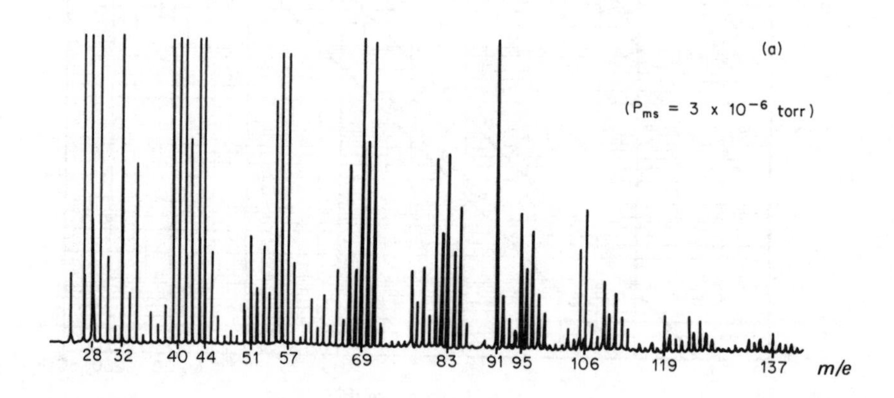

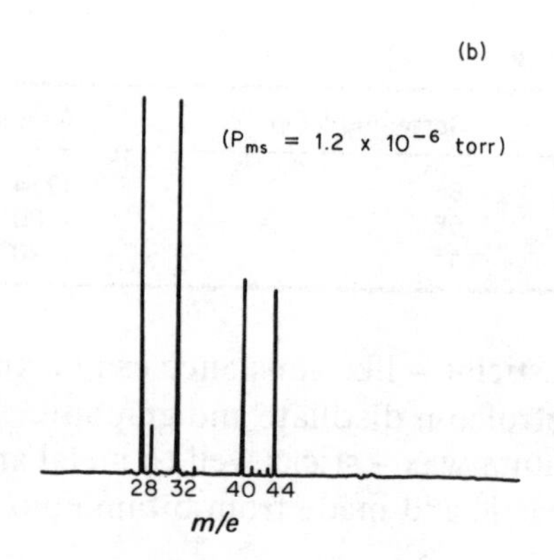

Fig. 1.12 Mass spectra of 75 l min^{-1} rotary pump, mineral oil 16, with pump exhaust connected via heated capillary to ion source of magnetic sector field instrument: (a) untrapped pump P_{ms} = 3 x 10^{-6} Torr; (b) with alumina trap P_{ms} = 1·2 x 10^{-6} Torr. Ion source background contained CO_2 (44) and CO (28) and trapped pump has added air-N_2 (28), O_2 (32) and Ar (40) after Holland (1971).

Shown in Fig. 1.12(a) is the mass spectrum of a rotary pump atmosphere resulting from decomposition processes both in the pump and the electron impact ion source; the reduction in backstreaming by the use of an alumina sorbent trap is shown in Fig. 1.12(b).

Corrosion inhibitors added to mineral oils can raise their vapour pressure and be selectively adsorbed to clean surfaces *in vacuo.*

Edwards mineral oil range (see also Table 1.31)
8A low viscosity for use in non-gas ballast pumps
15 medium viscosity for pumps > 2000 l/min (plain oil)
16 high vacuum oil for small gas-ballast pumps (plain oil)
17 high vacuum oil with corrosion inhibitors alternative to No. 15
18 high vacuum with corrosion inhibitors alternative to No. 16
29 for lubricating mechanical booster pumps.

TABLE 1.31 Vapour pressures of rotary pump oils, after Deville, Holland and Laurenson (1965)

Oil	Mol. wt. [average]	Viscosity cs (25°C)	Vapour pressure [Torr]			
			Temperature 50°C		Temperature 100°C	
			Manufacturer	Crystal*	Manufacturer	Crystal†
8A	410	65	5×10^{-5}	–	–	–
15	490	140	$4{\cdot}4 \times 10^{-5}$	9×10^{-4} to $4{\cdot}4 \times 10^{-4}$	3×10^{-3}	$1{\cdot}12 \times 10^{-4}$ to 6×10^{-5}
16	530	240	6×10^{-6}	$5{\cdot}6 \times 10^{-4}$ to $3{\cdot}87 \times 10^{-5}$	$4{\cdot}2 \times 10^{-4}$	–
17	500	170	$2{\cdot}3 \times 10^{-4}$	$4{\cdot}3 \times 10^{-4}$ to $9{\cdot}8 \times 10^{-5}$	$1{\cdot}4 \times 10^{-2}$	–
18	540	260	$1{\cdot}4 \times 10^{-4}$	$9{\cdot}6 \times 10^{-5}$ to $6{\cdot}5 \times 10^{-5}$	$1{\cdot}1 \times 10^{-2}$	$1{\cdot}54 \times 10^{-4}$ to $6{\cdot}65 \times 10^{-5}$

* Tube used as oil evaporator (value actually $a_e P$).
† Knudsen cell used as oil evaporator.

Vapour pressure and distillation. As the mineral oils used in rotary pumps are mixtures of molecules of different mass their evaporation rate *in vacuo* decreases as the light volatile fractions are emitted. This process occurs when a fluid is used in a pump and can be accelerated by gas-ballasting which expels light fractions. However fluid degradation enhances the backstreaming rate and although the ultimate pressure of a pump with a new oil charge decreases as the fluid is purified it finally reaches a value several orders above that of purified fluid.

Shown in Fig. 1.13 is the variation in condensation rate (proportional to evaporation rate) measured with a LN_2-cooled h.f. crystal microbalance above (1) a tubular evaporator and (2) a Knudsen cell; both contained mineral oil Edwards 18. There is an induction period of low evaporation rate (presumably during which surface impurities suppressing evaporation are removed) followed by emission of volatile components until the evaporation rate becomes constant. From evaporation/condenser geometry the v.p. of the oil

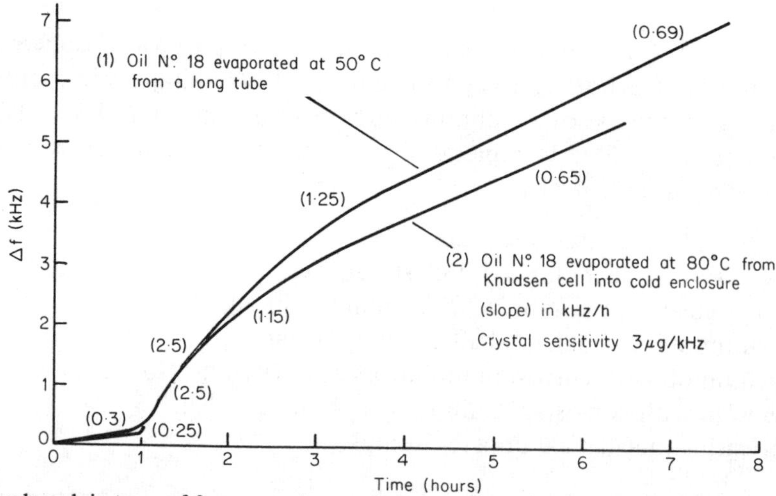

Fig. 1.13 Mass condensed, in terms of frequency change of h.f. (6 MHz) crystal, as a function of evaporation time for mineral oil 18, after Deville, Holland and Laurenson (1965).

can be calculated at a given temperature employing the Langmuir equation for free evaporation. One must use an average value for the molecular weight and as the latter must be estimated the v.p. is also an estimated value. Vapour pressures calculated for No. 18 and several other oils are given in Table 1.31 together with the manufacturer's values derived from the weight change of an effusion cell. The data presented shows that the v.p. of a fluid will depend on its previous history.

Backstreaming rates for different fluids – dependence on friction. The frictional degradation of organic fluids in a rotary pump results in a higher rate of backstreaming with limited influence of the v.p. of the parent oil. Shown in Table 1.32 are backstreaming rates measured by M. A. Baker in the writer's laboratory using a cooled h.f. crystal microbalance mounted on a two stage rotary pump. The pump was charged with fluids whose v.p. differed by at least 10^4 in magnitude, whereas the backstreaming rates are of comparable value. When the coefficient of friction of internal surfaces are reduced by use of additives or low friction materials the backstreaming rate is also reduced. Friction degradation would be reduced to a negligible level if the coefficient of friction of bearings and other contacting materials were less than 0·1 measured dry.

1.5.10 Hostile conditions

Oxygen. Mineral oil can explosively react with oxygen and when pumped a fire resistant fluid must be used in the rotary pump. Suitable fluids are estertritolyl phosphate or proprietary brands such as 'Cellulube 300'.* These fluids attack most elastomers and Viton A gaskets and valve seatings should be used. The inert fluorocarbons and fluorosilicones discussed below have been found useful when evacuating reactive gases and oxygen. Chlorotrifluoroethylene oils† can also be used because of their chemical inertness.

* A. Boake, Roberts and Co Ltd., and Amcel Co. Inc.–Celanese Corp. of America. Also 'Fyrquel' made by the Stauffer Chemical Co, New York, N.Y.

† Halocarbon Products Corp., Hackensack, N.J.

TABLE 1.32 Backstreaming rates for a 35 litre/s two stage mechanical pump (Edwards ED 35) when charged with different fluids, after M. A. Baker.

Oil or fluid	Backstreaming rate μ g cm^{-2} min^{-1}	Observations
Normal uninhibited mineral oil (Edwards 16 oil)	18	
Stripped mineral oil	18	
Mineral oil +1% MoS_2	4·7	
Mineral oil +1% Oleic acid	6·5	Oleic acid back-streaming evident
Stripped silicone fluid (DC 704)	10	Pump seized after $\frac{1}{2}$ h of operation
Polyphenyl ether (unstripped)	13	Difficult to start pump due to high viscosity
Apiezon C oil	15	

Acids. Leybold-Heraeus advise use of KS20 or Protelen (trade name) for exhausting corrosive vapours. The fluids are hygroscopic and form emulsions with water and a humid air gas-ballast cannot be used. The fluid pH should be checked with litmus paper to determine when it has become acid by exhaust gas reaction.
Radiation. Irradiation can degrade pump fluids. A radiation resistant oil is available from Shell – APL 731. This is a rust inhibited oil of medium viscosity suitable for radiation dosages up to 10^{18} neutrons cm^{-2}. The oil is an hydrocarbon which does not attack elastomers but rubber seals in a pump must be replaced by radiation resistant material.

1.5.11 Mineral oil lubricants for rotary seals

Apiezon J – a moderately viscous oil and Apiezon K – a very viscous oil can be used for gland seals. Their v.p.'s are given in Fig. 1.9. They have the following properties:

TABLE 1.33

	J	K
Density g/ml 20°C	0·918	0·916
Viscosity cS at 40°C	3330	5710
Av. molecular wt.	1130	1355
Thermal conductivity Wm^{-1}°C^{-1}	0·167	0·169
Specific heat 25°C joules/g	2·0	1·9

1.5.12 Mineral oil lubricants in dry stage pumps and compressors

Several types of rotary mechanical pump are in use which operate without oil seals, but invariably employ lubricated bearings. Typical examples are Roots blowers and axial flow compressors such as the turbomolecular (TM) pump. The Roots blower gives a large rate of displacement with a low compression ratio (backing to fineside pressure about 30 for air) whilst the TM typically gives a high compression ratio for O_2 and N_2 of between 10^6 to 10^9 but a low ratio in the region 10^2 to 10^3 for H_2. Blowers and TM pumps must be backed for exhausting to atmosphere and a rotary oil sealed pump is normally employed. The lubricated bearings in dry stage pumps are usually adjacent to their backing side. As degradation of the mineral oil lubricants can release hydrogen and light fragments from the rotary backing pump and dry pump bearings these can appear in the vacuum vessel depending on their respective compression ratios. Hydrogen is likely to have the highest partial pressure from back flow into the evacuated vessel, but when the dry stage pump is idle other oil components can backstream without the counter flow action of the pump.

Henning (1971) has shown (see Table 1.34) that the partial pressure of H_2 in a uhv-vessel exhausted by a TM pump is influenced by the grade of mineral oil used in the rotary pump. Not only is the residual atmosphere mainly composed of H_2 from this source but its pressure is increased depending on the oil chosen.

TABLE 1.34 Ultimate pressure and H_2 content of residual gas in a uhv-vessel exhausted by a turbomolecular pump using different mineral oils in the backing pump, after Henning (1971)

Oil test number	$P_u (\times 10^{-11}$ Torr)	% H_2
1	5·5	99·4
4	21·0	99·8
6	150	99·8

Holland (1972) has proposed using a fluorocarbonpolyether (discussed in section 1.5.16 below) as a lubricant in both the bearings of dry pumps and rotary backing pumps, because hydrogen is absent and the lightest degraded fragment CF has a mass of 31 a.m.u. which could not backstream readily. These fluids cannot be used with Al, Mg or their alloys if their abrading surfaces can reach high temperature (>300°C).

1.5.13 Esters of -phthalic, sebacic, and phosphoric acid

The work of C. R. Burch on petroleum distillates, discussed above, was followed by that of K. C. D. Hickman and his colleagues who investigated other oils and their behaviour in vapour stream pumps. Hickman and associates examined several esters which could be used in diffusion pumps and these are listed with their chemical formulae and trade names in Table 1.35 below; see also Fig. 1.8(b) and (c) for vapour pressure values. Although pumps with fractionating stages and purging facilities are now commonly employed their development was due to Hickman (1936), who recognized the need to prevent volatile and degraded fractions from reaching the high vacuum stage and if necessary to eject such components from the pump.

Esters are degraded by exposure to air whilst hot (tricresyl phosphate forms viscous tar-like deposits) and to water-vapour if exhausted in considerable quantities, e.g. butyl phthalate forms phthalic acid which is a solid crystalline substance. Latham et al. (1952) have described the use in diffusion pumps of tri-cresylphosphate (m- and p- mixed) and tri-xylenyl phosphate. Esters are used as plasticizers in plastics and elastomers and in contact with these materials can sometimes be absorbed causing swelling; fluorinated elastomers such as Viton A appear to be inert.

TABLE 1.35 Composition and properties of esters*

Name	Formula	Trade name*	Molecular weights	Molecular diam. (Å)	Specific gravity 25/25°C	Viscosity 26·7°C cS	B.P. (°C) at 0·5 Torr	Ultimate pressure 25°C/Bendix CVC GF.-26 pump/Torr		
n-dibutyl phthalate	$C_6H_4(COOC_4H_9)_2$	–	278·3	8·54	1·0435	14·4	135	$4 \cdot 10^{-5}$		
i-diamyl phthalate	$C_6H_4(COOC_5H_{11})_2$	Amoil†	306·2	8·90	–	–	–	–		
Di-2 ethylhexyl phthalate	$C_6H_4(COOC_8H_{17})_2$	Octoil	390·5	9·78	0·9827	51·5	183	$2 \cdot 10^{-7}$		
n-dibutyl sebacate	$C_8H_{16}(COOC_4H_9)_2$	–	314·3	9·25	–	–	–	–		
i-diamyl sebacate	$C_8H_{16}(COOC_5H_{11})_2$	Amoil S†	343·3	9·55	–	–	–	–		
Di-2 ethylhexyl sebacate	$C_8H_{16}(COOC_8H_{17})_2$	Octoil S	426·7	10·32	0·9122	18·2	199	$5 \cdot 10^{-8}$		
							B.P. (°C) at—Torr 1	10^{-3}	10^{-6}	V.P. at 25°C Torr
tri-m-cresyl phosphate	$(CH_3C_6H_4)_3PO_4$	–	368·36	–	–	–	~200	110	43-50	$\sim 5 \cdot 10^{-5}$
tri-p-cresyl phosphate	$(CH_3C_6H_4)_3PO_4$	–	368·36	–	–	–	212	116	52	$2 \cdot 10^{-5}$

* These fluids originally products of Distillation Products Inc. are now prepared by CVC Bendix Scientific Instruments & Equipment *Plant*. The table has been compiled from data published by the latter company and Dushman (1949).
† Not in current catalogue.
Shown in Fig. 1.8(c) is the v.p. as a function of temperature of a range of esters and other fluids, available from CVC Bendix Scientific Instruments and Equipment.

1.5.14 Polyphenyl ethers

Hickman (1960) first reported that polyphenyl ethers, developed as high temperature lubricants, could be used in diffusion pumps. A five ring ether (i.e. five phenyl groups linked by oxygen bonding) is currently used and this should have a mass of 446 a.m.u. for a mono-molecular fluid. The physical properties of two products are given in Table 1.38 as reported by their suppliers. A six ring ether is too viscous for the condensate on a water-cooled trap to return to the pump. Although the fluid is a high temperature lubricant it undergoes frictional degradation in a rotary vacuum pump and the backstreaming is not reduced compared with that from mineral oil.

TABLE 1.36 Properties of polyphenyl ether pump fluids

Fluid	Convalex-10*	Santovac 5†
Molecular wt.	454 (average)	446
Specific gravity 25°C/25°C	1·2	–
Density g/ml (20°C)	–	1·204
Viscosity cS (°C)	1000 (26·7) 363 (37·8)	16 (93·3) 2·1 (204)
Refractive index	1·630	1·6306 (N_D^{25})
Heat of vap. kcal/mole	25·3	21·97 (354–461°C)
v.p. Torr (25°C)	$2 \cdot 10^{-9}$	5×10^{-9}
b.p. (°C) at 0·5 Torr	275	275

* CVC Bendix, Scientific Instruments and Equipment Div.
† Monsanto Chemicals Ltd/Edwards Vacuum Components.
Note: Use of polyphenyl ethers as diffusion pump fluids is subject to US pat. 3,034,700.

The fluid can be oxidized if exposed whilst hot to air and in this respect is inferior to silicone. However, when heated *in vacuo* or exposed to electron bombardment it is more resistant to decomposition than silicones and particularly mineral oils. However it can form electrically conducting coatings when thermally or electron impact degraded, but adsorbed fragments are removed by baking. The mass spectrum of Santovac 5, from electron impact ionization in a magnetic sector field instrument, is shown in Fig. 1.14. The spectrum was prepared in the writer's laboratory by Cleaver and Fiveash (1970). The principal peaks at 446 and 447 a.m.u. coincide with those of the parent molecule and its isotope respectively. The relative peak heights of the fragmentation spectra remained unchanged for the fluid evaporation temperatures studied (up to 320°C) but raising the ionizing electron energy from 70 to 94 eV changed the relative peak heights. Convalex 10 gave a similar spectrum but with additional weak peaks at 470 and 471 a.m.u. and slightly different relative peak heights. Bultemann and Delgmann (1963) earlier reported similar spectra for polyphenyl ethers.

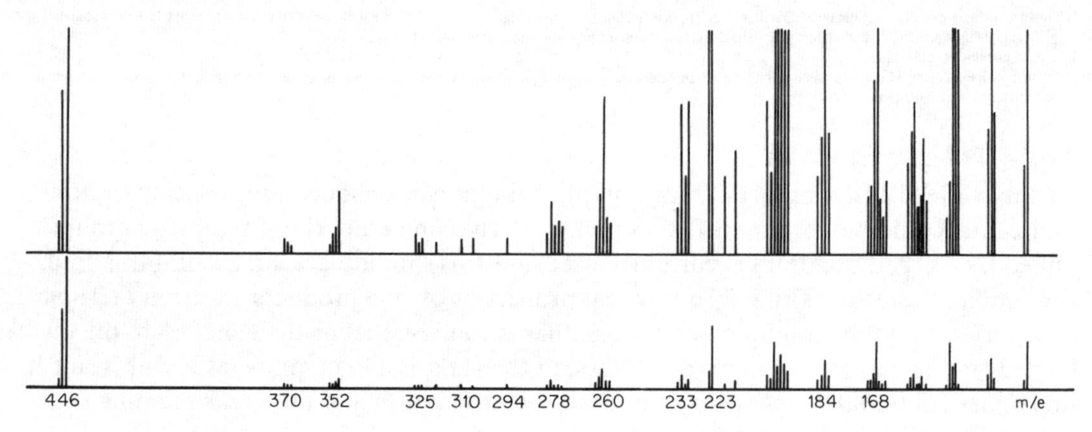

Fig. 1.14 Mass spectrum of Santovac 5 pump oil. Pressure 3×10^{-7} Torr, electron energy 70 eV, ion source temperature 170°C, after Cleaver and Fiveash (1970).

1.5.15 Silicones and related compounds

Production of silicone compounds commenced with work of F. S. Kipping about 1900. E. G. Rochow developed a process for producing alkyl chlorosilanes used for making many of the silicone products available. Silicone fluids were used in vapour stream pumps at the end of the last war. Initially dimethyl polysiloxanes fluids were employed and later methyl phenyl silicones of low vapour pressure were formulated. Silicone fluids in vapour stream pumps can be exposed whilst hot to air with negligible oxidation at temperatures up to 250°C. However, if either adsorbed molecules are or the parent fluid is electron bombarded insoluble polymer films are formed. These are electrically insulating and optically transparent or absorbing depending on the bombardment conditions; the presence of oxygen tends to produce silica deposits. Silicone molecules are chemisorbed to silicate glasses developing strongly hydrophobic surfaces if adsorbed hydroxyl groups are not present to shield the surface. Baking glass at ⩾250°C removes (OH)-group and silicone adsorption can proceed.

Diffusion pump fluids

Dimethylpolysiloxane fluids. These substances are based on compounds with the formula

$$(CH_3)_3SiO-[(CH_3)_2SiO]_n-Si(CH_3)_3.$$

The production method produces a range of molecular weights and pump fluids are fractional distillates to reduce the mass spread and volatility. Linear polymers for which $n \geqslant 5$ have sufficiently low vapour pressure at room temperature for use in vapour stream pumps, whilst providing adequate vapour pressure for moderate boiler temperatures i.e. ~175°C at $n = 5$. Silicone 702 with an average molecular weight of 530 corresponds to $n = 5$. Dushman (1949) has given v.p. values for linear polysiloxane fluids in relation to n, but his results suggest that n should equal twelve to obtain a v.p. characteristic similar to that of 702. This would give a molecular weight many times the average measured and it is concluded that the silicone studied must have contained appreciable light fractions.

Some vapour pressure characteristics of DC* 702 and 703 fluids are given in Tables 1.37 and 1.38. For economy, the widely used diffusion pump fluid is 702.

TABLE 1.37 B.P. values for linear dimethyl polysiloxanes

	Temperature °C v.p. Torr–0·01	0·1	0·5
DC 702	115	150	175
DC 703	145	180	205

The saturated v.p. of the fluid follows the equation

$$\log_{10}P = 10{\cdot}3 - \frac{4820}{T} \quad \text{where } P = \text{Torr},\ T = {}^\circ\text{K}.$$

Working in the writer's laboratory M. A. Baker (1971) has measured the v.p. of typical batches of fluids using GLC to determine the content of light ends, his results are in Table 1.38.

TABLE 1.38 Vapour pressure changes for silicone MS 702 with batch and free evaporation

	Batch no.	Initial v.p. (22°C) Torr	After 2 days *in vacuo* (22°C) Torr	% light ends
MS 702	(1)	7×10^{-6}	1×10^{-7}	0·1
MS 702	(2)	3×10^{-7}	1×10^{-7}	0·1
Defective	(3)	1×10^{-4}	1×10^{-5}	4·0

* Silicone fluids are produced by Dow Corning (designated DC) and Midland Silicones (designated MS). Although basically similar there can be small composition differences, e.g. in the presence of light fractions and additions to 'single molecule' type fluids, such as 704 or 705, to prevent crystallization.

TABLE 1.39 General properties of silicones in common use

Pump fluid	702	704	705
Average molecular weight	530	484	546
Pressure Torr at 15°C	10^{-6}	10^{-8}	10^{-10}
Specific gravity at 25°C	1·09	1·07	1·09
Viscosity at 25°C cS	45	40	175
Flash point °C	193	210	243
Boiling point* 0·5 Torr °C	180	215	245

* Typical operating temperature of pump boiler.

Tetra-methyl tetra-phenyl tri-siloxane (DC or MS 704). The mass spectrum is in Fig. 1.15

$$\log_{10} P = 11{\cdot}025 - \frac{5570}{T}, \quad \text{where } P = \text{Torr},\ T = {}^\circ\text{K},$$

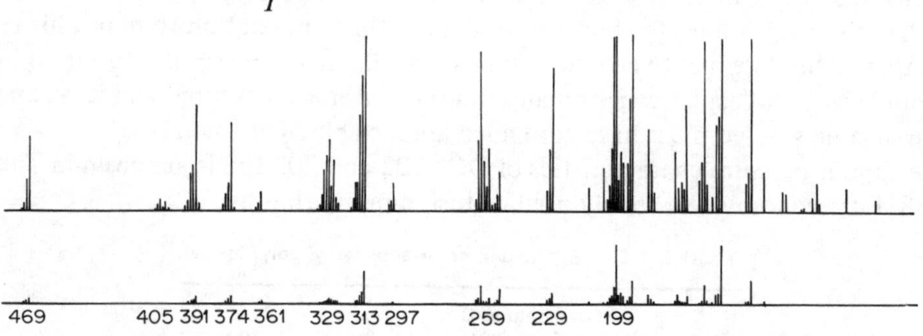

Fig. 1.15 Mass spectrum of Silicone MS 704. Pressure 1 x 10^{-7} Torr, electron energy 70 eV, ion source temp 150°C.

Principle ions in the mass spectrum of Silicone MS 704

Mass (amu)	Percentage major peak	Mass (amu)	Percentage major peak
499	< 1	313	5·5
483	< 1	259	36
469	7·7	229	20
405	2·3	213	2·5
391	10	199	100
329	7		

Tri-methyl penta-phenyl tri-siloxane (DC or MS 705). This is a uhv-fluid producing vacuum in the 10^{-11} Torr region with suitable traps and pump design. Shown in Fig. 1.16 is the v.p. of silicone 705 determined by Deville, Holland and Laurenson (1965) with that reported by Crawley et al. (1962).

The vapour pressure follows the relation

$$\log_{10} P = 12{\cdot}31 - \frac{6490}{T}, \quad \text{where } P = \text{Torr},\ T = {}^\circ\text{K}.$$

Silicones for lubrication and general use. Silicone fluids used in vapour pumps are unsuitable as lubricants for metals. Oxide formation on steel which reduces friction can be inhibited in presence of silicones which are very poor boundary lubricants. Tingle (1948)

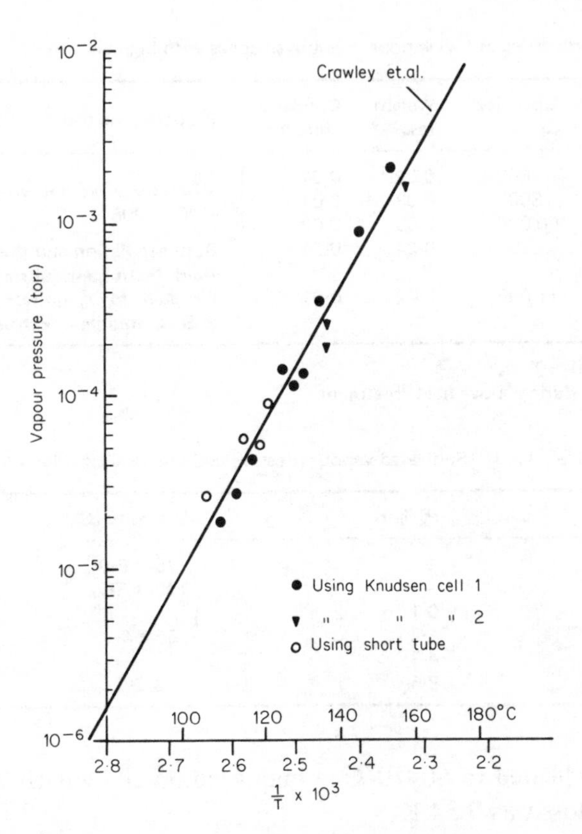

Fig. 1.16 Vapour pressure of MS 705 determined by Deville, Holland and Laurenson (1965) using a Knudsen cell with orifices 13·7 and 6·3 mm dia. and a tubular evaporator 13·7 mm diameter x 25 mm long. The full curve is from Crawley et al. (1962).

reported that silicones from 20 to 1000 cS (20°C) viscosity did not show much change in boundary friction between metals but a copolymer of dimethyl and diphenyl siloxane gave satisfactory results failing at 100°C. High viscosity fluids can operate as hydrodynamic lubricants for high temperature use and be superior to mineral oils, but if silicones are decomposed SiO_2-products can form with abrasive action.

Low v.p. fluids. A range of silicone fluids with various physical properties and low evaporation rate *in vacuo* have been prepared by the Dow Corning Corp. The polymer types used are those described above with viscosities appropriate to their application as shown in Table 1.40 (after Kummins 1970).

Silcodyne H (ICI) is a pure methyl chloro-phenyl polysiloxane fluid which is a good lubricant with a low rate of change of viscosity with temperature and oxidation and fire resistance. It can be used in vacuum pumps operated down to –50°C. Silcodyne H contains molecules with weights ranging from 800 to 6000; about 50% have molecular weights between 2500 and 3500. Thus the saturated vapour pressure values given in Table 1.41 only refer to equilibrium conditions. When free evaporation occurs *in vacuo* the light fraction will initially be preferentially removed.

TABLE 1.40 Physical properties of low vapour pressure silicones with high viscosities

Dow Corning fluid	Viscosity cS	Weight loss %*	Condensables %†	Property and use
Polydimethyl siloxane	600	0·02	0·01	Viscosity curve flat with temperature usable −40 to 205°C
	8000	0·03	0·01	
	50 000	0·05	0·02	
Polyphenyl methylsiloxane	800	0·04	0·04	Better radiation and thermal resistance than above fluid. Instrument lubricant. Base for grease
Poly (3,3,3-trifluoropropyl) methyl siloxane	11 000	0·02	0·01	Resistant to O_2-ignition. Excellent lubricant −40 to 205°C. Soluble in ketones and fluorocarbons only

* After 24 h at 125°C and 10^{-6} Torr.
† Collected on plate at 25°C during above heat treatment.

TABLE 1.41 Saturated vapour pressure and viscosity of Silcodyne H

°C	P_s Torr	Viscosity cS
−40	—	75–1150
0	—	130–170
50	0·1	50–54
150	1·2	
230	4·3	
315	8·5	2–2·5

When the fluid was heated to 60–70°C in high vacuum and volatiles condensed on an LN_2-trap the fluid weight loss was 0·34%.

Of available elastomers only Fluorosilicone and Viton have an operating range approaching that of Silcodyne H. Viton can be used up to 200°C in contact with the fluid. At 250°C Viton normally begins to embrittle and the rate is enhanced by contact with Silcodyne H. J. S. Olejniczak (1967) working in the writer's laboratory studied the behaviour of Silcodyne fluid H in a rotary pump (ED35) obtaining the results given in Table 1.42. A test was also made with Silcodyne M a chemically similar but more thermally stable fluid than H. This gave a Pirani gauge reading of 45 milliTorr after 15 min on gas-ballast stabilizing to 22 milliTorr and 0·7 milliTorr (Mcleod); after opening gas-ballast the Mcleod gauge stabilized

TABLE 1.42 Dependence of rotary pump ultimate pressure on pre-treatment of Silcodyne H

Silcodyne fluid	Duration (h)	Air-ballast Used	Air-ballast Not used	Pressure (milliTorr) Pirani	Pressure (milliTorr) Mcleod
H – as received	4		✓	9·5	4
	4	✓		9	1·2
	48		✓	8	0·4
	50	✓		9	3
H – 50% distilled *in vacuo*	21		✓	6	0·7
	21	✓		7	2·5
	45		✓	5	0·5
	50	✓		5	2·2

at 3 milliTorr. Fluid M was then distilled *in vacuo* with removal of 35% of fluid. Within 8 h the Pirani reading was 10 milliTorr, the Mcleod 0·5 milliTorr and a Penning gauge gave 3 milliTorr.

The distilled fluid must have given an ultimate pressure above its vapour pressure at the pump temperature because of frictional promoted degradation. Both the H and M fluids could be distilled to provide liquid which could be used in diffusion pumps; the distillate fractions had molecular weights of about 2400 which is several times that of silicone 705.

Fluorosilicone fluids

The lubricity of fluorosilicone fluids for steel moving on steel is claimed to be superior to that of silicone fluids and to be better than that of mineral oils at high loads and high speeds; they can be used in rotary vacuum pumps. It was found in the writer's laboratory that a fluid polymerized, as with silicones, when electron bombarded. Such fluids can be used from – 40 to 205°C. D.C. fluid (FS–1265) 300 cS is resistant to ignition in oxygen. The poly (3,3,3-trifluoropropyl) methyl siloxane is available from the Dow Corning Corp.

Greases

These are usually mixtures of a silicone compound with a silica filler. Formulations based on compounds discussed above can be used to give desired properties, e.g. lubrication, radiation resistance etc.

Vacuum sealents. Low volatility silicone compounds are available for sealing leaks and potting devices to be used in vacuum.

Optical properties

The optical constants n and k of silicones 704 and 705 have been determined in the vacuum ultraviolet by (1) Kerr et al. (1971) with the fluids in an open dish and reflectance measured as a function of incidence angle in the energy range 4–24·8 eV and by (2) Sowers et al. (1971) using critical angle measurements in the range 2–10·6 eV with the liquids in a closed cell with a sapphire window.

1.5.16 Perfluoropolyethers

Composition. A new class of fluids, lubricants and evaporants, based on the perfluoropolyethers have been extensively studied for vacuum use by Holland and his colleagues. The fluids are formed by u.v. irradiation of condensed hexafluoropropylene in the presence of oxygen with post-treatment in fluorine. They have the polymer form shown in Fig. 1.17 and the properties given in Table 1.43. As the atomic weight of fluorine is 19 and that of hydrogen 1 each [$-CF_2-$] chain unit has a mass of 50 a.m.u. compared with 14 for [$-CH_2-$]. Thus perfluoropolyether fluids usable in vacuum pumps have molecular masses of 2000 to 7000 a.m.u. compared with 350 to 500 a.m.u. for organic fluids. Perfluoropolyethers were first examined as a rotary pump lubricant and seal using a product developed by Montecatini Edison S.p.A. termed 'Fomblin'.* The fluid (Fomblin YR) studied was a

* Similar compounds with the trade name 'Krytox' are marketed by E. I. Du Pont De Nemours & Co.

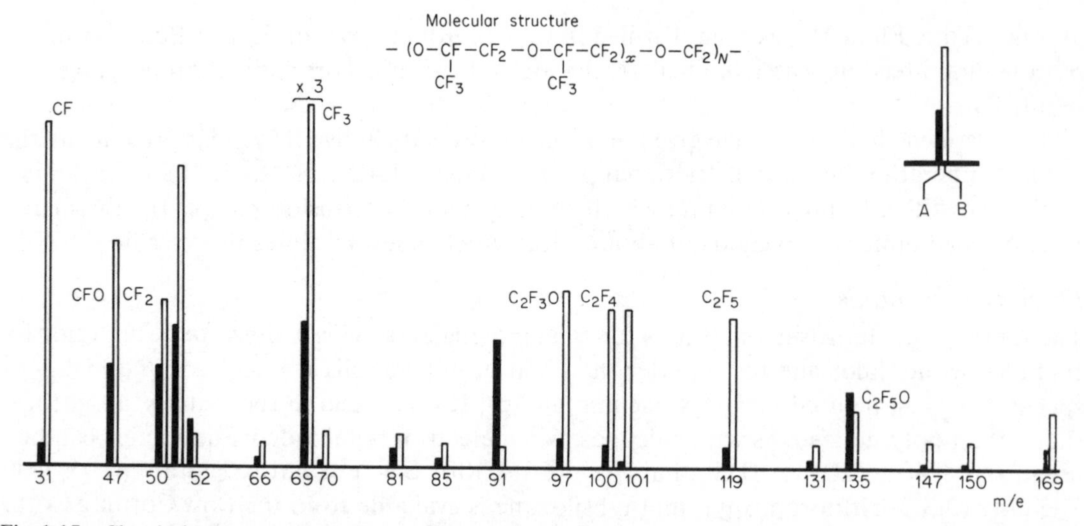

Fig. 1.17 Chemical structure and fragmentation spectra of perfluoropolyether (Fomblin). A. Atmosphere above vapour stream pump at a pressure of $2{\cdot}5 \times 10^{-6}$ Torr. B. Vapour admitted to ion source from fluid at 190°C. The mass range examined was limited to that shown, after Holland, Laurenson, Baker and Davis (1972).

TABLE 1.43 Physical properties of proprietary perfluoropolyethers

		'Fomblin'*			'Krytox'†		
Fluid type no.		YO4	Y25	YR	143AZ	143AY	143AD
Viscosity cS	−40°C	–	–	–	8000	–	–
	20°C	35	250	1–2000	–	–	–
	205°C	–	–	–	0·8	1·4	6·0
Density g/ml	20°C	1·87	1·90	1·91–92	(25°C-) 1·86	1·88	1·91
°C temp. range (sat. v.p. = 0·8 Torr)		–	–	–	143–185	210–217	‡
Surface tension dynes cm^{-1}	20°C	19	20	21	–	–	–
	27°C	–	–	–	16	17·7	19·3
Refractive index N_D	20°C	1·2945	1·300	1·304	–	–	–
	25°C	–	–	–	<1·3	<1·3	1·301

* Montecatini Edison SpA - YR is a distillate residue suitable for use in rotary pumps or blended to reduce its viscosity. The lower molecular weight fraction (Y16) distilled from Y25 can be used in a diffusion pump to give a performance comparable to silicone 702, but adjustment in boiler and condenser temperatures is necessary with some pumps for stable pumping.
† E. I. Du Pont De Nemours & Co.
‡ Decomposition before distillation complete.

distillation residue with an average molecular weight of ~6500 a.m.u. In use it gave similar backstreaming rates to conventional fluids, but it could be employed for pumping reactive compounds, such as fluorinated gases, and oxygen without reaction. Exposed to rubbing friction between steel components the fluid was partially degraded, as occurs with organic fluids, but the emitted components CF_2(50), CF_3(69), C_3F_3(93) are much heavier and can have different backstreaming behaviour. Perfluoropolyethers are thermally stable up to 350°C but the temperature limit is lowered in the presence of oxygen and oxide coated metals to about 300°C. The fluid can react with aluminium and magnesium and their alloys if their surfaces are exposed to mechanical shear.

It is possible to operate a rotary pump charged with a perfluoropolyether at high temperature thereby preventing condensation when evacuating vapours whose saturated pressure at the elevated temperature is higher than the exhaust pressure; one should use a distillate residue with a low vapour pressure at the working temperature. Thus gas-ballast would not be necessary when exhausting water vapour with a pump temperature $>100°C$. For a discussion of this work see Baker, Holland and Laurenson (1971). L. Laurenson has found that less viscous (lower molecular weight distillates) of the fluid can be used in large pumps to avoid a high starting torque.

Bearings and dry pumps. It has been shown earlier that mineral oil lubricants emit low molecular weight fragments and hydrogen when in contact with rubbing metal surfaces developing heat by friction. Holland (1972) has proposed that the perfluoropolyether is an ideal fluid for use in both turbomolecular pump bearings and in oil sealed rotary pumps used for backing. A perfluoropolyether cannot emit hydrogen from either source to backflow through the turbomolecular pump. Also backflow of any degraded fragments should be negligible during operation of such a turbomolecular pump because the compression ratio of typical pumps rises with the molecular weight of the gas. Fragments from a fluorocarbon oxide are greater in mass than those from a hydrocarbon, e.g. the smallest fragment CF has a mass of 31 a.m.u. The fluids should also be valuable for lubricating the bearings of Roots blowers and similar dry stage mechanical pumps.

Resistance to electron polymerization. When exposed to electron bombardment the polyfluoropolyether molecule is fragmented but *neither the liquid, vapour not condensed molecules of the perfluoropolyether can be polymerized when electron bombarded and insoluble material is not left in the oil,* see Holland, Laurenson, Baker and Davis (1972), and Holland, Laurenson and Baker (1972). Shown in Fig. 1.17 are the mass spectra of a Fomblin fluid obtained with a nude electron impact ion source exposed to (a) backstreaming vapour in a vessel evacuated by a diffusion and rotary pump charged with the fluid and without traps and (b) fluid directly evaporated into the ion source. The two fragmentation patterns are similar suggesting there is limited fluid degradation in the pump, i.e. vapour molecules are degraded in the ion source. (u.v. irradiation can cause breakdown but not further polymerization.) The absence of polymer growth separates the perfluoropolyether from the organic and silicone fluids in common vacuum use. *Note* fluorosilicones were found to polymerize when exposed to electron bombardment in the writer's laboratory. The polymerization resistance of the fluorocarbon oxide has important consequences for maintaining clean conditions in vacuum devices using electron beams; see for example the paper by Ambrose, Holland and Laurenson (1972) reporting prevention of specimen contamination in an electron microscope using Fomblin in both rotary and diffusion pumps. The perfluoropolyether could be degraded and made to form reactive gas (HF) when exposed to an r.f. discharge in hydrogen. Also, deposits have been observed in Penning gauge heads exposed to saturated Fomblin vapour.

Diffusion pump fluid. Several diffusion pumps of different size have been tested with a range of perfluoropolyether (Fomblin) fluids in the writer's laboratory. The work showed that pumping speeds and backing pressures comparable to those obtained with organic and silicone fluids in the same pumps could be obtained; results of the initial work giving a usable pumping performance was reported by Holland, Laurenson and Baker (1972). The perfluoropolyethers have a range of molecular weights after polymerization and must be fractionally distilled. The choice of distillate 'cut' or molecular weight spread determines both vacuum performance and fluid cost. Because of the nature and scale of current production the fluids are expensive, but only small charges are required for diffusion pumps in 100 l s^{-1}-speed range.* A mineral oil can be used in the rotary backing pump but it is essential to provide an effective trapping agent, e.g. alumina.

Greases. These are mixtures of a fluorocarbon oxide fluid and PTFE particles to enhance viscosity. Both oils and greases can be used for lubricating mechanisms and valve components operated *in vacuo.* They should be used only after careful test on aluminium or magnesium if high shear occurs, because of reaction hazard. Fomblin T greases are made by Montecatini Edison S.p.A. and Krytox 240 greases by E. I. Du Pont De Memours & Co.

1.5.17 Backstreaming characteristics of vacuum pumps

Manufacturers' catalogues sometimes list data on the backstreaming characteristics of pumps in terms of mass condensed per unit area in a given time. Apart from the problems of measuring the combined backstreaming rate of gas and vapour for well trapped pumps in the 10^{-9} Torr region (the surface impingement rate is equivalent to about a monomolecular layer per hour if the sticking coefficient is unity) the rate may vary with evacuation conditions and position of sensing head in the vessel. When other vapours are present from system desorption, e.g. H_2O, measurement is further complicated.

Effect on backstreaming rate of pipe conductance, gas flow and pressure. The rate of backstreaming from a rotary mechanical pump is limited by the conductance of the connecting pipelines as shown by Fig. 1.18 for a 900 l min^{-1} single stage pump charged with No. 16 mineral oil. The backstreaming rate is also reduced by raising the gas pressure and further reduced if the gas is in motion because of pumping, as then the emitted components must diffuse through a counter-flowing gas. Under certain conditions of pipe length, gas pressure and flow, the backstreaming rate in the vessel can be reduced to zero as shown in Fig. 1.19 for static and flowing gas conditions using a 75 l min^{-1} rotary pump with No. 16 mineral oil (Holland 1971).

Backstreaming from diffusion pumps. The backstreaming rate of a diffusion pump can change rapidly with pumping conditions and position above the pump inlet. Power and Crawley (1954) using a mechanical microbalance observed changes in the rate during pump warm-up and cool-down when 'lazy' vapour of low intensity enters the cold jets and more easily reaches the vessel by gas collision. These variations are easily observed with a LN_2-cooled h.f. crystal microbalance as shown in Fig. 1.20 after Baker (1968). The three peaks in the warm-up period appear to be related to the vapour feed to the three jet stages in the vertical column of the pump.

* The use of fluorocarbon oxide liquids in vacuum systems must be evaluated in stages. Molecular fragments did not appear to affect electron emission from bare tungsten emitters, but the fluorine released could poison oxide emitters. Holland et al. (1973) have described their behaviour in ion sources for accelerators.

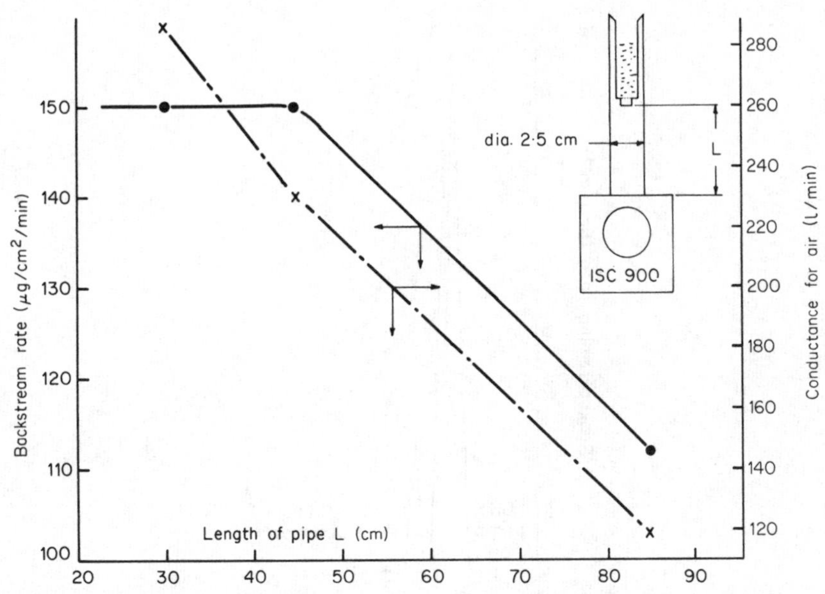

Fig. 1.18 Effects of pipe conductance on backstreaming rate of rotary pump (900 l/min) charged with mineral oil, and operating at pump ultimate pressure after Holland (1971).

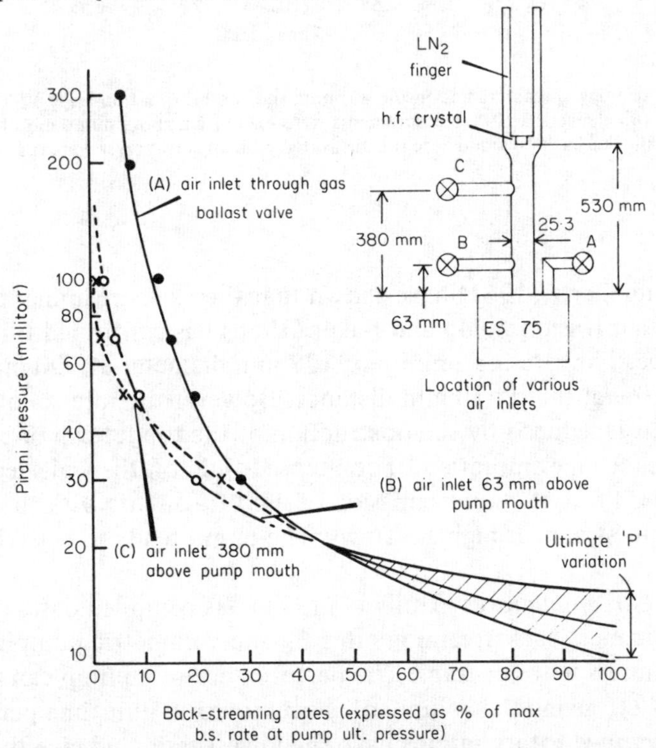

Fig. 1.19 Backstreaming rates as a function of pressure during gas flow into a 75 l/min rotary pump charged with No. 16 oil. Air inlets are in the pump line. Rates measured with a stationary gas in the same pressure range are plotted for comparison; the stationary gas pressure was raised by increasing the air-ballast flow to degrade the ultimate pressure (Holland, 1971).

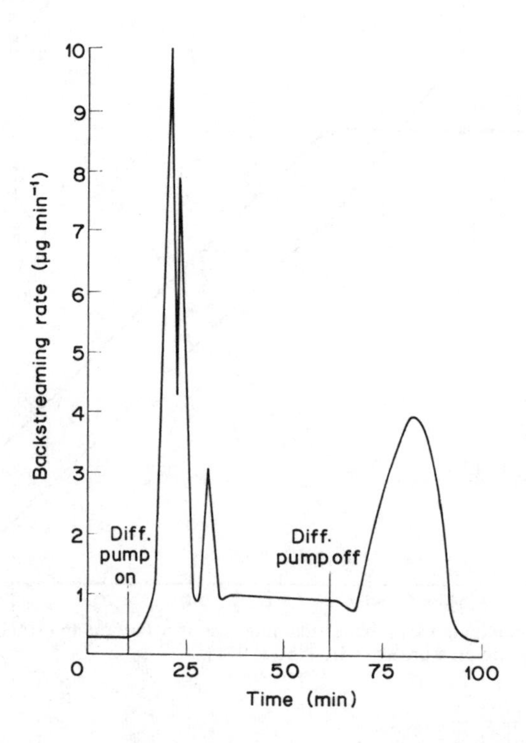

Fig. 1.20 Typical backstreaming characteristics during warming and cooling of a four-stage 12·7 cm diameter self-purifying diffusion pump (oil charge MS 704). Measurements were made 1 cm above pump mouth and mid-way between jet assembly and pump wall. The backstreaming rate was measured with an active crystal area of ~1 cm² (Baker, 1968).

Hablanian and Steinhertz (1961) have shown that the backstreaming rate falls of rapidly from the mouth of a diffusion pump and Baker (1968) has confirmed this with an h.f. crystal microbalance. The rate for a pump of 127 mm diameter at 100 mm from the pump mouth was 10% of that at 10 mm distance. However the rate is not uniform across the pump orifice but is reduced by the obstruction of the top jet cap or cooled guard cap. Backstreaming arises from evaporation of condensed fluid on the walls, evaporation from droplets condensed on the jet stages, emission of light fragments with the vapour and disruption of the vapour stream at high gas flows. The pump tends to behave as an annular emitting source.

Comparison of backstreaming rates. Holland (1971) has compiled data, reproduced in Table 1.44 giving the mass backstreaming rates Q_b and volumetric pumping speeds S of a range of pumps with and without traps. The performance of a pump can then be judged in terms of the ratio S/Q_b giving the volume of gas removed in unit time per unit of backstreaming rate. *Untrapped* rotary pumps have the lowest ratio and give the highest rates of contamination. Of course they operate in a different pressure region to vapour stream pumps.

TABLE 1.44 Showing number of litres of gas pumped per microgram of backstreaming products

Pump/°C	Fluid/trap	(Air S† (l/min)	Q_b max (μg/min) Total	S/Q_b (l/μg)	P (ultimate) 20°C (Torr)	VP 20°C (Torr)
Rotary pumps					*Pirani*	
Single stage/57 (ES 75)	Mineral (16)/none	75	180	0·42	10^{-2}	2×10^{-8}
Single stage/50	Mineral (16)/none	900	750	1·2	5×10^{-3}	2×10^{-8}
Two stage/60 (ED 35)	Mineral (16)/none	35	90	0·39	5×10^{-3}	2×10^{-8}
Two stage/60 (ED 35)	Mineral (16)/alumina trap	35	<0·1	>350	$< 10^{-3}$	–
Two stage/60 (ED 35)	Apiezon C/none	35	75	0·47	5×10^{-3}	1×10^{-8}
Two stage/60 (ED 35)	–Polyphenyl ether/none	35	75	0·47	–	5×10^{-10}
*Diffusion pumps**					*Ion gauge*	
EO4 + Guard ring	Silicone 704/none	36 000	2500	14·4	5×10^{-7}	9×10^{-9}
	Silicone 704/chevron 15°C	14 400	10	1440	5×10^{-8}	
EO4 + Guard ring	Silicone 705/none	36 000	1800	20·0	5×10^{-8}	5×10^{-10}
	Silicone 705/chevron 15°C	14 400	1	14 400	5×10^{-10}	
EO4 + Guard ring	‡Polyphenyl ether/none	36 000	1600	22·5	5×10^{-10}	3×10^{-10}
	‡Polyphenyl/chevron 10°C	14 400	1	14 400	2×10^{-10}	
EO4 + Guard ring	Apiezon C/none	36 000	1800	20·0	5×10^{-7}	1×10^{-8}
Diffusion pump (22·5 cm diam.)	Silicone 704 No Guard ring/none	90 000	$2·8 \times 10^5$	0·32	–	–
Diffusion pump (22·5 cm diam.)	Silicone 704 + Guard ring /none	90 000	2500	36·0	–	–

* All pumps are water cooled fractionating with self-purifying ejector stage and guard ring above top jet.
† Note. The diffusion pump speeds are given in l/min for backstreaming comparison.
‡ Santovac 5.

In the uhv-region the backstreaming rate can decrease to less than a monomolecular layer per hour. Holland, Laurenson and Priestland (1963) detected such low growth rates by electron induced polymerization on the anode of an electron gun measuring the deposit thickness by multiple beam interferometry. They showed that oil condensed on a trap could badly contaminate a vessel during bakeout to accelerate degassing.

REFERENCES

AMBROSE, B. K., HOLLAND, L., and LAURENSON, L. (1972) *J. Microscopy*, **96**, 389–391.
BAKER, M. A. (1968) *J. Phys. E. Scient. Instrum.*, **1**, 774–776.
BAKER, M. A., HOLLAND, L., and LAURENSON, L. (1971) *Vacuum*, **21**, 479–481.
BLEARS, J. (1944), *Nature*, **154**, 20.
BLEARS, J. (1947) *Proc. Roy. Soc. Lond. A*, **188**, 62–76.
BULTEMANN, H. J., and DELGMANN, L. (1963) *Trans. Europaisches Symp. Vacuum*, DAGV, 213–222 (Rudolf A. Lang, Verlag-Esch (Taunus)).
BURCH, C. R. (1928) *Nature*, **122**, 729.
BURCH, C. R. (1929) *Proc. Roy. Soc. Lond. A*, **123**, 271.
BURROWS, G. (1946) *J. Soc. Chem. Ind.*, **65**, 360.
BURROWS, G. (1960) 'Molecular distillation' (Oxford University Press).
CLEAVER, J. S., and FIVEASH, P. N. (1970) *Vacuum*, **20**, 49–54.
CRAWLEY, D. J., TOLMIE, E., and HUNTRESS, A. (1962) *Trans. 9th Nat. Vac. Symp. A.V.S.*, 399–403 (Macmillan and Co.).

DEVILLE, J. P., HOLLAND, L., and LAURENSON, L. (1965) *Trans. 3rd Int. Vacuum Congress,* 153–160 (Pergamon Press, Oxford).
DUSHMAN, S. (1949) 'Scientific foundations of vacuum techniques', Chapter 5 (J. Wiley Inc., New York). (Note 2nd edn., Lafferty, J. M., 1962, omits most of the v.p. data on oils and greases.)
ESPE, W. (1968) 'Materials of high vacuum technology', 3, 'Auxiliary materials' (Pergamon Press, Oxford).
HABLANIAN, M. H., and STEINHERTZ, H. A. (1961) *2nd Int. Vacuum Congress, A.V.S.,* **1**, 33–41 (Pergamon Press, Oxford).
HENNING, J. (1971) *Vacuum,* **21**, 523-526.
HICKMAN, K. C. D. (1936) *J. Franklin Inst.,* **221**, 215, 383.
HICKMAN, K. C. D. (1960) *Nature,* **187**, 405.
HICKMAN, K. C. D. (1961) *Trans. 8th A.V.S. Ann. Symp.,* **1**, 307-314.
HOCKLY, D. A., and BULL, C. S. (1954) *Vacuum,* **4**, 40–47.
HOLLAND, L. (1971) *Vacuum,* **21**, 45-54.
HOLLAND, L. (1972) *Vacuum,* **22**, 234.
HOLLAND, L., LAURENSON, L., and BAKER, P. N. (1972) *Vacuum,* **22**, 315-319.
HOLLAND, L., LAURENSON, L., BAKER, P. N., and DAVIS, H. J. (1972) *Nature,* **238**, 36-37.
HOLLAND, L., LAURENSON, L., and DEVILLE, J. P. (1965) *Nature,* **206**, 883-884.
HOLLAND, L., LAURENSON, L., HURLEY, R. E., and WILLIAMS, K. (1973) *Nucl. Instrum. Meth.,* **111**, 555-560.
HOLLAND, L., LAURENSON, L., and PRIESTLAND, C. R. D. (1963) *Rev. Sci. Instrum.,* **34**, 377-382.
KERR, G. D., WILLIAMS, M. W., BIRKHOFF, R. D., and PAINTER, L. R. (1971) *J. Appl. Phys.,* **42**, 4258–4261.
KUMMINS, J. S. (1970) *Research/Development,* **21**, 46–48.
LATHAM, D., POWER, B. D., and DENNIS, N. T. M. (1952) *Vacuum,* **2**, 33–49.
LEYBOLD HERAEUS GMBH, 'Vacuum technology: Its foundations, formulae and tables', HV 152, Section H–B.
NÖLLER, H. G. (1966) 'Theory of vacuum diffusion pumps', 'Handbook of vacuum physics', **1**, Part 6 (Pergamon Press, Oxford).
POWER, B D., and CRAWLEY, D. J. (1954) *Vacuum,* **4**, 415–437.
REICH, G. (1957) *Z. Angew Phys.,* **9**, 23–29.
SOWERS, B. L., WILLIAMS, M. W., HAMM, R. N., and ARAKAWA, E. T. (1971) *J. Appl. Phys,* **42**, 4252–4257.
SMITH, H. R., (1959) *AVS Symp. Trans.,* 140.
TINGLE, E. D. (1948), Ph.D. Dissertation Cambridge. (See p. 192 of Part 1 and general discussion in I (1958) and II (1964) on silicones in Bowden, F. P., and Tabor, D., 'The friction and lubrication of solids', Oxford University Press.)

2 – Vacuum Equipment

This section provides information on currently available vacuum pumps, instrumentation and systems. It was compiled from information supplied by manufacturers in response to a questionnaire mailed to them. In some cases the details were extracted from brochures and data sheets forwarded by the manufacturers. The names and addresses of the manufacturers listed are given at the end of the section.

Every effort has been made to ensure that the information provided is comprehensive and accurate but neither the editors nor the publishers can be held responsible for any inconvenience which may be caused by errors or omissions. If such mistakes do exist we hope that users of the manual will bring them to our attention so that they may be rectified in future editions.

2.1 Vacuum pumps, valves and accessories

2.1.1 Cryopumps

Pumping by a surface cooled with liquid helium to 4·2 K or lower (e.g. 2·3 K) by reducing the pressure above the helium with a mechanical rotary pump.

Andonian Cryogenics

Model no. 0-11-2000: cryopump with a pumping speed for air of 2000 litre/second without baffling.

Model No. 300: ultra high vacuum cryopump.

A wide range of accessories for cryogenic work is available.

Balzers

	Model number		
	KRY 1000 U	KRY 2000 U	KRY 5000
Pumping speed in l/s for air	1000	2000	5000
Helium required (litre) for cooling the He container from room temperature to 4·2 K	1–2	2–3	5·5
Cooling time (min) to 4·2 K	15	30	15
Capacity (cm^3) of the container	300	900	2500
Duration (hour) of one He filling at 4·2 K	12	22	50

Balzers *cont'd*

	Model number		
	KRY 1000 U	KRY 2000 U	KRY 5000
Helium required (cm^3) for cooling the condensation surface from 4·2 K to 2·3 K	100	200	800
Duration of He filling at 2·3 K (hour)	6	8	32
Nitrogen required for cooling the LN_2 container and the radiation shield (litre)	1–2	3–4	9
LN_2 required when pump is operating (litre/hour)	0·5	0·7	1·9

2.1.2 Getter-ion pumps

A class of pumps in which an evaporated metal film sorbs active gas and inert gas is removed by ionizing the gas and implanting or burying extracted ions in the metal coating.

Evaporation getter-ion pump

Granville–Phillips

Electro Ion Vacuum Pump Series 220:
Baffled pumping speed for nitrogen: 1600 litre/second in 6 inch size
3200 litre/second in 8 inch size
Bakeout temperature (max): 300°C
Sublimator: large area, resistance heated BeO insulated sublimator giving sublimation rate of titanium of up to 0·02 g/hour.

A water-cooled evaporation getter-ion pump (no magnetic field) capable of producing ultimate pressures in the 10^{-10} Torr region.

Orbitron getter-ion pump

Cenco

Model no.	Pumping speed (litre/second)	Intake port I.D. in mm
* 93840-002	300	83
† 93840-004	500	135
‡ 93840-006	900	178

Nominal sizes in inch are *2, †4 and ‡6.

Water-cooled stainless steel cylindrical body with orbitron arrangement of tungsten filament electron emitter and central anode loaded with titanium. No magnet.

Radial electric field pump

Edwards High Vacuum

Pumping speed for nitrogen is 300 litre/second at 10^{-7} Torr, and provides an ultimate vacuum of better than 10^{-10} Torr. Bakeable to 450°C, An orbitron-type of getter-ion pump for which the filament current is 5 to 6 A at 6 V and the filament bias is + 60 V relative to earth. Anode voltage = 5 kV, anode current = 30 to 50 mA. Cooling water flow rate = 0·5 litre/minute. This pump is of vacuum brazed, stainless steel construction, with an external diameter of 114 mm.

2.1.3 Magnetless ion pumps

Veeco-Andar

Magnetless ion appendage pumps use a combination of ionization and sublimation for their pumping action and can also operate as vacuum gauges.

The pumps are lightweight, operate from 10^{-3} Torr to below 10^{-9} Torr and are bakeable to 425°C. (Models HV 002, 005, 006, 007, 008.)

2.1.4 Non-evaporable getter pumps

S.A.E.S. Getters

A unique series of non-evaporable getter cartridge pumps. A cartridge containing the getter material (16% Al, 84% Zr alloy) is heated to temperatures up to 750°C which enhances the diffusion into the bulk of the getter of surface sorbed atoms of active gases.

Model no.	Nominal pumping speed (litre/second)	Capacity (litre-Torr)	Cartridge heater characteristics			
			Activation to 150°C		Operation at 400°C	
			V	A	V	A
* NP2	2§	4§	50	1·25	17·5	0·65
NP10	10	34	92	1·2	32	0·6
† AP5G6	1·2	14	27	1·0	16	0·6
AP10G6	1·6	34	38	0·95	25	0·55
AP10GP	6·5	34	38	0·95	25	0·55
MAP10	1·6	34	74	1·05	25	0·55
‡ GP50W	50	140	55	4·7	16	2·2
GP100W	100	260	80	5·5	32	3·0
GP200W	200	1000	190	4·7	65	2·2
GP500W	500	2400	110	18·0	35	8·0

* NP series: small pumps comprising only cartridge and heater.
† AP series: small pumps in glass envelopes with feed throughs and a heat reflecting aluminium film.
‡ GPW series: large pumps constructed of metal suitable for use on standard uhv assemblies.
§ Values given are *not* for air but for carbon monoxide (CO) with the getter cartridge at 400°C.
This firm also supplies a wide range of getters.

For sputter-ion pumps *see* Section 2.1.8.

2.1.5 Positive displacement pumps

A positive displacement pump is defined as 'a mechanical pump in which the gas is drawn in, compressed and discharged by means of periodic variations in the volume of the working chamber' (BS 2951: Part 1, 1959). Classes of such pump are (a) piston pump, (b) rotary pump, (c) rotary piston pump, (d) sliding vane rotary pump, (e) liquid ring pump, (f) rough vacuum pumps, including those which are non or partially lubricated.

MANUFACTURER:	**Alcatel**								
Class of pump	Sliding vane rotary pumps →								
Manufacturer's model no.	Z1012	Z1030	Z1060	Z2004	Z2007	Z2012	Z2030	Z2060	Z2100
No. of stages	1	1	1	2	2	2	2	2	2
Gas ballast provision: Yes/No	Yes	Yes	Yes	Yes	Yes	Yes	Yes	Yes	Yes
Ultimate total pressure in Torr (with gas ballast off)	5×10^{-3}	4×10^{-3}	4×10^{-3}	10^{-4}	10^{-4}	10^{-4}	10^{-4}	10^{-4}	10^{-4}
Pumping speed in litre/min at 760 Torr	300	535	1000	75	165	300	500	900	2500
Pump rotational speed, rev/min	1800	1450	1000	1500	1800	1800	1200	900	1800

Remarks: Also supply H-series (hermetically sealed) pumps types 2004H, 2012H and 2030H and CHEMIST series of pumps types 1030C and 2030C.
Oil filtering systems for the pumping of contaminated circuits are available, also oil mist eliminators and molecular sieve filters.

MANUFACTURER:	**Alley Compressors**		
Class of pump	Rotary pumps →		
Manufacturer's model no.	8/50X/63	12X/63	M300
No. of stages	1	1	1
Gas ballast provision: Yes/No	Yes	Yes	Yes
Ultimate total pressure in Torr (with gas ballast off)	4×10^{-3}	4×10^{-3}	4×10^{-3}
Pumping speed in litre/min at 760 Torr	2180	4650	10100
Pump rotational speed, rev/min	450	400	250

Remarks: Water-cooled, oil-filled pumps.

MANUFACTURER:	**Atlas Copco**						
Class of pump	Rough vacuum piston pumps →						
Manufacturer's model no.	FE20	FE40	FE50	KE20	KE40	KE60	BE40
No. of stages	1 (1-cylinder)	1 (1-cylinder)	1 (1-cylinder)	1 (2-cylinder)	1 (2-cylinder)	1 (2-cylinder)	1 (2-cylinder)
Gas ballast provision: Yes/No	No	No	No	No	No	No	No
Maximum vacuum (%)	90	92	94	92	92	92	88
Pumping speed in litre/min at 760 Torr	118	183	253	460	1100	1640	11 100
Pump rotational speed, rev/min	1400	1400	1400	1000	1000	750	1460

MANUFACTURER:	**Atlas Copco** *cont'd*		
Class of pump	Rough vacuum piston pumps →		
Manufacturer's model no.	BE60	DA20	EA50
No. of stages	1 (4-cylinder)	1 (2-cylinder)	1 (2-cylinder)
Gas ballast provision: Yes/No	No	No	No
Maximum vacuum (%)	90	92	92
Pumping speed in litre/min at 760 Torr	22 200	29 300	43 600
Pump rotational speed, rev/min	1460	970	730

Remarks: DA20 and EA50 are water-cooled pumps; others are air-cooled.

MANUFACTURER:	Charles Austen Pumps							
Class of pump	Rough vacuum pumps →							
Manufacturer's model no.	LVM8	LVM9	Dymax Mk. I	Dymax Mk. II	T.M.6	Capex Mk. I	Capex Mk. II‡	M.361
No. of stages	1	1	1	1	1	1	1	1
Gas ballast provision: Yes/No	No	No	No	No	No	No	No	No
Ultimate total pressure in Torr (with gas ballast off)	150	200	380	450	500	380	380	230
Pumping speed in litre/min at 760 Torr	1	0·2	5·3	3·1†	4	17	6	14
Pump rotational speed, rev/min								

Remarks:

* Oil-free diaphragm or membrane sampling pump; output can be piped away or re-circulated; also act as means of providing pressure above atmospheric.

† Dymax Mk. IIA (intended for air pollution sampling) has a speed of 1·5 litre/min.

‡ Capex Mk. IIFP is the same as Capex Mk. II but mounted on a flame-proof motor.

MANUFACTURER:	Charles Austen Pumps *cont'd*		Balzers				
Class of pump	Rough vacuum pumps*		Sliding vane rotary pumps →				
Manufacturer's model no.	F.65	2F	Medvak MP1	Mavak MP6	UNO 004	UNO 008	UNO 012
No. of stages	1	2 in parallel §	1	1	1	1	1
Gas ballast provision: Yes/No	No	No	No	No	Yes	Yes	Yes
Ultimate total pressure in Torr (with gas ballast off)	150	150	7·6	12	5×10^{-2}	5×10^{-2}	5×10^{-2}
Pumping speed in litre/min at 760 Torr	33	40	16·7	100	66	132	198
Pump rotational speed, rev/min			2800	1400	1400	1400	1400

Remarks:

§ Duplex 2 in series; also Duplex 4 in parallel and Duplex 4 in series are available.

Range of accessories is provided for assembling pumps within vacuum systems.

MANUFACTURER:	Balzers *cont'd*								
Class of pump	Sliding vane rotary pumps →								
Manufacturer's model no.	UNO 035	BA 056	BA 091	UNO 100	UNO 170	BA 181	BA 361	DUO 004	DUO 008
No. of stages	1	1	1	1	1	1	1	2	2
Gas ballast provision: Yes/No	Yes	Yes	Yes	Yes	Yes	Yes	Yes	Yes	Yes
Ultimate total pressure in Torr (with gas ballast off)	5×10^{-2}	5×10^{-2}	5×10^{-2}	5×10^{-2}	5×10^{-2}	5×10^{-2}	5×10^{-2}	10^{-2}	10^{-2}
Pumping speed in litre/min at 760 Torr	580	920	1500	1584	2580	3000	6000	66	132
Pump rotational speed, rev/min	585	475	365	420	350	325	230	1400	1400

Remarks: Wide range of accessories provided: models BA 056, BA 091, BA 181 and BA 361 are water-cooled and provided with oil regenerating equipment.

MANUFACTURER:	Balzers *cont'd*			
Class of pump	Sliding vane rotary pumps →			
Manufacturer's model no.	DUO 012	DUO 035	DUO 100	DUO 170
No. of stages	2	2	2	2
Gas ballast provision: Yes/No	Yes	Yes	Yes	Yes
Ultimate total pressure in Torr (with gas ballast off)	10^{-2}	5×10^{-3}	5×10^{-3}	5×10^{-3}
Pumping speed in litre/min at 760 Torr	198	618	1620	2820
Pump rotational speed, rev/min	1400	560	430	400

MANUFACTURER:	Bingham-Willamette								
Class of pump	Liquid ring pumps →								
Manufacturer's model no.	2G	3G	4G	5G	6G	8G	10G	12G	14G
No. of stages	1	1	1	1	1	1	1	1	1
Gas ballast provision: Yes/No	No	No	No	No	No	No	No	No	No
Pumping speed in litre/min at 125 Torr	7750	14 600	22 800	29 400	48 100	70 800	127 000	178 000	268 000
Pump rotational speed, rev/min	1800	1400	1200	900	660	520	450	360	295

Remarks: Rotational speeds given are maximum; lower speeds can be used with sacrifice of pumping speed. Pumping speeds are higher at higher pressures. For example, type 2G at 1800 rev/min has a pumping speed of 8660 litre/min at 250 Torr

MANUFACTURER:	Bosch and Noltes Apparaten		Brizio Basi	
Class of pump	Sliding vane rotary rough vacuum pumps		Piston pumps →	
Manufacturer's model no.	BNA B60	BNA B100	3C	3C
No. of stages	2	2	1	2
Gas ballast provision: Yes/No	No	No	No	No
Ultimate total pressure in Torr (with gas ballast off)	50	50	10	4×10^{-1}
Pumping speed in litre/min at at 760 Torr	270	345	1000	
Pump rotational speed, rev/min	1450	1450		

Remarks: Oil-filling is by a drip-feed lubricator. Pumps also usable as compressors to 1·5 atm continuously or to 3 atm intermittently.

MANUFACTURER:	Burckhardt										
Class of pump	Rough vacuum piston pumps; vertical one-stage slide-valve vacuum pumps, type VA →										
Manufacturer's model no.	100/65*	125/80	160/100	200/125	250/160	280/160	320/200	360/200	400/250	450/250	500/320
No. of stages	1	1	1	1	1	1	1	1	1	1	1
Gas ballast provision: Yes/No	No	No	No	No	No	No	No	No	No	No	No
Ultimate total pressure in Torr (with gas ballast off)	4–2	4–2	4–2	4–2	4–2	4–2	5–3	5–3	5–3	5–3	6–4
Pumping speed in litre/min at 760 Torr	425	835	1500	2660	4600	5800	8800	11 300	15 500	19 500	24 800
Pump rotational speed, rev/min	480	480	420	380	330	330	310	310	275	275	220

Remarks: * Model numbers consist of three groups of digits. The first, single digit indicates the number of cylinders (where there is no first digit the pump is single-cylinder). The second group gives the diameter in mm of the piston; the third gives the stroke in mm of the piston. Models 2 x 450/250, 2 x 500/320, 3 x 450/250, 2 x 560/320 and 3 x 560/320 of these VA type pumps will provide pumping speeds at 760 Torr equal to the speed of the single cylinder pump multiplied by the number of cylinders.

MANUFACTURER:	Burckhardt *cont'd*						
Class of pump	Rough vacuum piston pumps; vertical one-stage plate-valve vacuum pumps, type VD →						
Manufacturer's model no.	560/320	500/250*	2X500/250	3X500/250	630/320	2X630/320	3X630/320
No. of stages	1	1	1	1	1	1	1
Gas ballast provision: Yes/No	No	No	No	No	No	No	No
Ultimate total pressure in Torr (with gas ballast off)	6–4	40	40	40	40	40	40
Pumping speed in litre/min at 760 Torr	31 000	33 000	66 000	99 000	52 500	105 000	157 500
Pump rotational speed, rev/min	220	380	380	380	380	380	380

Remarks: * Model numbers consist of three groups of digits. The first, single digit indicates the number of cylinders (where there is no first digit the pump is single-cylinder). The second group gives the diameter in mm of the piston; the third gives the stroke in mm of the piston.

MANUFACTURER:	**Burckhardt** *cont'd*								
Class of pump	Piston pumps; vertical two-stage sliding valve vacuum pumps, type V2A →								
Manufacturer's model no.	100/65*	125/80	160/100	200/125	250/160	280/160	320/200	360/200	400/250
No. of stages	2	2	2	2	2	2	2	2	2
Gas ballast provision: Yes/No	No	No	No	No	No	No	No	No	No
Ultimate total pressure in Torr (with gas ballast off)	0·2–0·04	0·2–0·04	0·2–0·04	0·2–0·04	0·2–0·04	0·2–0·04	0·2–0·05	0·2–0·05	0·2–0·05
Pumping speed in litre/min at 760 Torr	425	835	1500	2670	4600	5850	8900	11 250	15 500
Pump rotational speed, rev/min	480	480	420	380	330	330	310	310	275

Remarks: * Model numbers consist of three groups of digits. The first, single digit indicates the number of cylinders (where there is no first digit the pump is single-cylinder). The second group gives the diameter in mm of the piston; the third gives the stroke in mm of the piston. Also available in the V2A series are models 2X450/220, 2X500/320, 3X450/320, 2X560/320, 3X500/320 and 3X560/320. These give two or three times the pumping speeds obtainable with single cylinder first stage pumps.

MANUFACTURER:	**Burckhardt** *cont'd*					
Class of pump	Piston pumps; vertical two-stage sliding valve vacuum pumps, type V2A			Piston pumps; vertical three-stage slide-valve and plate-valve pumps, type V3A		
Manufacturer's model no.	450/250	500/320	560/320	2X450/250*	2X500/320	2X560/320
No. of stages	2	2	2	3	3	3
Gas ballast provision: Yes/No	No	No	No	No	No	No
Ultimate total pressure in Torr (with gas ballast off)	0·2–0·05	0·2–0·05	0·2–0·05	0·1–0·02	0·1–0·02	0·1–0·02
Pumping speed in litre/min at 760 Torr	19 500	24 700	31 000	39 000	49 000	62 000
Pump rotational speed, rev/min	275	220	220	275	220	220

Remarks: * Model numbers conist of three groups of digits. The first, single digit indicates the number of cylinders (where there is no first digit the pump is single-cylinder). The second group gives the diameter in mm of the piston; the third gives the stroke in mm of the piston.

MANUFACTURER:	**Burckhardt** *cont'd*							
Class of pump	Piston pumps; vertical two-stage slide-valve and plate-valve vacuum pumps type V2D →							
Manufacturer's model no.	250/160*	280/160	320/200	360/200	400/250	450/250	500/320	560/320
No. of stages	2	2	2	2	2	2	2	2
Gas ballast provision: Yes/No	No	No	No	No	No	No	No	No
Ultimate total pressure in Torr (with gas ballast off)	0·8–0·4	0·8–0·4	0·8–0·4	0·8–0·4	0·8–0·4	0·8–0·4	0·8–0·4	0·8–0·4
Pumping speed in litre/min at 760 Torr	4250	5350	8250	10 500	14 250	18 000	22 500	28 000
Pump rotational speed, rev/min	360	360	340	340	300	300	240	240

Remarks: * Model numbers consist of three groups of digits. The first, single digit indicates the number of cylinders (where there is no first digit the pump is single-cylinder). The second group gives the diameter in mm of the piston; the third gives the stroke in mm of the piston. Also available in the V2D series are models 2X450/250, 2X500/320 and 2X560/320. These models give two or three times the pumping speed obtainable with single-cylinder first stage pumps.

MANUFACTURER:	**Burckhardt** *cont'd*							
Class of pump	Piston pumps; horizontal slide-valve vacuum pumps types VB (1-stage) and V2B (2-stage) →							
Manufacturer's model no.	630/400*	700/400	800/500	900/500	630/400	700/400	800/500	900/500
No. of stages	1	1	1	1	2	2	2	2
Gas ballast provision: Yes/No	No	No	No	No	No	No	No	No
Ultimate total pressure in Torr (with gas ballast off)	6–4	6–4	6–4	6–4	0·2–0·05	0·2–0·05	0·2–0·05	0·2–0·05
Pumping speed in litre/min at 760 Torr	45 000	56 500	72 000	92 000	45 000	56 500	72 000	92 000
Pump rotational speed, rev/min	200	200	160	160	200	200	160	160

Remarks: * Model numbers consist of three groups of digits. The first, single digit indicates the number of cylinders (where there is no first digit the pump is single-cylinder). The second group gives the diameter in mm of the piston; the third gives the stroke in mm of the piston.

MANUFACTURER:	**Burckhardt** *cont'd*							
Class of pump	Piston pumps; horizontal slide-valve and plate valve vacuum pumps: 2-stage type V2C; 3-stage type V3C							
Manufacturer's model no.	630/400*	700/400	800/500	900/500	630/400	700/400	800/500	900/500
No. of stages	2	2	2	2	3	3	3	3
Gas ballast provision: Yes/No	No	No	No	No	No	No	No	No
Ultimate total pressure in Torr (with gas ballast off)	0·8–0·4	0·8–8·4	0·8–0·4	0·8–0·4	0·1–0·02	0·1–0·02	0·1–0·02	0·1–0·02
Pumping speed in litre/min at 760 Torr	38 000	47 000	63 000	80 000	45 000	55 500	72 000	92 000
Pump rotational speed, rev/min	200	200	160	160	200	200	160	160

Remarks: * Model numbers consist of three groups of digits. The first, single digit indicates the number of cylinders (where there is no first digit the pump is single-cylinder). The second group gives the diameter in mm of the piston; the third gives the stroke in mm of the piston.

MANUFACTURER:	**Burckhardt** *cont'd*								
Class of pump	Rough vacuum liquid ring pumps ⟶								
Manufacturer's model no.	PC 220/95	PC 220/140	PC 220/200	PC 225/250	PC 340/230	PC 340/300	PC 340/410	PC 460/380	PC 460/580
No. of stages	1	1	1	1	1	1	1	1	1
Ultimate total pressure in Torr*	30	30	30	30	35	35	35	35	35
Pumping speed in litre/min at 760 Torr	1920	3000	3700	5400	7700	11 000	14 500	14 700	23 400
Pump rotational speed, rev/min	1460	1460	1460	1460	960	960	960	730	730

Remarks: * The admissible service vacuum with cooling water inlet temperature of 12°C. Above types are for direct coupling to normal electric motors. A wide range of accessories such as water separators, coolers, condensers, valves, is available.

MANUFACTURER:	**Burckhardt** *cont'd*								
Class of pump	Rough vacuum liquid ring pumps →								
Manufacturer's model no.	PC 600/600	PC 600/750	P2J 220/95	P2J 220/140	P2J 220/200	P2J 225/250	P2J 340/230	P2J 340/300	P2J 340/410
No. of stages	1	1	2	2	2	2	2	2	2
Ultimate total pressure in Torr*	35	35	15	15	15	15	15	15	15
Pumping speed in litre/min at 760 Torr	47 500	62 000	1840	2400	3100	4350	6700	9300	12 000
Pump rotational speed, rev/min	580	580	1460	1460	1460	1460	960	960	960

Remarks: * The admissible service vacuum with cooling water inlet temperature of 12°C. Above types are for direct coupling to normal electric motors. A wide range of accessories such as water separators, coolers, condensers, valves is available.

MANUFACTURER:	**Burckhardt** *cont'd*					
Class of pump	Rough vacuum liquid ring pumps →					
Manufacturer's model no.	PG 135/30	PG 135/45	PG 135/70	P2G 135/30	P2G 135/45	P2G 135/70
No. of stages	1	1	1	2	2	2
Ultimate total pressure in Torr*	30	30	30	15	15	15
Pumping speed in litre/min at 760 Torr	520	780	1220	530	750	1100
Pump rotational speed, rev/min	2850	2850	2850	2850	2850	2850

Remarks: * The admissible service vacuum with cooling water inlet temperature of 12°C. Above types have flanged-on electric motor. A wide range of accessories such as water separators, coolers, condensers, valves is available.

MANUFACTURER:	**Burckhardt** *cont'd*				
Class of pump	Rough vacuum liquid ring pumps →				
Manufacturer's model no.	PMH 122	PMH 124	PMH 136	PMH 146	PMH 156
No. of stages	1	1	1	1	1
Ultimate total pressure in Torr*	30	30	30	30	30
Pumping speed in litre/min at 760 Torr	350	835	1170	1500	1840
Pump rotational speed, rev/min					

Remarks: *The admissible service vacuum with cooling water inlet temperature of 12°C. Monoblock pumps of compact construction.

MANUFACTURER:	**Cenco**								
Class of pump	Sliding vane rotary pumps →								
Manufacturer's model no.	Hyvac 300	Hyvac 225	Hyvac 150	Hyvac 45	Hyvac 28	Hyvac 14	Hyvac 7	Hyvac 6	Hyvac 4
No. of stages	2	2	2	2	2	2	2	2	2
Gas ballast provision: Yes/No	Yes	Yes	Yes	Yes	Yes	Yes	Yes	Yes	Yes
Ultimate total pressure in Torr (with gas ballast off)	5×10^{-5}	5×10^{-5}	5×10^{-5}	5×10^{-5}	5×10^{-5}	10^{-4}	10^{-4}	10^{-4}	5×10^{-5}
Pumping speed in litre/min at 760 Torr	3000	2250	1500	450	280	140	79	65	48
Pump rotational speed, rev/min	465	465	465	550	525	525	525	440	325

Remarks: A wide range of accessories is available.

MANUFACTURER:	Cenco *cont'd*							
Class of pump	Sliding vane rotary pump →							
Manufacturer's model no.	Hyvac 2	Hyvac	Hyvac	Hypervac 25	Hypervac 23	Megavac	Hyvac 45	Hyvac 28
No. of stages	2	2	2	2	2	2	1	1
Gas ballast provision: Yes/No	Opt.	No	No	No	No	No	Yes	Yes
Ultimate total pressure in Torr (with gas ballast off)	10^{-4}	3×10^{-4}	3×10^{-4}	10^{-4}	5×10^{-3}	10^{-4}	5×10^{-3}	5×10^{-3}
Pumping speed in litre/min at 760 Torr	25	20	10	264	240	57	450	280
Pump rotational speed, rev/min	530	580	510	570	510	600	565	525
Remarks:	A wide range of accessories is available.							

MANUFACTURER:	Cenco *cont'd*					Dawson, McDonald and Dawson	
Class of pump	Sliding vane rotary pumps →					Rough vacuum pumps	
Manufacturer's model no.	Hyvac 14	Hyvac 7	Pressovac*	Pressovac†	Blower-suction pump	D range	2D range
No. of stages	1	1	1	1	1	1	2
Gas ballast provision: Yes/No	Yes	Yes	No	No	No	No	No
Ultimate total pressure in Torr (with gas ballast off)	5×10^{-3}	10^{-2}	$1{\cdot}5 \times 10^{-2}$	$1{\cdot}5 \times 10^{-2}$	3 in. Hg.	120	380
Pumping speed in litre/min at at 760 Torr	140	79	35	35	35	70	70
Pump rotational speed, rev/min	525	525	600	600	–	1425	1425
Remarks:	* Vacuum only. † Vacuum and pressure pump.					A diaphragm type of pump of which output can be piped away or re-circulated. No oil filling used. Motor drive is to customer's requirements.	

MANUFACTURER:	Dresser Industries							
Class of pump	Liquid ring pumps →							
Manufacturer's model no.	2BD-202	2BD-212	2BA-212	2BA-222	2BA-232	2BA-312	2BA-322	2BA-532
No. of stages	1	1	1	1	1	1	1	1
Gas ballast provision: Yes/No	No	No	No	No	No	No	No	No
Ultimate total pressure in Torr (with gas ballast off)	50*	50	50	50	50	50	50	50
Pumping speed in litre/min at 760 Torr	1840	3170	4050	6200	8200	10 900	15 300	20 600
Pump rotational speed, rev/min	1750	1750	1750	1750	1750	880	880	700

Remarks: * Given as 28 inch Hg vacuum

MANUFACTURER:	Dresser Industries *cont'd*						
Class of pump	Liquid ring pumps →						
Manufacturer's model no.	2BA-552	2BA-662	2BA-722	2BA-762	2BA-822	2BB-862	2BB-962
No. of stages	1	1	1	1	1	1	1
Gas ballast provision: Yes/No	No	No	No	No	No	No	No
Ultimate total pressure in Torr (with gas ballast off)	50*	50	50	50	50	50	50
Pumping speed in litre/min at at 760 Torr	28 000	41 500	58 000	78 000	107 000	134 000	218 000
Pump rotational speed, rev/min	700	585	500	420	360	322	256

Remarks: * Given as 28 inch Hg vacuum

MANUFACTURER:	Edwards High Vacuum				
Class of pump	Oil lubricated vacuum/compressor pumps			Oil-free vacuum/compressor pumps	
Manufacturer's model no.	EB3A*	RB5†	EB10*	ECB1†	RCB2‡
No. of stages	1	1	1	1	1
Gas ballast provision: Yes/No	No	No	No	No	No
Ultimate total pressure in Torr (with gas ballast off)	<100	<125	<50	<125	<200
Pumping speed in litre/min at 760 Torr	78	160	290	28	62
Pump rotational speed, rev/min					

Remarks:

* Provides continuous pressure at 18 lbf/sq. in.
† Provides continuous pressure at 15 lbf/sq. in.
‡ Provides continuous pressure at 12 lbf/sq. in.
A wide range of accessories for these vacuum/compressor pumps is available.

MANUFACTURER:	Edwards High Vacuum *cont'd*								
Class of pump	Rotary pumps →								
Manufacturer's model no.	EDM2	ISP30C	ES50	ED50	ES100	ED100	ES200	ED200	ES330
No. of stages	2	1	1	2	1	2	1	2	1
Gas ballast provision: Yes/No	Yes	No	Yes	Yes	Yes	Yes	Yes	Yes	Yes
Ultimate total pressure in Torr (with gas ballast off)	7×10^{-4}	10^{-1}	10^{-2}	5×10^{-4}	5×10^{-3}	2×10^{-4}	5×10^{-3}	10^{-4}	5×10^{-3}
Pumping speed in litre/min at 760 Torr	33, 40*	30	53	53	100	100	190	190	334
Pump rotational speed, rev/min	1425,† 1725								

Remarks:

* The first figure is the pumping speed when the supply is at 50 Hz and the second when the supply is at 60 Hz (U.S.A. electrical supply mains practice). EDM2 is a direct drive rotary vacuum pump.
† The first figure is for a supply at 50 Hz, the second for a supply at 60 Hz.
A complete range of fore-line traps, manifolds and filters is available.

MANUFACTURER:	Edwards High Vacuum *cont'd*							Officine Galileo
Class of pump	Rotary pumps ⟶							Sliding vane rotary pumps
Manufacturer's model no.	ED330	1SC450B	ED660	1SC900	ES2000	ES4000	ES7500	V2d
No. of stages	2	1	2	1	1	1	1	2
Gas ballast provision: Yes/No	Yes	Yes	Yes	Yes	Yes	Yes	Yes	Yes
Ultimate total pressure in Torr (with gas ballast off)	10^{-4}	5×10^{-3}	10^{-4}	5×10^{-3}	5×10^{-3}	5×10^{-3}	5×10^{-3}	2×10^{-4}
Pumping speed in litre/min at 760 Torr	334	555	233	1110	2100	4200	7800	1000
Pump rotational speed, rev/min								510
Remarks:	Oil mist and inlet dust filters are available.							Utilizes a 1·5 kVA three-phase electric motor and average oil capacity is 1·3 litre.

MANUFACTURER:	Gast Manufacturing					
Class of pump	Rough vacuum sliding vane rotary pumps ⟶					
Manufacturer's model no.	0406	0211	0322	0522	0822	1022
No. of stages	1	1	1	1	1	1
Gas ballast provision: Yes/No	No	No	No	No	No	No
Maximum vacuum in inches of mercury	25* 15	27 20	26 26	27 27	27 27	27 27
Pumping speed in litre/min at 760 Torr	16	37	71	113	203	280
Pump rotational speed, rev/min	1725	1725	1725	1725	1725	1725

Remarks: * In each case the top figure is for intermittent and the lower figure is for continuous vacuum. Standard accessories supplied for these oil-lubricated models are foot support assemblies, oiler-filters and exhaust mufflers. For models 0322 and 0522 also siphon oilers and for models 0822 and 1022, constant level oilers. Direct drive models.

MANUFACTURER:	Gast Manufacturing *cont'd*											
Class of pump	Rough vacuum sliding vane rotary pumps →											
Manufacturer's model no.	0240	0440	0740	1550	3040	0465	0765	1065	2065	2565	4565	4565
No. of stages	1	1	1	1	1	1	1	1	1	1	1	1
Gas ballast provision: Yes/No	No	No	No	No	No	No	No	No	No	No	No	No
Maximum vacuum in inches of mercury	26* 15	26 15	26 10	25 15	27 15	28 28	28 28	28 28	28 28	28 28	27 25	27 20
Pumping speed in litre/min at 760 Torr	54	113	167	326	880	113	167	235	480	595	1100	1415
Pump rotational speed, rev/min	1725	1725	1725	1440	1325	1725	1725	1725	1725	1725	1350	1725

Remarks: * In each case the top figure is for intermittent and the lower figure is for continuous vacuum. Standard accessories supplied for these oil-lubricated models are: for models 0240, 0440 and 0740, oilers, mufflers and drive coupling assemblies; for model 1550, oiler muffler and fan pulley; for model 3040, heavy duty lubricator, muffler and pulley; for models 0465, 0765, 1065, 2065 and 2565, heavy duty lubricators, mufflers and drive coupling assemblies; for model 4565, heavy duty lubricator, muffler, filter and drive coupling assembly. Separate drive models.

MANUFACTURER:	Gast Manufacturing *cont'd*			
Class of pump	Rough vacuum sliding vane rotary pumps →			
Manufacturer's model no.	10 x 1040	10 x 1040	11 x 1740	11 x 1740
No. of stages	dual chamber	dual chamber	dual chamber	dual chamber
Gas ballast provision: Yes/No	No	No	No	No
Maximum vacuum in inches of mercury	20	20	20	20
Pumping speed in litre/min at 760 Torr	198 x 198	254 x 254	198 x 312	254 x 396
Pump rotational speed, rev/min	950	1200	800	1000

Remarks: Also useable as compressors to 15 psi. Accessories for 10 x 1040 and 11 x 1740 are oilers, intake filters and mufflers on both chambers. Oil-lubricated models with separate drive.

MANUFACTURER:	Gast Manufacturing *cont'd*								
Class of pump	Rough vacuum sliding vane rotary pumps →								
Manufacturer's model no.	0406	0531†	1031†	0211	1531†	0322	0522	0822	1022
No. of stages	1	1	1	1	1	1	1	1	1
Gas ballast provision: Yes/No	No	No	No	No	No	No	No	No	No
Maximum vacuum in inches of mercury	20* / 10	20 / 20	20 / 15	20 / 20	20 / 15	25 / 25	26 / 26	26 / 26	26 / 26
Pumping speed in litre/min at 760 Torr	14	17	31	31	42	71	113	204	280
Pump rotational speed, rev/min	1725	3450	3450	1725	3450	1725	1725	1725	1725

Remarks:

Motor-mounted *oil-less* models.
* In each case the top figure is for intermittent and the lower figure is for continuous vacuum.
Accessories in form of foot support assemblies, filters and mufflers are furnished.
† Miniature extremely compact pumps.

MANUFACTURER:	Gast Manufacturing *cont'd*											
Class of pump	Rough vacuum sliding vane rotary pumps →											
Manufacturer's model no.	0330	0330	0630	1030	1533	0240	0440	0740	1550	1550	3040	3040
No. of stages	1	1	1	1	1	1	1	1	1	1	1	1
Gas ballast provision: Yes/No	No	No	No	No	No	No	No	No	No	No	No	No
Maximum vacuum in inches of mercury	20* / 20	22 / 20	23 / 20	24 / 20	20 / 15	24 / 20	24 / 20	24 / 20	24 / 20	20 / 20	25 / 20	20 / 15
Pumping speed in litre/min at 760 Torr	9·9	19	17	28	42	54	113	167	326	412	710	880
Pump rotational speed, rev/min	3450	6000	3450	3450	3450	1725	1725	1725	1440	1725	1060	1325

Remarks:

Separate drive *oil-less* models.
* In each case the top figure is for intermittent and the lower figure is for continuous vacuum.
Standard accessories supplied: for model 0330, hose nipples; for 0630 and 1030, filters and mufflers; for 1533, filter and exhaust filter; for 0240, 0440 and 0740, inlet filters and exhaust mufflers; for 1550 and 3040, fans, fan pulleys, inlet filters and exhaust filters.

MANUFACTURER:	**Gast Manufacturing** *cont'd*							
Class of pump	Rough vacuum rotary piston pump →							
Manufacturer's model no.	4VSF	1VSF	4VCF	1VBF	1VAF	VAB	VBB	VCD
No. of stages	2	2	1	1	1	1	1	1
Gas ballast provision: Yes/No	No	No	No	No	No	No	No	No
Maximum vacuum in inches of mercury	29	29	27·5	27·5	27·5	15	15	15
Pumping speed in litre/min at 760 Torr	113	85	142	91	57	34	68	133
Pump rotational speed, rev/min	1725	1725	1725	1725	1725			

Remarks: Motor-mounted direct-drive models; accessories comprise: carrying handles, foot support assemblies, intake filters and vacuum relief valves.

MANUFACTURER:	**Apparatebau Gauting**						**Gelman Instrument**
Class of pump	Rough vacuum pumps →						Rough vacuum piston pump
Manufacturer's model no.	SDP1	SDP2	SDP4	SDP8	SDP14	SDP25	13152 (Little Giant)
No. of stages	1	1	1	1	1	1	1
Gas ballast provision: Yes/No	No	No	No	No	No	No	No
Ultimate total pressure in Torr (with gas ballast off)	76 (90% Vac)	76	76	38 (95% Vac)	38	38	Maximum vacuum given as 22 inches of mercury
Pumping speed in litre/min at 760 Torr	16·5	33	66	132	233	415	25·5
Pump rotational speed, rev/min							1725
Remarks:	Oil-less 'dry running' rough vacuum pumps; also act as compressors to approx. 20 psi. Accessories supplied: handles, air filters, pressure balances, over-pressure valves, dial manometer.						As a compressor provides 60 psi. A pressure/vacuum pump with a hard graphite piston ring which is self lubricating without oil. Also supply a range of membrane filters (made from polymerized plastic).

MANUFACTURER:	General Engineering									
Class of pump	Rotary piston pumps →									
Manufacturer's model no.	GHS3	GHS6	GHS12	GHS18	GKS13	GKS27	GKS47	GKD30	GDK110	GKDH130
No. of stages	1	1	1	1	1	1	1	1	1	1
Gas ballast provision: Yes/No	Yes	Yes	Yes	Yes	Yes	Yes	Yes	Yes	Yes	Yes
Ultimate total pressure in Torr (with gas ballast off)	5×10^{-3}	5×10^{-3}	5×10^{-3}	5×10^{-3}	10^{-2}	10^{-2}	10^{-2}	10^{-2}	10^{-2}	10^{-2}
Pumping speed in litre/min at 760 Torr	85	170	340	510	340	710	1200	780	2800	3200
Pump rotational speed, rev/min										
Remarks:	Water vapour tolerance for all these models is 30 Torr.									

MANUFACTURER:	General Engineering *cont'd*								
Class of pump	Rotary piston pumps →								
Manufacturer's model no.	GKD220	GKD780	GT80	GKT150A	GKT300A	GKT500	GHD3	GHD6	GHD9
No. of stages	1	1	1	1	1	1	2	2	2
Gas ballast provision: Yes/No	Yes	Yes	Yes	Yes	Yes	Yes	Yes	Yes	Yes
Ultimate total pressure in Torr (with gas ballast off)	10^{-2}	10^{-2}	10^{-2}	10^{-2}	10^{-2}	10^{-2}	5×10^{-4}	5×10^{-4}	5×10^{-4}
Pumping speed in litre/min at 760 Torr	5600	18 500	2100	3900	7800	12 800	85	170	255
Pump rotational speed, rev/min									
Remarks:	Water vapour tolerance for all these models is 30 Torr.								

MANUFACTURER:	**General Engineering** *cont'd*							
Class of pump	Rotary piston pumps		Sliding vane rotary pumps			Liquid ring pumps ⟶		
Manufacturer's model no.	GKC15	GKTC21	GSV4	GDV4	GDV9	GLRS13	GLRS30	GLRS55
No. of stages	2	2	1	2	2	1	1	1
Gas ballast provision: Yes/No	Yes	Yes	Yes	Yes	Yes			
Ultimate total pressure in Torr (with gas ballast off)	5×10^{-4}	5×10^{-4}	10^{-2}	5×10^{-4}	5×10^{-4}	120 working ultimate	120 working ultimate	120 working ultimate
Pumping speed in litre/min at 760 Torr	400	525	54	51	125	235	540	910
Pump rotational speed, rev/min								

Remarks: Water vapour tolerance 30 Torr. Data given is with water at 15°C as sealing fluid: performance varies with temperature and other fluids can be used. Working ultimate means without cavitation.

MANUFACTURER:	**General Engineering** *cont'd*								
Class of pump	Liquid ring pumps ⟶								
Manufacturer's model no.	GLRS100	GLRS145	GLRS175	GLRS250	GLRS340	GLRS420	GLRS475	GLRS700	GLRS1000
No. of stages	1	1	1	1	1	1	1	1	1
Gas ballast provision: Yes/No									
Ultimate total pressure in Torr (with gas ballast off)	120 working ultimate ⟶								
Pumping speed in litre/min at 760 Torr	1750	2450	2830	4200	5600	7000	8000	12 600	16 600
Pump rotational speed, rev/min									

Remarks: Data given is with water at 15°C as sealing fluid. Peformance varies with temperature and other fluids can be used. Working ultimate means without cavitation.

MANUFACTURER:	**General Engineering** *cont'd*						
Class of pump	Liquid ring pumps →						
Manufacturer's model no.	GLRS870	GLRS1250	GLRS1550	GLRS2000	GLRS2500	GLRS3000	GLRS3750
No. of stages	1	1	1	1	1	1	1
Gas ballast provision: Yes/No							
Ultimate total pressure in Torr (with gas ballast off)	120 working ultimate →						
Pumping speed in litre/min at 760 Torr	13 400	21 000	26000	33 500	42 000	52 000	62 500
Pump rotational speed, rev/min							

Remarks: Data given is with water at 15°C as sealing fluid. Performance varies with temperature and other fluids can be used. Working ultimate means without cavitation.

MANUFACTURER:	**General Engineering** *cont'd*								
Class of pump	Liquid ring pumps →								
Manufacturer's model no.	GLRD4	GLRD14	GLRD33	GLRD55	GLRD108	GLRD150	GLRD170	GLRD290	GLRD360
No. of stages	2	2	2	2	2	2	2	2	2
Gas ballast provision: Yes/No									
Ultimate total pressure in Torr (with gas ballast off)	25 working ultimate →								
Pumping speed in litre/min at 760 Torr	74	235	550	935	1800	2350	2550	4750	5700
Pump rotational speed, rev/min									

Remarks: Data given is with water at 15°C as sealing fluid. Performance varies with temperature and other fluids can be used. Working ultimate means without cavitation.

MANUFACTURER:	General Engineering *cont'd*								
Class of pump	Liquid ring pumps →								
Manufacturer's model no.	GLRD450	GLRD520	GLRD775	GLRD880	GLRD1350	GLRD1700	GLRD2100	GLRD2700	GLRD3200
No. of stages	2	2	2	2	2	2	2	2	2
Gas ballast provision: Yes/No									
Ultimate total pressure in Torr (with gas ballast off)	25 working ultimate →								
Pumping speed in litre/min at 760 Torr	7000	9000	13 000	14 600	21 000	24 500	34 000	40 000	50 000
Pump rotational speed, rev/min									
Remarks:	Data given is with water at 15°C as sealing fluid. Performance varies with temperature and other fluids can be used. Working ultimate means without cavitation.								

MANUFACTURER:	Hick Hargreaves	Ingersoll Rand						Jigtool High Vacuum	
Class of pump	A wide range of pumps, including sliding vane and liquid ring types is supplied.	Rough vacuum pumps →						Sliding vane pumps	
Manufacturer's model no.		V23A	V235	V244	V255	7V	15V	SE 45; 65; 100	SEV 65; 100
No. of stages		1 or 2	1 or 2	1 or 2	1 or 2	1 or 2	1 or 2	2	
Gas ballast provision: Yes/No		No	No	No	No	No	No	Yes	
Ultimate total pressure in Torr (with gas ballast off)		25 (single-stage) and 9 (two-stage)						10^{-2}	
Pumping speed in litre/min at 760 Torr								Model number denotes speed	
Pump rotational speed, rev/min		850	850	850	850	850	850	500(45) 660(65) 700(100)	
Remarks:		Suitable only for dry duty. Lubricated cylinders.						Incorporate VACPAK vacuum cartridge allowing replacement of all working parts in event of damage.	

MANUFACTURER:	Kinney Vacuum								
Class of pump	Rotary piston pumps →								
Manufacturer's model no.	KC-2	KC-3	KC-5	KC-8	KC-15	KTC-21	KTC-60	KTC-75	KTC-112
No. of stages	2	2	2	2	2	2	2	2	2
Gas ballast provision: Yes/No	Yes	Yes	Yes	Yes	Yes	Yes	Yes	Yes	Yes
Ultimate total pressure in Torr (with gas ballast off)	2×10^{-4}	2×10^{-4}	2×10^{-4}	2×10^{-4}	2×10^{-4}	2×10^{-4}	2×10^{-4}	2×10^{-4}	2×10^{-4}
Pumping speed in litre/min at 760 Torr	50	78	115	188	360	510	1380	1680	2770
Pump rotational speed, rev/min	755	1135	630	1000	525	1735	960	1200	1060

Remarks: Models KC-2 and KTC-21 are air-cooled; models KTC-60 to KTC-112 are water-cooled.

MANUFACTURER:	Kinney Vacuum *cont'd*										
Class of pump	Rotary piston pumps →										
Manufacturer's model no.	KS-13	KD-30	KDH-35	KDH-80	KDH-130	KDH-150	KD-780	KD-850	KT-150	KT-300	KT-500
No. of stages	1	1	1	1	1	1	1	1	1	1	1
Gas ballast provision: Yes/No	Yes	Yes	Yes	Yes	Yes	Yes	Yes	Yes	Yes	Yes	Yes
Ultimate total pressure in Torr (with gas ballast off)	10^{-2}	10^{-2}	10^{-2}	10^{-2}	10^{-2}	10^{-2}	10^{-2}	10^{-2}	10^{-2}	10^{-2}	10^{-2}
Pumping speed in litre/min at 760 Torr	283	750	1700	2130	3400	3960	18 650	20 500	3680	7500	13 380
Pump rotational speed, rev/min	460	523	440	555	535	650	400	444	1060	880	845

Remarks: Models KS-13 and KD-30 are air-cooled, the others are water-cooled.

MANUFACTURER:	**Kinney Vacuum** *cont'd*							
Class of pump	Rotary piston	Liquid ring pumps →						
Manufacturer's model no.	KT-850	KLR-360	KLR-700	KLR-1050	KLR-2600	KLRC-75	KLRC-125	KLRC-300
No. of stages	1	1	1	1	1	2	2	2
Gas ballast provision: Yes/No	Yes	No	No	No	No	Yes	Yes	Yes
Ultimate total pressure in Torr (with gas ballast off)	10^{-2}	10^2	10^2	10^2	10^2	20	20	20
Pumping speed in litre/min at 760 Torr	22 000	10 380	20 400	31 130	73 500	1420	2400	4250
Pump rotational speed, rev/min	614	1750	1750	1150	870	1750	1750	1750

MANUFACTURER:	**Kinney Vacuum** *cont'd*						
Class of pump	Liquid ring pumps →						
Manufacturer's model no.	KLRC-375	KLRC-525	KLRC-775	KLRC-950	KLRC-1500	KLRC-2100	KLRC-2
No. of stages	2	2	2	2	2	2	2
Gas ballast provision: Yes/No	Yes	Yes	Yes	Yes	Yes	Yes	Yes
Ultimate total pressure in Torr (with gas ballast off)	20	20	20	20	20	20	50
Pumping speed in litre/min at 760 Torr	5650	9300	13 600	17 000	24 000	28 300	45 (at 250 Torr)
Pump rotational speed, rev/min	1750	1750	1150	1150	870	870	1750

MANUFACTURER:	**Kinney Vacuum** *cont'd*				
Class of pump	Liquid ring pumps →				
Manufacturer's model no.	KLRC-3	KLRC-11	KLRC-25	KLRC-40	KLRC-200
No. of stages	2	2	2	2	2
Gas ballast provision: Yes/No	Yes	Yes	Yes	Yes	Yes
Ultimate total pressure in Torr (with gas ballast off)	50	50	25	20	20
Pumping speed in litre/min at 760 Torr	88 (at 250 Torr)	310 (at 250 Torr)	740 (at 250 Torr)	1220 (at 250 Torr)	5380 (at 250 Torr)
Pump rotational speed, rev/min	1750	1750	1750	1750	1750

MANUFACTURER:	**Kraissl**										
Class of pump	Rough vacuum rotary piston pumps →					Rough vacuum sliding vane rotary pumps →					
Manufacturer's model no.	21-3	21-5	21-6	21-6A	21-6B	25-5	25-6	25-6A	25-6B	25-9	25-11
No. of stages	1	1	1	1	1	1	1	1	1	1	1
Gas ballast provision: Yes/No	No	No	No	No	No	No	No	No	No	No	No
Ultimate total pressure in Torr	125	125	125	125	125	50	50	50	50	50	50
Pumping speed in litre/min at 760 Torr	28	77	113	142	190	99	142	170	254	425*	990*
Pump rotational speed, rev/min	1800	1800	1800	1800	1800	1800	1800	1800	1800	1800	1800

Remarks:

* Intermittent service; slower speeds or water-cooled for continuous duty.
Class 21 and 25 series are radiant cooled, former makes use of rollers carried in the recesses of the rotary piston and latter series makes use of blades as similar displacement elements. Automatic system of oil lubrication. Also available are a class 25 WJS series of water-cooled pumps and a class 25-FCF series of fan-cooled pumps with integrated class 26 series air filter.

MANUFACTURER:	Leybold-Heraeus						
Class of pump	Sliding vane rotary pumps →						
Manufacturer's model no.	MINNI	S2A	S4A	S8A	S16A	S30A	S60A
No. of stages	1	1	1	1	1	1	1
Gas ballast provision: Yes/No	Yes	Yes	Yes	Yes	Yes	Yes	Yes
Ultimate pressure in Torr (with gas ballast off)	< 1	2×10^{-2}	2×10^{-2}	2×10^{-2}	2×10^{-2}	2×10^{-2}	2×10^{-2}
Pumping speed in litre/min at 760 Torr	15	54	107	166	335	640	1260
Pump rotational speed, rev/min	1500	1500	1500	1500	1500	1500	1500

Remarks: Wide range of accessories is available for all pumps.

MANUFACTURER:	Leybold-Heraeus *cont'd*					
Class of pump	Sliding vane rotary pumps →					
Manufacturer's model no.	D2A	D4A	D8A	D16A	D30A	D60A
No. of stages	2	2	2	2	2	2
Gas ballast provision: Yes/No	Yes	Yes	Yes	Yes	Yes	Yes
Ultimate pressure in Torr (with gas balast off)	2×10^{-4}	2×10^{-4}	2×10^{-4}	2×10^{-4}	2×10^{-4}	2×10^{-4}
Pumping speed in litre/min at 760 Torr	54	107	166	335	640	1260
Pump rotational speed, rev/min	1500	1500	1500	1500	1500	1500

Remarks: Wide range of accessories is available for all pumps.

MANUFACTURER:	Leybold-Heraeus *cont'd*				
Class of pump	Rotary piston pumps →				
Manufacturer's model no.	S60	S100	S200	S400	S800
No. of stages	1	1	1	1	1
Gas ballast provision: Yes/No	Yes	Yes	Yes	Yes	Yes
Ultimate pressure in Torr (with gas ballast off)	2×10^{-2}	2×10^{-2}	5×10^{-2}	5×10^{-2}	5×10^{-2}
Pumping speed in litre/min at 760 Torr	1000	1666	3333	6666	13 333
Pump rotational speed, rev/min	480	800	800	400	400

Remarks: Multiple coupled two-stage vacuum pumps (Duplex type) all driven by a common 1450 rev/min motor are in series Duplex D5 pumps: 4-fold (4 x 83·3 litre/min); 6-fold (6 x 83·3 litre/min) and 12-fold (12 x 83·3 litre/min) are available. Ultimate total pressure is approx. 10^{-2} Torr whereas ultimate partial pressure is approx. 2×10^{-5} Torr. Wide range of accessories is available for all pumps. All E models are monoblock pumps.

MANUFACTURER:	Leybold-Heraeus *cont'd*							
Class of pump	Rotary piston pumps →							
Manufacturer's model no.	E38	E75	E150	E250	DK25	DK50	DK100	DK200
No. of stages	1	1	1	1	2	2	2	2
Gas ballast provision: Yes/No	Yes	Yes	Yes	Yes	Yes	Yes	Yes	Yes
Ultimate pressure in Torr (with gas ballast off)	2×10^{-2}	2×10^{-2}	2×10^{-2}	2×10^{-2}	10^{-4}	10^{-4}	10^{-4}	10^{-4}
Pumping speed in litre/min at 760 Torr	633	1250	2500	4170	417	834	1666	3333
Pump rotational speed, rev/min	1500	1500	1500	1500	1500	1500	1500	1500

MANUFACTURER:	**Leybold-Heraeus** *cont'd*				
Class of pump	Roots tandem pumps: combination of a single-stage mechanical rotary pump with a Roots pump				
Manufacturer's model no.	RUTA 60	RUTA 100	RUTA 200	RUTA 400	RUTA 800
No. of stages	S60 with E106	S100 with E111	S200 with E116	S400 with E126	S800 with E136
Gas ballast provision: Yes/No	Yes	Yes	Yes	Yes	Yes
Ultimate total pressure in Torr (with gas ballast off)	2×10^{-3}	2×10^{-3}	5×10^{-3}	5×10^{-3}	5×10^{-3}
Pumping speed in litre/min at 760 Torr	917	1750	3170	6330	13 000
Pump rotational speed, rev/min					

MANUFACTURER:	**Ch. Lumpp**								
Class of pump	Cylinder pumps: single cylinder with horizontal reciprocating piston								
Manufacturer's model no.*	120-70	160-70	180-100	200-150	250-150	280-150	320-200	360-200	400-200
No. of stages	1	1	1	1	1	1	1	1	1
Gas ballast provision: Yes/No	No	No	No	No	No	No	No	No	No
Ultimate total pressure in Torr (with gas ballast off) approx.	3	3	3	3	3	3	3	3	3
Pumping speed in litre/min at 760 Torr	400	700	1500	2200	3200	4200	5300	7000	9000
Pump rotational speed, rev/min									

Remarks: * Pumps of type H^3 are horizontal with one or more pistons; pumps are designated by the bore in mm of the cylinder, the traverse in mm of the piston and the number of cylinders (one in cases above), e.g. PVH^3–280/150 bicylindrical = horizontal vacuum pump, bore 280 mm, traverse 150 mm – 2 cylinders.

MANUFACTURER:	Ch. Lumpp *cont'd*							
Class of pump	Cylinder pumps: single cylinder with horizontal reciprocating piston		Two cylinder pumps ⟶					
Manufacturer's model no.*	450-200	500-300	120-70	160-70	180-100	250-150 ——	280-150	360-200
No. of stages	1	1	2 cylinders	2 cylinders	2 cylinders	2 cylinders	2 cylinders	2 cylinders
Gas ballast provision: Yes/No	No	No	No	No	No	No	No	No
Ultimate total pressure in Torr (with gas ballast off) approx.	3	3	2	2	2	2	2	2
Pumping speed in litre/min at 760 Torr	12 000	15 500	400	800	1500	3500	4500	8000
Pump rotational speed, rev/min								

Remarks: * Pumps of type H^3 are horizontal with one or more pistons; pumps are designated by the bore in mm of the cylinder, the traverse in mm of the piston and the number of cylinders (one in cases above), e.g. PVH^3–280/150 bicylindrical = horizontal vacuum pump, bore 280 mm, traverse 150 mm – 2 cylinders.

MANUFACTURER:	Ch. Lumpp *cont'd*		N.G.N.					
Class of pump	Two cylinder pumps		Rotary piston pumps ⟶					
Manufacturer's model no.	450-200	500-300	PSR1	PSR2	PSR6	PSR12	PD2	505A
No. of stages	2 cylinders	2 cylinders	1	1	1	1	2	1
Gas ballast provision: Yes/No	No	No	Yes	Yes	Yes	Yes	Yes	Yes
Ultimate total pressure in Torr (with gas ballast off)	2	2	5×10^{-5}	5×10^{-5}	5×10^{-5}	5×10^{-5}	2×10^{-6}	10^{-4}
Pumping speed in litre/min at 760 Torr	12 300	17 000	45	85	221	370	85	44
Pump rotational speed, rev/min			870	870	770	770	870	870

Remarks: (N.G.N.) Accessories supplied: electromagnetic automatic air admittance valves, non-return valves, cone fittings, gaskets, O-rings, oils and greases.

MANUFACTURER:	Nash Engineering											
Class of pump	Rough vacuum liquid ring pumps →											
Manufacturer's model no.	571	572	573	574	671	672	673	674	H-3	H-4	H-5	H-6
No. of stages	1	1	1	1	1	1	1	1	1	1	1	1
Gas ballast provision: Yes/No	No	No	No	No	No	No	No	No	No	No	No	No
Ultimate total pressure in Torr* (with gas ballast off)												
Pumping speed in litre/min at 380 Torr	312	425	600	1000	530	765	1160	1380	1840	3820	5660	8750
Pump rotational speed, rev/min	1750	1750	1750	1750	1750	1750	1750	1750	1750	1750	1150	870

Remarks:

* The H-series of pumps have ultimate (shut-off) vacuum of 27 inches to 28 inches of mercury, depending on the pump size and speed. Also available are units of greater pumping capacity in the new type CL vacuum pumps. A wide range of compressors is also furnished.

MANUFACTURER:	Nash Engineering *cont'd*			
Class of pump	Rough vacuum liquid ring pumps			
Manufacturer's model no.	L-2	L-3	L-4	L-5
No. of stages	1	1	1	1
Gas ballast provision: Yes/No	No	No	No	No
Ultimate total pressure in Torr* (with gas ballast off)				
Pumping speed in litre/min at 380 Torr	1840	3960	5660	9200
Pump rotational speed, rev/min	1750	1750	1150	870

MANUFACTURER:	Nash Engineering *cont'd*				
Class of pump	Rough vacuum liquid ring pumps with Nash air ejector pump connected				
Manufacturer's model no.	CL-203	CL-403	CL-703	CL-1003	CL-2003
No. of stages	2*	2*	2*	2*	2*
Gas ballast provision: Yes/No	No	No	No	No	No
Ultimate total pressure in Torr (with gas ballast off)	19	19	19	19	19
Pumping speed in litre/min at 10 x ultimate pressure	2120	4520	6800	11 000	15 600
Pump rotational speed, rev/min	1750	1750	1750	1750	1750

Remarks: * Air-ejector pump forms 1 stage and rotary pump the 2nd ('backing') stage.

MANUFACTURER:	Neuman and Esser									
Class of pump	Rough vacuum piston pumps (reciprocating piston in horizontal cylinder) →									
Manufacturer's model no.	IVK4	IVK5	IVK6	IVK7	IVK8	IVK9	IVK10	IVK11	2VK4	2VK5
No. of stages	1	1	1	1	1	1	1	1	2	2
Gas ballast provision: Yes/No	No	No	No	No	No	No	No	No	No	No
Ultimate total pressure in Torr (with gas ballast off)	6	5	5	4	4	3·5	3	3	5×10^{-1}	5×10^{-1}
Pumping speed in litre/min at 760 Torr	4500	6250	10 000	15 000	20 000	30 000	45 000	66 500	4500	6250
Pump rotational speed, rev/min	180	130	150	145	135	125	122	104	180	130

MANUFACTURER:	Neuman and Esser *cont'd*					
Class of pump	Rough vacuum piston pump (reciprocating piston in horizontal cylinder)					
Manufacturer's model no.	2VK6	2VK7	2VK8	2VK9	2VK10	2VK11
No. of stages	2	2	2	2	2	2
Gas ballast provision: Yes/No	No	No	No	No	No	No
Ultimate total pressure in Torr (with gas ballast off)	4×10^{-1}	4×10^{-1}	3×10^{-1}	3×10^{-1}	3×10^{-1}	3×10^{-1}
Pumping speed in litre/min at 760 Torr	10 000	15 000	20 000	30 000	45 000	66 500
Pump rotational speed, rev/min	150	145	135	125	122	104

MANUFACTURER:	New Jersey Machine							
Class of pump	Rough vacuum sliding vane rotary pump →							
Manufacturer's model no.	CAC	CA	CB	CC	CD	CE	CG	ACG
No. of stages	1	1	1	1	1	1	1	1
Gas ballast provision: Yes/No	No	No	No	No	No	No	No	No
Ultimate vacuum in inches of mercury	{20* 15	15 10	15 10	15 10	15 10	15 10	15 10	27·5 20
Pumping speed in litre/min at 760 Torr	76	110	164	186	208	262	326	450
Pump rotational speed, rev/min								

Remarks: * Upper figure for intermittent operation, lower figure for continuous operation. Pumps also act as compressors. All air-cooled. Pump is integral with motor. Circulating lubrication system includes an oil separation device.

MANUFACTURER:	Northey Rotary Compressors								
Class of pump	Rough vacuum lubricated* or oil-free rotary pumps with two rotors, having some similarities to Roots pumps.								
Manufacturer's model no.	82	83	101	102	162	163	165	167	1652
No. of stages	1	1	1	1	1	1	1	1	1
Gas ballast provision: Yes/No	No	No	No	No	No	No	No	No	No
Ultimate total pressure in Torr (with gas ballast off)	100	100	25	25	100† 25	100† 25	100† 25	100† 25	100† 25
Pumping speed in litre/min at 760 Torr	200	270	425	710	1250	1640	2500	3500	5000
Pump rotational speed, rev/min	1450	1450	1450	1450	1450	1450	1450	1450	1450

Remarks:

* Lubricated models (oil-drip feed from oil-metering device) have same performance as oil-free pumps except that the ultimate pressures for the oil-free water-cooled models are at 50 Torr whereas for the lubricated models the values are 25 Torr.

† Air-cooled (denoted by VAL) and water-cooled (denoted by VWL) versions are available. In all cases, the ultimate pressure provided by the air-cooled type is 100 Torr or 26 inch Hg and by the water-cooled models it is 25 Torr or 29 inch Hg.

MANUFACTURER:	Northey Rotary Compressors *cont'd*							
Class of pump	Lubricated* or oil-free rotary pumps with two rotors, having some similarities to Roots pumps.				Lubricated two-stage water-cooled rough vacuum pumps			
Manufacturer's model no.	1672	365	367	3610	101TVWL	102TVWL	162TVWL	163TVWL
No. of stages	1	1	1	1	2	2	2	2
Gas ballast provision: Yes/No	No	No	No	No	No	No	No	No
Ultimate total pressure in torr (with gas ballast off)	100† 25	25	25	25	5×10^{-1}	5×10^{-1}	5×10^{-1}	5×10^{-1}
Pumping speed in litre/min at 760 Torr	7100	8500	12 700	17 000	333	540	1030	1350
Pump rotational speed, rev/min	1450	960	960	960	1450	1450	1450	1450

MANUFACTURER:	Northey Rotary Compressors *cont'd*						
Class of pump	Lubricated two-stage water-cooled rough vacuum pumps →						
Manufacturer's model no.	165TVWL	167TVWL	1652/165TVWL	1672/167TVWL	365/365TVWL	375/365TVWL	3610/365TVWL
No. of stages	2	2	2	2	2	2	2
Gas ballast provision: Yes/No	No	No	No	No	No	No	No
Ultimate total pressure in Torr (with gas ballast off)	5×10^{-1}	5×10^{-1}	5×10^{-1}	5×10^{-1}	5×10^{-1}	5×10^{-1}	5×10^{-1}
Pumping speed in litre/min at 760 Torr	2080	2800	3900	5600	7100	10 700	14 200
Pump rotational speed, rev/min	1450	1450	1450	1450	1450	1450	1450

MANUFACTURER:	Northey Rotary Compressors *cont'd*						
Class of pump	Oil-free two-stage water-cooled rough vacuum pumps →						
Manufacturer's model no.	101TVWD	102TVWD	162TVWD	163TVWD	165TVWD	167TVWD	1652/165TVWD
No. of stages	2	2	2	2	2	2	2
Gas ballast provision: Yes/No	No	No	No	No	No	No	No
Ultimate total pressure in Torr (with gas ballast off)	1	1	1	1	1	1	1
Pumping speed in litre/min at 760 Torr	285	450	780	1030	1630	2260	3250
Pump rotational speed, rev/min	1450	1450	1450	1450	1450	1450	1450

MANUFACTURER:	**Northey Rotary Compressors** *cont'd*			
Class of pump	Oil-free two-stage water-cooled rough vacuum pumps →			
Manufacturer's model no.	1672/167TVWD	365/365TVWD	367/365TVWD	3610/365TVWD
No. of stages	2	2	2	2
Gas ballast provision: Yes/No	No	No	No	No
Ultimate total pressure in Torr (with gas ballast off)	1	1	1	1
Pumping speed in litre/min at 760 Torr	4500	6000	8800	11 700
Pump rotational speed, rev/min	1450	1450	1450	1450

MANUFACTURER:	**Pennwalt/Stokes**					
Class of pump	Rotary piston pumps →					
Manufacturer's model no.	149H	212H	412H	612H	212MB	412MB
No. of stages	1	1	1	1	2	2
Gas ballast provision: Yes/No	Yes	Yes	Yes	Yes	Yes	Yes
Ultimate total pressure in Torr (with gas ballast off)	1×10^{-2}	1×10^{-2}	1×10^{-2}	1×10^{-2}	5×10^{-4}	5×10^{-4}
Pumping speed in litre/min at 760 Torr	1840	3960	7400	14 800	5370	8800
Pump rotational speed, rev/min	490	500	500	500	3200 + 500	4950 + 500

Remarks: Provide also exhaust filters, flexible connectors, butterfly valves, gate valves, poppet valves, small angle valves and ball valves.

MANUFACTURER:	Precision Scientific								Red Point
Class of pump	Sliding vane rotary pumps →								Sliding vane rotary pump
Manufacturer's model no.	S-35*	D-25	D-75	D-150	D-300	D-500	D-1000	D-1500	69BF
No. of stages	1	2	2	2	2	2	2	2	1
Gas ballast provision: Yes/No	Yes	Yes	Yes	Yes	Yes	Yes	Yes	Yes	No
Ultimate total pressure in Torr (with gas ballast off)	2×10^{-2}	10^{-4}	10^{-4}	10^{-4}	10^{-4}	10^{-4}	10^{-4}	10^{-4}	2×10^{-2}
Pumping speed in litre/min at 760 Torr	35	25	75	150	300	500	1000	1500	100
Pump rotational speed, rev/min	1725	1725	1725	1725	1725	1725	1725	1725	690
Remarks:	* Model PV-35 provides pressure up to 15 lb/sq. in.								Provided with intake and exhaust check valves, large oil reservoir with forced oil circulation, intake filter, exhaust filter (oil-mist separator) and dial gauge. Also available: vapour condensing cold traps, freon refrigerator (–40°C).

MANUFACTURER:	Sargent-Welch Scientific								
Class of pump	Sliding vane rotary pumps →								
Manufacturer's model no.	1400	1405H	1405B	1402	1376	1397	1375	1398M	1398H
No. of stages	2	2	2	2	2	2	2	2	2
Gas ballast provision: Yes/No	Optional	Optional	Optional	Yes	Yes	Yes	Yes	Yes	Yes
Ultimate total pressure in Torr (with gas ballast off)	1×10^{-4}	6×10^{-5}	1×10^{-4}	1×10^{-4}	1×10^{-4}	1×10^{-4}	1×10^{-4}	1×10^{-4}	1×10^{-4}
Pumping speed in litre/min at 760 Torr	25	35	60	160	300	500	1000	1500	2000
Pump rotational speed, rev/min	580	340	525	525	525	400	335	245	340
Remarks:	Model 1392 is a combined mechanical-rotary/diffusion pump unit. Pumping speed at 10^{-2} Pa is 600 litre/min: ultimate total pressure obtainable is 10^{-4} Pa. Backing pump is model 1400.								

MANUFACTURER:	Sargent-Welch Scientific *cont'd*						
Class of pump	Sliding vane rotary pumps →						
Manufacturer's model no.	1410	1403	1399	1404	8805	8810	8815
No. of stages	1	1	1	1	2	2	2
Gas ballast provision: Yes/No	No	No	Yes	No	Yes	Yes	Yes
Ultimate total pressure in Torr (with gas ballast off)	2×10^{-2}	5×10^{-3}	$<2 \times 10^{-2}$	2×10^{-2}	10^{-4}	10^{-4}	10^{-4}
Pumping speed in litre/min at 760 Torr	21	100	35	35	50	100	150
Pump rotational speed, rev/min	450	375	750	300	1725	1725	1725
Remarks:					Direct drive pumps of 'DirecTorr' class: pump is directly coupled to electric-motor		

MANUFACTURER:	Sihi						
Class of pump	Rough vacuum liquid ring pumps →						
Manufacturer's model no.	LOHE20103	LOHE20107	LPHE40106	LPHE40411	LPHE40516	LOHE05501	LOHE25003
No. of stages	1	1	1	1	1	1	1
Ultimate total pressure in Torr	110	110	110	110	110	25	25
Pumping speed in litre/min at 760 Torr	400	920	1670	3000	4300	102	400
Pump rotational speed, rev/min	3400	3400	1740	1740	1740	1700	3400

MANUFACTURER:	Sihi *cont'd*					
Class of pump	Rough vacuum liquid ring pumps →					
Manufacturer's model no.	LRVE25007	LOHE25309	LPHE45008	LPHE45311	LPHE45316	GOVE3211, LOHE25007 or LRVE25007
No. of stages	1	1	1	1	1	2*
Ultimate total pressure in torr	25	25	25	25	25	4-5
Pumping speed in litre/min at 760 Torr	800	800	1830	2250	2840	370
Pump rotational speed, rev/min	3400	3500	1740	1740	1740	2800

Remarks:

* Second stage is an air ejector pump so system comprises an air ejector pump backed by a liquid ring pump.

MANUFACTURER:	**Sihi** *cont'd*					
Class of pump	Rough vacuum liquid ring pumps →					
Manufacturer's model no.	GOVE3212 LOHE25309	GOVE4011 LPHE45008	GPVE4012 LPHE45316	DSW080 022 030 + LPHE45 404	DSW080 033 030 + LPHE45 404	DSW080 033 030 + LPHE45 406
No. of stages	2*	2*	2*	2‡	2‡	2‡
Ultimate total pressure in Torr	4-5	4-5	3-4	4	4	4
Pumping speed in litre/min at 760 Torr	550	1170	2000†	2840§	3300§	4350§
Pump rotational speed, rev/min	2900	1450	1450	1450	1450	1450

Remarks:
* Second stage is an air ejector pump so system comprises an air ejector pump backed by a liquid ring pump.
† Over the pressure range from 15 to 110 Torr.
‡ Second stage is a steam injector pump, type DSW so system comprises a steam injector pump backed by a liquid ring pump.
§ At a pressure of 8 Torr at which pumping speed is a maximum.

MANUFACTURER:	**Sihi** *cont'd*				
Class of pump	Rough vacuum liquid ring pumps →				
Manufacturer's model no.	DSW080 044 030 + LPHE45406	DSW080 044 030 + LPHE45408	LPH11055	LRMA00602	LRMA10604
No. of stages	2‡	2‡	1	1	1
Ultimate total pressure in Torr	4	4	80	<150	<80
Pumping speed in litre/min at 760 Torr	5100§	5900§	180 000	136	400
Pump rotational speed, rev/min	1450	1450	4700		

Remarks: 'Vac block' liquid ring vacuum pumps designed with the pump unit close coupled to a standard electric motor.

MANUFACTURER:	Sihi *cont'd*				
Class of pump	Rough vacuum liquid ring pumps			Liquid-ring pump in conjunction with gas-ejector pump	
Manufacturer's model no.	LOH05501	LOH25003*	LPH11535†	GOV3211* with LOH25007	GPV10012* with LPH7630
No. of stages	2	2	2	2: gas-ejector pump backed by liquid-ring pump	2: gas-ejector pump backed by liquid-ring pump
Ultimate total pressure in Torr	45	25	25	4	3
Pumping speed in litre/min at 760 Torr	75	300	150 000	275	12 700
Pump rotational speed, rev/min	1450	2800	335–475		
Remarks:	* The smallest pump in the range supplied capable of an ultimate pressure of 25 Torr; † The largest pump in this range of 2-stage liquid-ring pumps. There are 25 different sizes between the smallest and the largest of these liquid ring pumps. Accessories available include separators for operating liquids.			GOV and GPV models are gas-ejector pumps: * there are 17 different sizes between the smallest pump GOV3211 and the largest pump GPV10012. LOH and LPH models are liquid-ring pumps. The maximum allowable backing pressure for these gas-ejector pumps is 40 Torr.	

MANUFACTURER:	Teltron	Varian	
Class of pump	Sliding vane rotary pumps	Carbon vane rotary pump; oil-less, clean operation.	GASP roughing pump operated by compressed air
Manufacturer's model no.	TEL505b	931-5001	942-1000
No. of stages	1	1	1
Gas ballast provision: Yes/No	Yes	No	No
Ultimate total pressure in Torr (with gas ballast off)	$<5 \times 10^{-2}$	250 (continuous) 125 (intermittent)	125
Pumping speed in litre/min at 760 Torr	16·7	700	
Pump rotational speed, rev/min	1400	1060	
Remarks:	Pump is mounted on a die-cast base accommodating motor-starter, protection devices, and solenoid-operated air-admittance valve. Designed for educational work to form a compatible pumping station with other accessories.	Vanes are carbon-graphite, self-sealing and self-adjusting	Operated by compressed air at 70–100 lb/sq. inch, this pump is designed to rough-pump a vacuum system to about 125 Torr before bringing into operation sorption pumps to reduce the pressure into the sub-torr region.

MANUFACTURER:	Veeco								
Class of pump	Sliding vane rotary pumps →								
Manufacturer's model no.	1410	1400	1404	1399	1405	1405B	1403	1376	1397
No. of stages	1	2	1	1	2	2	1	2	2
Gas ballast provision: Yes/No									
Ultimate total pressure in Torr (with gas ballast off)									
Pumping speed in litre/min at 760 Torr	21	25	33·4	35	35	60	100	300	500
Pump rotational speed, rev/min									

MANUFACTURER:	Veeco *cont'd*				J. M. Voith				
Class of pump	Sliding vane rotary pumps →				Liquid ring pumps →				
Manufacturer's model no.	1375	1398	1398	1402	2	3	3a	4	5k
No. of stages	2	2	2	2	1	1	1	1	1
Gas ballast provision: Yes/No					No	No	No	No	No
Ultimate total pressure in Torr (with gas ballast off)									
Pumping speed in litre/min at 760 Torr	1000	1500	2000	150	3600	4100	6500	9500	13 800
Pump rotational speed,* rev/min					1450	1450	960	960	730

Remarks:

* Maximum speed; lower speeds also used.

MANUFACTURER:	J. M. Voith *cont'd*								
Class of pump	Liquid ring pumps →								
Manufacturer's model no.	6	7k	8k	9	10	11	12k	2E	3E
No. of stages	1	1	1	1	1	1	1	1	1
Gas ballast provision: Yes/No	No	No	No	No	No	No	No	No	No
Ultimate total pressure in Torr (with gas ballast off)									
Pumping speed in litre/min at 760 Torr	17 200	25 000	33 500	43 500	44 500	62 000	106 000	5000	7300
Pump rotational speed,* rev/min	730	585	585	480	480	480	300	1450	1450

Remarks: * Maximum speed; lower speeds also used.

MANUFACTURER:	J. M. Voith *cont'd*								
Class of pump	Liquid ring pumps →								
Manufacturer's model no.	4E*	5E*	6E*	7E	8E	9E*	10E	11E*	12E
No. of stages	1	1	1	1	1	1	1	1	1
Gas ballast provision: Yes/No	No	No	No	No	No	No	No	No	No
Ultimate total pressure in Torr (with gas ballast off)									
Pumping speed in litre/min at 760 Torr	10 000	15 000	22 500	30 500	45 000	63 000	77 000	88 000	138 000
Pump rotational speed,† rev/min	960	960	730	730	667	535	535	455	363

Remarks: * Calculated values.
† Maximum speed; lower speeds also used.

MANUFACTURER:	J. M. Voith *cont'd*		Westinghouse		
Class of pump	Liquid ring pumps		Rough vacuum rotary pumps		
Manufacturer's model no.	14E	15E*	VC1	VB2	VC3
No. of stages	1	1	1	1	1
Gas ballast provision: Yes/No	No	No	No	No	No
Ultimate total pressure in Torr (with gas ballast off)			Given for all types as degree of vacuum attainable = 95%		
Pumping speed in litre/min at 760 Torr	170 000	270 000	3420	4830	8715
Pump rotational speed†, rev/min	324	270	1500 max.	1500 max.	1500 max.
Remarks:	*Calculated values. † Maximum speed; lower speeds also used.		Vacuum-compressor pumps for use as vacuum brakes on rolling stock.		

2.1.6 Roots pumps

MANUFACTURER:	Alcatel		Balzers			
Manufacturer's model no.	Series MIV	Series MAV	MTP200	MTP400	MTP1000	MTP2000
No. of stages						
Maximum pumping speed in litre/min			3700	6700	18 000	34 000
Pump rotational speed, rev/min						
Remarks:	For pumping speeds below 2·8 x 10^{-1} litres/min motor under vacuum (no shaft seal); pumps above this speed have shaft seal.		Wide range of accessories available; also standard combinations of Roots pumps with backing pumps and condensers.			

MANUFACTURER:	Balzers *cont'd*					Brizio Basi
Manufacturer's model no.	MTP4000	MTP5000	MTP7500	MTP12000	MTP50000	BRV
No. of stages						1
Maximum pumping speed in litre/min	67 000	84 000	127 000	192 000	810 000	1 333 000
Pump rotational speed, rev/min						1600

Remarks: Wide range of accessories available; also standard combinations of Roots pumps with backing pumps and condensers

MANUFACTURER:	Dresser Industries							
Manufacturer's model no.	1020	1026	1215	1220	1224	1232	1419	1424
No. of stages	1	1	1	1	1	1	1	1
Maximum pumping speed in litre/min	90 000*	126 000†	78 000‡	104 000‡	130 000*	180 000†	116 000‡	147 000‡
Pump rotational speed, rev/min	1750	1750	1450	1450	1450	1450	1250	1250

Remarks:
* At 20 inch Hg. vac.
† At 15 inch Hg. vac.
‡ At 22 inch Hg. vac.

MANUFACTURER;	Dresser Industries *cont'd*						
Manufacturer's model no.	1428	1432	1537	1832	1836	1842	1847
No. of stages	1	1	1	1	1	1	1
Maximum pumping speed in litre/min	180 000*	214 000†	247 000†	252 000‡	297 000*	360 000†	405 000†
Pump rotational speed, rev/min	1250	1250	1250	970	970	970	970

Remarks:
* At 20 inch Hg. vac.
† At 15 inch Hg. vac.
‡ At 22 inch Hg. vac.

MANUFACTURER:	Edwards High Vacuum						
Manufacturer's model no.	ER100	ER200	ER600	ER1000	ER2000	ER2500	ER5000
No. of stages							
Maximum pumping speed in litres/min	88	180	600	800	1650	2500	4800
Pump rotational speed, rev/min	2850	2850	2850	2150	1430	1430	1430
Remarks:	All pumping speeds are for air at 0·1 Torr.						

MANUFACTURER:	General Engineering						Hick Hargreaves
Manufacturer's model no.	AGMB130	AGMB300	AGMB570	HGMB750	HGMB1300	HGMB5000	A wide range of Roots pumps is supplied.
No. of stages							
Maximum pumping speed in litre/min	3300*	6800†	2100‡; 3900§	34 000‖; 37 000¶	73 500**; 76 500††	115 000‡‡	
Pump rotational speed, rev/min							

Remarks:

* At 10^{-1} Torr: used with rotary pump 6K518 forms combination model no. MBG1; with rotary pump 6KD30 forms combination model no. MBG2 with a higher Roots cut in pressure.
† At 10^{-1} Torr: used with rotary pump 6KD30 forms combination model no. MBG3 ; with rotary pump 6T80 forms combination model no. MBG4.
‡ At 760 Torr, with rotary pump 6T80 forming combination model no. MBG5.
§ At 760 Torr, with rotary pump 6KT150A forming combination model no. MBG6.
‖ At 10^{-1} Torr: used with rotary pump 6KT150A forming combination model no. MBG7.
¶ At 10^{-1} Torr with rotary pump 6KT300 forming combination model no. MBG8.
** At 10^{-1} Torr: used with rotary pump 6KT300 forming combination model no. MBG9.
†† At 10^{-1} Torr: used with rotary pump 6KT500 forming combination model no. MBG10.
‡‡ At 10^{-1} Torr: used with rotary pump 6KT500 forming combination model no. MBG11.
A wide range of Roots pumps is supplied with displacement up to 570 000 litre/min, and they are useable with a wide variety of backing pumps; the more popular standard types are listed above.

MANUFACTURER:	Kinney Vacuum						
Manufacturer's model no.	KMBD-150	KMBD-400	KMBD-540	KMBD-850	KMBD-1600	KMBD-1601	KMBD-1602
No. of stages	3*	3†	2‡	2§	2‖	2¶	2**
Maximum pumping speed in litre/min	3980	7500	10 150	18 400	34 000	28 400	26 250
Pump rotational speed, rev/min	3600	3600	3600	3600	3600	3000	2700

Remarks:

Allowable differential pressure for all models is 380 Torr = 5 x 10^4 Pa (0·5 atm).
* Backed by rotary piston pump model no KC-15; models KTC-21, KD-30 and KDH-65 are also used.
† Backed by rotary piston pump model no. KTC-21; models KD-30, KDH-65, KDH-130, KT-150 are also used.
‡ Backed by rotary piston pump model no. KD-30; models KDH-80, KT-150 and KT-300 are also used.
§ Backed by rotary piston pump model no. KDH-80; models KT-150 and KT-300 are also used.
‖ Backed by rotary piston pump model no. KT-150; models KT-300 and KT-500 are also used.
¶ Backed by rotary piston pump model no. KT-150; models KT-300 and KT-500 are also used.
** Backed by rotary piston pump model no. KT-150; models KT-300 and KT-500 are also used.
Pumping speed obtainable depends on capacity of backing rotary pump.

MANUFACTURER:	Kinney Vacuum *cont'd*						
Manufacturer's model no.	KMBD-2000	KMBD-2700	KMBD-3200	KMBD-4000	KMBD-5400	KMBV-5700	KMBV-11000
No. of stages	2*	2†	2‡	2§	2‖	2¶	2**
Maximum pumping speed in litre/min	44 000	51 000	68 000	84 900	119 000	144 000	280 000
Pump rotational speed, rev/min	3600	3600	3600	3600	3600	3600	3600

Remarks:

Allowable differential pressure for all models is 380 Torr = 5 x 10^4 Pa (0·5 atm).
* Backed by rotary piston pump model no. KT-150, models KT-300 and KT-500 are also used.
† Backed by rotary piston pump model no. KT-150; models KT-300, KT-500 and KD-850 are also used.
‡ Backed by rotary piston pump model no. KT-300; models KT-500 and KD-850 are also used.
§ Backed by rotary piston pump model no. KT-300; models KT-500 and KD-850 are also used.
‖ Backed by rotary piston pump model no. KT-500; model KD-850 also used.
¶ Backed by rotary piston pump model no. KT-500.
** Backed by rotary piston pump model no. KD-850.

MANUFACTURER:	Leybold-Heraeus					
Manufacturer's model no.	RUVAC WS1000	RUVAC WS2000	WS152	WS252	RUVACZ05	R6000
No. of stages	1	1	2	2	2	1
Maximum pumping speed in litre/min	15 300	29 200	2106	3500	1500	100 000
Pump rotational speed, rev/min	3000	3000	3000	3000		3000

Remarks: WS series is without shaft leadthrough (canned motor); Z05 has shaft leadthrough; R series is without shaft leadthrough (vacuum jacket); RA series has shaft leadthrough.

MANUFACTURER:	Leybold-Heraeus *cont'd*					
Manufacturer's model no.	RUVAC RA300	RUVAC RA500	RUVAC RA700	RUVAC9001	RUVAC13 000	RUVAC WA150
No. of stages	1	1	1	1	1	1
Maximum pumping speed in litre/min	50 000	83 000	117 000	153 000	210 000	2334
Pump rotational speed, rev/min	3000	1000	3000	1500	2000	

Remarks: WA series has shaft leadthrough; WS series is without shaft leadthrough (canned motor).

MANUFACTURER:	Leybold-Heraeus *cont'd*				
Manufacturer's model no.	RUVAC WA250	RUVAC WA500	RUVAC WA1000	RUVAC WA2000	RUVAC WS150
No. of stages	1	1	1	1	1
Maximum pumping speed litre/min	3830	8000	15 500	30 800	2160
Pump rotational speed, rev/min	3000	3000	3000	3000	3000

Remarks: WA series has shaft leadthrough; WS series is without shaft leadthrough (canned motor).

MANUFACTURER:	Leybold-Heraeus *cont'd*		Pennwalt/Stokes						
Manufacturer's model no.	RUVAC WS250	RUVAC WS500	1730	1731	1721	1722	1723	1724	1725
No. of stages	1	1	2	2	2	2	2	2	2
Maximum pumping speed in litre/min	3500	7300	14 600	14 600	37 000	37 000	109 000	81 500	154 000
Pump rotational speed, rev/min	3000	3000	3600	3600	1700	1700	1450	1450	1450

MANUFACTURER:	Sihi	
Manufacturer's model no.	RPH200*	RPH1200†
No. of stages	1	1
Maximum pumping speed in litre/min	3700	193 000
Pump rotational speed, rev/min	2900	1450

Remarks:
* The smallest Roots pump in this range.
† The largest Roots pump in this range.
There are 6 different sizes between the smallest and largest of these liquid ring pumps. Accessories available include pressure control units, manual operating and automatic process control devices.

2.1.7 Sorption pumps

A pump in which a renewable porous substance of large surface area (e.g. molecular sieve material) is used, usually at low temperature.

MANUFACTURER:	AEI Scientific Apparatus		Aero Vac
Manufacturer's model no.	SP3	SP150	22 HP111
Sorbent material used	Zeolite	Zeolite	Linde molecular sieve 5A
Mass of sorbent (gram)	1360	150	1200
Normal operating temperature, °C	−196	−196	−196 (liquid nitrogen)
Remarks:	Normal working range down to 10^{-4} Torr, increased to 10^{-7} Torr by backing with a rotary pump.		Three models having different connections to the vacuum system are available.

MANUFACTURER:	Cooke	Edwards High Vacuum
Manufacturer's model no.	Standard or special types supplied	EZ500
Sorbent material used	Zeolite 5A	Aluminium calcium silicate
Mass of sorbent (gram)	Dependent upon size of pump requested	500
Normal operating temperature, °C	–196 (liquid nitrogen)	–196 (liquid nitrogen)

MANUFACTURER:	Ion Equipment	Leybold-Heraeus	
Manufacturer's model no.	SP-11*	ASP20	ASP32
Sorbent material used	Molecular sieve 5A	Zeolite 13X	Zeolite 13X
Mass of sorbent (gram)	1350	400	2000
Normal operating temperature, °C	–196	–196 (liquid nitrogen)	–196 (liquid nitrogen)

Remarks:

* The standard pump is SP-11, the double pump is SP-12 and the triple pump is SP-13
Recommended maximum pumping capacities are: SP-11, 40 litre; SP-12, 110 litre and SP-13, 220 litre. Bakeout temperature for all sizes is 250°C. The pump body and cooling fins are made of aluminium, the pump neck and connecting flange of type 304 stainless steel. A range of accessories is available.

MANUFACTURER:	RIBER		Torr Vacuum Products			UHV
Manufacturer's model no.	PA10	PA50	CSP100	CSP200	CSP400	Made to order
Sorbent material used	Zeolite	Zeolite	Zeolite 5A	Zeolite 5A	Zeolite 5A	
Mass of sorbent (gram)	1200	16 000	910	1810	3630	
Normal operating temperature, °C			–196 (liquid nitrogen)	–196 (liquid nitrogen)	–196 (liquid nitrogen)	

Results:

Normal working pressure range: 760 to 10^{-4} Torr.

MANUFACTURER:	Ultek		Vacuum Generators	
Manufacturer's model no.	236-1500	236-1000	MSS100	MSS200
Sorbent material used	5A zeolite	5A zeolite	Molecular sieve 5A	Molecular sieve 5A
Mass of sorbent (gram)	900	900	600	1200
Normal operating temperature, °C	–196	–196	–196	–196
Remarks:	Working pressure range normally from 760 → 10^{-2} Torr. Bakeout temperature: 300°C. A special construction utilizing copper finned heat exchanger tubes assuring that no sorbent material is more than 0·5 inch from a liquid-nitrogen cooled surface.		Single MSS100 used on systems of up to 20 litre volume. Single MSS200 used on systems of up to 40 litre volume. Two MSS100 operated in sequence on systems of up to 100 litre volume. Two MSS200 operated in sequence on systems of up to 250 litre volume. Normal working pressure range from 760 to 10^{-2} Torr.	

MANUFACTURER:	Varian			Veeco
Manufacturer's model no.	949-0010	949-0020	949-0030	SP500
Sorbent material used	Zeolite 5A	Zeolite 5A	Zeolite 5A	Zeolite molecular sieve
Mass of sorbent (gram)	1400	2400	3500	–
Normal operating temperature, °C	–196	–196	–196	–
Remarks:				Bakeable to 250°C.

2.1.8 Sputter-ion pumps

MANUFACTURER:	AEI Scientific Apparatus					
Manufacturer's model no.	P10	P25	P60	P120	P300	P500
Pump mouth internal diameter (mm)	25·4	50·8 x 101·6	114·3	114·3	114·3	152·4
Pumping speed for air in litre/second. (Pressure at which measured in Torr)	10 (10^{-6})	24 (10^{-6})	50 (10^{-6})	120 (10^{-6})	300 (10^{-6})	450 (10^{-6})
Operating voltage (kV)	3	5	5	5	5	5
Magnetic flux density (tesla)	0·145	0·19	0·2	0·2	0·2	0·2

Remarks: Working pressure for P10 is 10^{-10} Torr; for the other models it is between 10^{-5} and below 10^{-12} Torr. For all models the maximum starting pressure is 10^{-2} Torr and the bakeout temperature 450°C.

MANUFACTURER:	Aero Vac					
Manufacturer's model no.	22TP152	22TP202	22TP250*	22TP275†	22TP300‡	22TP410§
Pump mouth internal diameter (mm)	25	100	100	100	200	200
Pumping speed for air in litre/second. (Pressure at which measured in Torr)	8 (10^{-7})	25 (10^{-7})	100 (10^{-7})	250 (10^{-7})	500 (10^{-7})	1200 (10^{-7})
Operating voltage (kV)	5·2 (max.)	5·2 (max.)	5·2 (max).	5·2 (max.)	5·2 (max.)	5·2 (max.)
Magnetic flux density (tesla)						

Remarks: Triode sputter-ion pumps of all-stainless-steel construction with Alnico magnets, bakeable to 400°C with magnet in place. Pumping speed for inert gases up to 30% of pumping speed for air.

* Model 22TP254 is a 'nude' version.
† Model 22TP279 is a 'nude' version.
‡ 22TP300 is a welded version; model 22TP301 is a flanged version where contamination is probable so requires occasional cleaning.
§ 22TP410 is a welded version; model 22TP411 is a flanged version where contamination is probable so requires occasional cleaning.

MANUFACTURER:	Cenco		Edwards High Vacuum	
Manufacturer's model no.	93800-002*	93800-003	EP1-30	EP1-125
Pump mouth internal diameter (mm)	19		80	152·4
Pumping speed for air in litre/second at 10^{-6} Torr	8–10	125	30*	125*
Operating voltage (kV)	3·9	3·9	12 (max.)	12 (max.)
Magnetic flux density (tesla)	0·14	0·14	0·13 to 0·15	0·13 to 0·15
Remarks:	* Data given is for ion pump system (93218) which has max. bakeout temperature of 450°C without magnet and 300°C with magnet.		* Pumping speed for nitrogen	

MANUFACTURER:	Ion Equipment					
Manufacturer's model no.	IP-011	IP-020	IP-025	IP-050	IP-100	IP-150
Pump mouth internal diameter (mm)	38	38	50	152	152	152
Pumping speed for air in litre/second at 10^{-5} Torr	11	20	25	50	100	150
Operating voltage (kV)	4·75	4·75	4·75	4·75	4·75	4·75
Magnetic flux density (tesla)	High strength Ferrites used	High strength Ferrites used	High strength Ferrites used	High strength Ferrites used	High strength Ferrites used	High strength Ferrites used
Remarks:	Known as 'magnetic ion pumps': to each IP (ion pump) number there corresponds a NP (noble pump) number. Temperature limits are 500°C for stainless steel 304 body and 250°C for ferrite magnet slabs.					

MANUFACTURER:	Ion Equipment			Leybold-Heraeus		
Manufacturer's model no.	IP-200	IP-400	IP-002S*	IZ02*	IZ1*	IZ120E
Pump mouth internal diameter (mm)	152	152	19†	27	46	
Pumping speed for air in litre/second at 10^{-5} Torr	200	400	2	0·2	1·0	120
Operating voltage (kv)	4·75	4·75	3·5	4 (max)	4 (max)	7·5 (max)
Magnetic flux density (tesla)	High strength Ferrites used	High strength Ferrites used	0·12	0·1	0·1	0·15

Remarks:

* IP-002S: stainless steel tubulation; IP-002C: copper tubulation; IP-002G: glass tubulation; IP-002F: flanged tubulation.
† Outside diameter of pump with stainless steel tubulation.

* Appendage pumps designed for flangeless connection to vacuum chambers or systems. Directly welded or brazed to the system.

MANUFACTURER:	Leybold-Heraeus *cont'd*							
Manufacturer's model no.	IZ8	IZ50	IZ80	IZ120	IZ230	IZ340	IZ450	IZ700
Pump mouth internal diameter (mm)	35	50	80	120	230	340	450	700
Pumping speed for air in litre/second at 10^{-6} Torr	8	50	80	120	230	340	450	700
Operating voltage (kV)	4	4	7·5	7·5	7·5	7·5	7·5	7·5
Magnetic flux density (tesla)	0·09	0·14	0·2	0·15	0·15	0·15	0·15	0·15

MANUFACTURER:	RIBER								
Manufacturer's model no.	PI1	PI3	PI11	PI25	PI50	PI100	PI200	PI400	PI600
Pump mouth internal diameter (mm)	10	18	38	38	150	150	150	150	150
Pumping speed for air in litre/second at 10^{-6} Torr	1	3	11	25	50	100	200	400	600
Operating voltage (kV)	3·2	3·2	4·75	4·75	4·75	4·75	4·75	4·75	4·75
Magnetic flux density (tesla)									

Remarks: Working pressure range: 10^{-4} to 10^{-10} Torr. Maximum bakeout temperatures: with magnet, 200°C; without magnet, 400°C.

MANUFACTURER:	RIBER *cont'd*			Ultek			
Manufacturer's model no.	PI1000	PI2200	PI4000	202-0100	202-0500	202-0800	202-2000
Pump mouth internal diameter (mm)	250	450	450	19*	25†	38†	38†
Pumping speed for air in litre/second at 10^{-6} Torr	1000	2200	4000	1	5	11	20
Operating voltage (kV)	4·75	4·75	4·75	4·75	3·0	5·0	5·0
Magnetic flux density (tesla)					High strength Ferrite slabs used‡	High strength Ferrite slabs used‡	High strength Ferrite slabs used‡

Remarks: (Ultek) Range of pumps known as D.I. (differential ion), stated to give very favourable pumping speeds for the inert gases. Bakeable to 500°C or 400°C with magnet in position.

* Outside diameter of stainless steel tubulation; other sizes are available with copper or glass tubulation. A full range of accessories for these pumps is available; also 'modular ion pump assemblies'.

† A main flange 'Curvac' seal of given size.

‡ Temperature limit: 250°C.

MANUFACTURER:	**Ultek** *cont'd*			
Manufacturer's model no.	202-2500	206-0505	206-1000	206-1500
Pump mouth internal diameter (mm)	50†	152†	152†	152†
Pumping speed for air in litre/second at 10^{-6} Torr	25	50	100	150
Operating voltage (kV)	5·0	4·75	4·75	4·75
Magnetic flux density (tesla)	High strength Ferrite slabs used‡	High strength Ferrite slabs used‡	High strength Ferrite slabs used‡	High strength Ferrite slabs used‡

Remarks:
† A main flange 'Curvac' seal of given size.
‡ Temperature limit: 250°C.

MANUFACTURER:	**Ultek**				
Manufacturer's model no.	206-2000	206-4000	206-5000	206-6000	206-7100
Pump mouth internal diameter (mm)	152*	152*	152*	152*	254†
Pumping speed for air in litre/second at 10^{-6} Torr	200	400	500	600	1200
Operating voltage (kV)	4·75	4·75	4·75	4·75	4·75
Magnetic flux density (tesla)	High strength Ferrite slabs used‡	High strength Ferrite slabs used‡	High strength Ferrite slabs used‡	High strength Ferrite slabs used‡	High strength Ferrite slabs used‡

Remarks:
* A main flange 'Curvac' seal of given size.
† A 10 inch wire seal on main flange.
‡ Temperature limit: 250°C.
A full range of accessories for these pumps is available.

MANUFACTURER:	**Varian**					
Manufacturer's model no.	913-0036*	913-0037†	913-0038‡	913-0039*	913-0040†	913-0041‡
Pump mouth internal diameter (mm)	6·3	8	8	6·3	8	8
Pumping speed for air in litre/second at 10^{-7} Torr	0·2	0·2	0·2	0·2	0·2	0·2
Operating voltage (kV)	3·5	3·5	3·5	3·5	3·5	3·5
Magnetic flux density (tesla)	0·125/0·1	0·125/0·1	0·125/0·1	0·125/0·1	0·125/0·1	0·125/0·1

Remarks:
* Glass tubulation.
† Copper.
‡ Stainless steel.

MANUFACTURER:	**Varian** *cont'd*					
Manufacturer's model no.	913-0032‡	913-0034*	913-0035†	913-5000§	911-5001‖	911-5027¶
Pump mouth internal diameter (mm)	16·6	15·8	8	16·6	38	38
Pumping speed for air in litre/second at 10^{-7} Torr	2	2	2	2	8	11
Operating voltage (kV)	3·5	3·5	3·5	3·5	3·5	7·5
Magnetic flux density (tesla)	0·12	0·12	0·12	0·12	0·12	0·12

Remarks:
§ Mini flange.
‖ Tee style.
¶ Water-cooled, high throughput.

MANUFACTURER:	Varian					
Manufacturer's model no.	911-5030*	911-5000	911-5032*	911-5034*	912-7000	912-7006*
Pump mouth internal diameter (mm)	38	38	72	108	152	152
Pumping speed for air in litre/second at 10^{-7} Torr	20	8	30	60	140	110
Operating voltage (kV)	5	3·5	5	5	7·5	5
Magnetic flux density (tesla)	0·12	0·12	0·12	0·12	0·12	0·12

Remarks: * Triode.

MANUFACTURER:	Varian *cont'd*					
Manufacturer's model no.	912-7008	912-7014*	912-7016	912-7022*	912-7024	912-7030*
Pump mouth internal diameter (mm)	152	152	152	152	275	275
Pumping speed for air in litre/second at 10^{-7} Torr	270	220	500	400	1000	800
Operating voltage (kV)	7·5	5	7·5	5	7·5	5
Magnetic flux density (tesla)	0·12	0·12	0·12	0·12	0·12	0·12

MANUFACTURER:	Veeco							
Manufacturer's model no.	PD101	PD150	PD201	PD251	PD301	PD401	PD601	PD1201
Pump mouth internal diameter (mm)								
Pumping speed for air in litre/second	11	50	100	150	200	400	600	1200
Operating voltage (kV)								
Magnetic flux density (tesla)								

MANUFACTURER:	Veeco							
Manufacturer's model no.	PN103 PG102	PN152 PG151	PN203 PG203	PN252 PG251	PN303 PG302	PN404 PG403	PN603 PG602	PN1201 PG1200
Pump mouth internal diameter (mm)								
Pumping speed for air in litre/second	11	50	100	150	200	400	600	1200
Operating voltage (kV)								
Magnetic flux density (tesla)								
Remarks:	Noble gas magnetic ion pumps. PN series are diode ion pumps, PG series are triode ion pumps.							

MANUFACTURER:	Veeco					
Manufacturer's model no.	MI-20	MI30	MI75	MI150	MI225	MI300
Pump mouth internal diameter (mm)	38·1 Main flange	60·3 Main flange	152·4 Main flange	152·4 Main flange	152·4 Main flange	152·4 Main flange
Pumping speed for air in litre/second	20	30	75	150	225	300
Operating voltage (kV)						
Magnetic flux density (tesla)						
Remarks:	Magnets bakeable to 250°C, pump elements to 400°C.					

MANUFACTURER:	Veeco *cont'd*			
Manufacturer's model no.	MI600	MI900		TG500
Pump mouth internal diameter (mm)	152·4 Main flange	152·4 Main flange		
Pumping speed for air in litre/second	600	900		500 (2730 internal)
Operating voltage (kV)				
Magnetic flux density (tesla)				
Remarks:				Combines titanium sublimation and sputtering.

2.1.9 Steam ejector pumps

Hick Hargreaves

A wide range of steam ejector and air and steam eductor pumps is supplied.

Jet Vac

Single stage ejector Model number	Suction port I.D. (mm)	Steam consumption (kg per hour)
S10	38	20
S20	50	30
S30	75	50
S40	100	70
S50	150	100
S60	150	115

Number of steam ejector stages	Operating pressure (Torr)	Ultimate pressure (Torr)
1	76	40
2	10 to 100	5
3	2 to 20	1
4	0·2 to 3	5×10^{-2} to 10^{-1}
5	0·3	5×10^{-3} to 10^{-2}

Pumps are made to order for large-scale rough vacuum plant, chiefly used in chemical engineering industry.

Leybold-Heraeus

WDS steam ejector pump (glass): ultimate total pressure 2 Torr, pumping speed for air at 4 Torr is 33·3 litre/minute.

Penberthy Division, Houdaille Industries

U series of ejectors give greatest operating efficiency in pressure range from 150 to 300 Torr; L series give greatest operating efficiency in pressure range from 75 to 150 Torr; 2NC series of two-stage non-condensing ejectors perform efficiently in pressure range from 12 to 75 Torr.

Model no.	L-1 U-1	L-2 U-2	L-3 U-3	L-4 U-4	L-5 U-5	L-6 U-6	L-7 U-7	L-8 U-8	L-9 U-9
Steam consumption in lb/hour	85	125	195	270	370	480	610	755	910

Model no.	L-10 U-10	L-11 U-11	L-12 U-12	L-13 U-13	L-14 U-14	L-15 U-15	L-16 U-16	L-17 U-17	L-18 U-18
Steam consumption in lb/hour	1090	1280	1480	1820	2190	2580	3030	3780	4710

Note: 1 lb = 0·454 kg.

Model no.	2NC1	2NC2	2NC3	2NC4	2NC5	2NC6	2NC7	2NC8	2NC9
Steam consumption in lb/hour	106	160	240	330	450	590	740	920	1110

Model no.	2NC10	2NC11	2NC12	2NC13	2NC14	2NC15	2NC16	2NC17	2NC18
Steam consumption in lb/hour	1320	1525	1800	2200	2660	3140	3660	4600	5700

A wide range of two-stage condensing ejectors is also available in units for suction pressure ranges from 12·5 to 125 Torr with motive steam pressures of 80 to 200 lb/sq. in. in 20 lb/sq. in. increments and cooling water supply temperatures not above 32°C.

Model no.	1C16	2C16	2C26	3C16	3C26	3C36	4C16	4C26	4C36
Max. air load (lb/hour)	20	20	35	20	35	50	20	35	50
Motive steam (lb/hour)	106	114	154	120	160	230	135	175	245

Model no.	4C46	5C16	5C26	5C36	5C46	5C58	6C16	6C26	6C36
Max. air load (lb/hour)	75	20	35	50	75	100	20	35	50
Motive steam (lb/hour)	320	155	195	265	340	440	175	215	285

Model no.	6C46	6C58	6C68	7C28	7C38	7C48	7C510	7C610	7C710
Max. air load (lb/hour)	75	100	130	35	50	75	100	130	170
Motive steam (lb/hour)	360	460	570	240	310	385	485	595	725

Model no.	8C28	8C38	8C48	8C510	8C610	8C710	8C810	9C28	9C38
Max. air load (lb/hour)	35	50	75	100	130	170	210	35	50
Motive steam (lb/hour)	265	335	410	510	620	750	895	295	365

Note: 1 lb = 0·454 kg.

Model no.	9C48	9C510	9C610	9C712	9C812	9C912	10C310	10C410	10C510
Max. air load (lb/hour)	75	100	130	170	210	250	50	75	100
Motive steam (lb/hour)	440	540	650	780	925	1080	400	475	575

Model no.	10C610	10C712	10C812	10C912	10C1012	11C310	11C410	11C510	11C610
Max. air load (lb/hour)	130	170	210	250	300	50	75	100	130
Motive steam (lb/hour)	685	815	960	1115	1295	435	510	610	720

Model no.	11C712	11C812	11C912	11C1014	11C1114	12C310	12C410	12C510	12C610
Max. air load (lb/hour)	170	210	250	300	350	50	75	100	130
Motive steam (lb/hour)	850	995	1150	1330	1520	470	545	645	755

Model no.	12C712	12C812	12C912	12C1014	12C1114	12C1214	13C312	13C412	13C512
Max. air load (lb/hour)	170	210	250	300	350	400	50	75	100
Motive steam (lb/hour)	885	1030	1185	1365	1555	1755	535	610	710

Model no.	13C612	13C712	13C812	13C912	13C1014	13C1114	13C1218	13C1318	14C312
Max. air load (lb/hour)	130	170	210	250	300	350	400	500	50
Motive steam (lb/hour)	790	950	1095	1250	1430	1620	1820	2160	605

Model no.	14C412	14C512	14C612	14C712	14C812	14C912	14C1014	14C1114	14C1218
Max. air load (lb/hour)	75	100	130	170	210	250	300	350	400
Motive steam (lb/hour)	680	780	890	1020	1165	1320	1500	1690	1890

Model no.	14C1318	14C1418	15C514	15C614	15C714	15C814	15C914	15C1014	15C1114
Max. air load (lb/hour)	500	600	100	130	170	210	250	300	350
Motive steam (lb/hour)	2230	2600	855	965	1095	1240	1395	1575	1765

Note: 1 lb = 0·454 kg.

Model no.	15C1218	15C1318	15C1418	15C1518	16C514	16C614	16C714	16C814
Max. air load (lb/hour)	400	500	600	700	100	130	170	210
Motive steam (lb/hour)	1965	2305	2675	3065	935	1045	1175	1320

Model no.	16C914	16C1014	16C1114	16C1218	16C1318	16C1418	16C1518	16C1618
Max. air load (lb/hour)	250	300	350	400	500	600	700	820
Motive steam (lb/hour)	1475	1655	1845	2045	2385	2755	3145	3595

Model no.	17C518	17C618	17C718	17C818	17C918	17C1018	17C1118	17C1218
Max. air load (lb/hour)	100	130	170	210	250	300	350	400
Motive steam (lb/hour)	1080	1190	1320	1465	1620	1800	1990	2190

Model no.	17C1318	17C1418	17C1518	17C1618	18C518	18C618	18C718	18C818
Max. air load (lb/hour)	500	600	700	820	100	130	170	210
Motive steam (lb/hour)	2530	2900	3290	3740	1250	1360	1490	1635

Model no.	18C918	18C1018	18C1118	18C1218	18C1318	18C1418	18C1518	18C1618
Max. air load (lb/hour)	250	300	350	400	500	600	700	820
Motive steam (lb/hour)	1790	1970	2160	2360	2700	3070	3460	3910

Note: 1 lb = 0·454 kg.

Steam ejector pump series 1A and 20A are also recommended for producing vacuum.

These ejectors can be operated with either high pressure steam or air as the motive fluid. For the full tabulated details needed to use a 1A or 2A ejector for producing rough vacua on the large engineering scale, reference should be made to this firm's Catalogue No. 67, in which series 1A is Bulletin 100 and series 20A is Bulletin 200.

Schutte and Koerting

Range of steam jet exhausters and compressors which make use of either steam or compressed air as the motive force. Standard units have suction and discharge port diameters in inch of $\frac{1}{2}$, $\frac{3}{4}$, 1, $1\frac{1}{2}$, 2, $2\frac{1}{2}$, 3, 4, 5, 6, 8 and 10. The $\frac{1}{2}$ and $\frac{3}{4}$ inch sizes are all bronze, the larger sizes have iron bodies, tails and handwheels with bronze nozzles, spindles and stuffing boxes. Bronze and iron exhausters can be used for saturated steam at pressures up to

250 lb/sq. inch. For higher steam pressures or for superheated steam, steel body exhausters with stainless steel nozzles and spindles are used. Dimensions and technical data are given in the firm's Bulletin 4E.

Range of steam jet vacuum pumps capable of producing ultimate pressures of $2{\cdot}5 \times 10^{-2}$ Torr is described in Bulletins 5E, 5H2 and 5H3.

For handling corrosive vapours, the SK hard lead steam jet exhausters are used.

SK hard lead steam jet exhausters

Size (inch)	Steam inlet port diameter (inch)	Suction and discharge port diameters (inch)	Suction capacity (ft^3*/min)
1	$\frac{3}{8}$	1	16
$1\frac{1}{2}$	$\frac{1}{2}$	$1\frac{1}{2}$	33
2	$\frac{3}{4}$	2	66
$2\frac{1}{2}$	1	$2\frac{1}{2}$	100
3	$1\frac{1}{4}$	3	200
4	$1\frac{1}{2}$	4	300
5	2	5	450
6	2	6	600

* 1 ft^3 = 28·3 litre.

SK steam jet exhausters made of Pyrex glass

Size (inch)	Steam inlet port diameter (inch)	Suction and discharge port diameters (inch)	Suction capacity (ft^3/min)
$\frac{1}{2}$	$\frac{3}{8}$	$\frac{1}{2}$	4
$\frac{3}{4}$	$\frac{3}{8}$	$\frac{3}{4}$	8
1	$\frac{1}{2}$	1	16
$1\frac{1}{2}$	$\frac{3}{4}$	$1\frac{1}{2}$	33
2	1	2	66
$2\frac{1}{2}$	$1\frac{1}{2}$	$2\frac{1}{2}$	100

SK steam jet laboratory vacuum pump, made in $\frac{1}{4}$ inch connection size only.

Operating with a steam pressure of 50 lb/sq. inch a vacuum of 24 inch of mercury (nominally 150 Torr) is produced in a vessel of volume 5 litre in a time of just over 7 seconds.

SK single-nozzle water jet exhausters (Type 484) are supplied in (a) cast iron with bronze nozzle, (b) bronze or (c) stainless steel. Other materials on special order supplied are Monel, Evedur, Haveg, lead, fibreglass and glass. Suction and discharge port diameters are in sizes (inch) of $\frac{1}{2}$, $\frac{3}{4}$, 1, $1\frac{1}{2}$, 2, $2\frac{1}{2}$ and 3.

An ultimate pressure of about 5×10^{-2} Torr is obtainable.

SK multi-nozzle water jet exhausters in 13 standard sizes (suction port diameters ranging from 1 to 10 inch). Types 488 and 489 are made in (a) bronze with phenolic plastic nozzles (sizes 1 and 2 only), (b) cast iron with phenolic plastic nozzles; alternatively in stainless steel or Haveg and lined with abrasion- and corrosion-resistant materials such as rubber, carbon and Penton.

Barometric hydro-vac exhausters (Type 483) in eight standard sizes with suction connections from 10 to 36 inch diameter. Generally fabricated from steel.

SK Type 7010 ejector-venturi exhauster in sizes up to 12 inch diameter suction port made in a variety of materials – metal and plastics.

SK hydro-steam vacuum units consist of a tank (available in steel, in rubber-lined steel or in polyester fibreglass) used as a discharge and water-storage receptacle and as a chamber from which non-condensable gases are released, a flange-mounted integral centrifugal pump with motor, a multi-nozzle water jet exhauster and one or more steam jet vacuum pumps. For example, a five-stage unit (water jet exhauster and four steam jet pumps) will provide an ultimate pressure of 3×10^{-2} Torr.

Wiegand Apparatebau

Five-stage steam-ejector vacuum pump: suction capacity at 10^5 Pa = 300 kg/hour; ultimate pressure = 1 Torr with time of 3 minutes for the evacuation of a tank of volume 135 m^3.
Multi-stage steam-ejector vacuum pump: suction capacity at 120 Torr = 1100 kg/hour; ultimate pressure = 2×10^{-1} Torr.

Other steam-ejector vacuum pump units are constructed for special performances and purposes such as freeze-drying, steel degassing, vacuum distillation (e.g. in an oil refinery), condensation plants (e.g. in the production of pharmaceutical products), de-salination plants, crystallization plants etc.

2.1.10 Titanium sublimation pumps

In a sublimation pump, titanium (or, in a few cases, zirconium) is evaporated in a chamber at low pressure (usually, $<10^{-2}$ Torr) so that the titanium is deposited on the chamber walls which are at ambient temperature, or cooled with water or liquid nitrogen. The pumping action is then due to gettering of the gas by the active titanium film produced.

The titanium charge is supported on a heater which is raised in temperature by the passage of electric current or the charge in a suitable container is heated by electron bombardment.

The pumping speed obtainable will depend upon the surface area of the deposited film, the temperature of the substrate (usually the chamber wall) on which the film is deposited, the rate of evaporation of the titanium and the prevailing pressure. Approximately for 1 cm^2 of deposited film surface the pumping speed is 0·5 to 1·0 litre/second if the substrate is uncooled, 1 to 1·5 litre/second if the substrate is water-cooled and 2 to 2·5 litre/second if the substrate is cooled by liquid nitrogen.

With a suitable disposition of the titanium sublimator in a chamber, the pumping speed is limited in some cases by the access port area to the chamber, but with ready access, can be typically 1000 litre/second or more.

AEI Scientific Apparatus

Model TS4: 4 filaments, each with a one gram loading giving a speed for air of 1000 litre/second, the evaporation times are manually controlled, cooling is by water or liquid nitrogen. Line voltage 240 V, heater voltage 6 V, heater current 50 A.

Model TS6: as above but with a speed for air of 5000 litre/second.

In both pumps the sublimation surface is thermally insulated from the vacuum envelope, ensuring minimum consumption of liquid nitrogen.

Airco Temescal

Model no. RFP1, with electron-bombardment heated titanium wire of 1·6 mm dia. which is unwound *in vacuo* from a reel containing 105 g of this wire. The sublimation rate is up to 1 g/hour.

Model no. RFP-35, with electron-bombardment heated titanium rod of 25 mm dia. x 180 mm length bottom fed. The sublimation rate is up to 35 g/hour. Pumping speed is conductance limited but system can be designed for speeds of several thousands of litre/second.

Model no. RFP-500 and other special models capable of subliming kilogramme quantities of titanium per hour.

Balzers

Ti-sublimation pump model no. USP1 (control unit is no. USS2).

The sublimator consists of three evaporators for which the operating voltage is 3 to 6 and the operating current is 30–50 A.

Granville-Phillips

Ti-sublimation pump model no. 287: total charge of titanium is 20 g approx. Water-cooled, bakeable stainless steel pump chambers are available which give pumping speeds for nitrogen of 1600 litre/second and 3200 litre/second.

Also available for sublimation pumping are new spiral titanium filaments made of 85% Ti, 15% Mo alloy wire, having a filament life of 8–30 hours, depending on the system size.

Ion Equipment

The Combo-Vac pumps combine sublimation and sputter-ion pumping in one unit.

Model no. COV-500, water-cooled, contains a replaceable titanium sublimator and a replaceable 25 litre/second sputter-ion pump (so-called 'noble ion pump'). The pumping speed for nitrogen is 500 litre/second through a port fitted with a 20 mm O.D. x 15 mm I.D. metal gasketed flange.

Model no. COV-1000, water-cooled, contains a 200 litre/second noble ion pump, a cryo-shroud and a titanium sublimator. The pumping speed for nitrogen is 1000 litre/second.

Leybold-Heraeus

The titanium sublimation pump V400 is mainly intended as a booster pump in combination with sputter-ion pumps. In combination with model IZ230 it has a pumping speed for air in the high and ultra-high vacuum range of 620 litre/second; V400's own speed is rated at 400 litre/second. Two hairpin shaped filaments (one is a reserve) are of tungsten wrapped with titanium wire (weight of titanium 4·5 grams per filament). Bakeable to 400°C the pump is either water or LN_2 cooled. The duration of evaporation is continuously variable between 2-60 minutes.

RIBER

Model no 302, of which the sublimator consists of four directly electrically-heated filaments each loaded with 1·6 g of titanium (heater voltage max.: 11; heater current max.: 55 A).

Model no. SB301, with a charge of 30 g of titanium which is heated by electron bombardment.

Ultek

Model no. 224-1800, which has a sublimator in the form of a 40 g rod of titanium which is heated by electron bombardment to give a sublimation rate of 0·4 g/hour.

Model no. 214-0400, which has a sublimator comprising four directly electrically-heated filaments each loaded with 3·4 g of titanium (heater voltage max.: 8; heater current max.: 55 A). Sublimation rate is 1 g/hour max.

Also supply 'BoostiVac' D-I pumps which are combinations of an ion pump and a titanium sublimation pump. Models are:

No. 210-1500 D-I Boosti-Vac pump. Pumping speed in litre/second: air, 25; argon, 5, helium 7 at 10^{-6} Torr.

No. 210-15600 conventional BoostiVac pump. Pumping speed in litre/second: air, 25; argon, 0·3; helium, 2·5 at 10^{-6} Torr.

The Cryo-BoostiVac D-I pump which can be operated as:

a 150 litre/second D-I pump without cooling;

a 660 litre/second BoostiVac sublimation pump in which the cryo-shroud is water-cooled;

a 1000 litre/second Cryo-BoostiVac pump in which the cryo-shroud is cooled with liquid nitrogen.

Vacuum Generators

Model no. ST2 pump with three titanium/molybdenum alloy hair-pin filaments where the pump body may be manufactured in any required configuration.

Varian

Model no.	916-0017	916-0006
No. of filaments	3	3
Loading in gram/filament	1·2	5
Power supply	Model no. 922-0052/32	Model no. 922-0039/41
Line voltage	230 V/115 V	240 V/115 V
Heater voltage	0–6 V	450–750 watt
Heater current	0–50 A	
Evaporation times	0–4 min	0–12 min

These titanium sublimation pumps can be supplied in various containers which give various speeds and use either water or liquid nitrogen cooling.

Veeco

'E' beam equipment

Type	*Model*
Control unit 115 or 230 volt	PS220
10 amps max.	FH270
'E' beam cartridge with 2 anodes on $3\frac{3}{8}$ inch OD flange	

Filament type equipment

Control unit 115 volt	PS201
Filament type holder for 4 filaments on $3\frac{3}{8}$ inch OD flange	FH253
Filament type holder for 3 filaments on $2\frac{2}{4}$ inch OD flange	FH254

Sublimation chambers

Model no.	SC250	SC251	SC255	SC257	SC252	SC254
Type	Closed end	Closed end	In line	In line	Tee	Tee
Pumping speed	754 l/s	125 l/s	950 l/s	1250 l/s	440 l/s	750 l/s

Model no.	SC261	SC262	SC264	SC265	SC266	SC267	SC268
Type	Internal	Cylinder	Cylinder	Baseplate	Baseplate	Baseplate	Baseplate
Pumping speed	700 l/s	1450 l/s	3200 l/s	3800 l/s	7800 l/s	3800 l/s	7800 l/s

Torgett getter ion pump, Model TG500, is a combination titanium sublimation and sputter ion pump with a speed of 500 litres/second.

2.1.11 Turbomolecular pumps

A class of pumps derived from the axial flow compressor used in turbines. The gas is displaced axially by rotating blades mounted on discs alternating with fixed bladed discs. A horizontal rotor shaft drive is used with bearings and backing connections at the rotor ends, or a vertical shaft drive with a single base bearing and backing connection.

Airco Temescal

Model 614: Pump mouth internal dia.: 280 mm; rotor speed: 13 700 rev/min; pumping speed for air: 650 litre/second; operating pressure range: 10^{-1} to 10^{-10} Torr; maximum permissible backing pressure: 10 Pa; pumping speed of recommended backing pump: 600 litre/minute.
Models 9016 (9000 litre/second), 814 (800 litre/second) and 514 (500 litre/second).

Balzers

Model no.	Pumping speed for air (litre/second)	Ultimate pressure (torr)	Rotor speed (rev/min)
TVP250	70	$< 5 \times 10^{-11}$ * $< 10^{-9}$ †	30 000
THP1000	280	$< 10^{-7}$ ‡	18 000
TVP2500	700	$< 5 \times 10^{-1}$ * $< 5 \times 10^{-10}$ †	12 000
TVP6000	1700	$< 5 \times 10^{-11}$ * $< 5 \times 10^{-10}$ †	12 000
TVP15000	4250	$< 5 \times 10^{-11}$ * $< 5 \times 10^{-10}$ †	6000

* Backed by two-stage rotary vane pump and a turbomolecular pump or a diffusion pump.
† Backed by a two-stage rotary vane pump.
‡ Backed by a single-stage rotary vane pump.

Turbomolecular pumping units consisting of the turbomolecular pump with the appropriate backing pumps are supplied complete with necessary control units, ultra-high vacuum flanges and cooling water connections.

Leybold-Heraeus

Model no.	Pumping speed for N_2 (litre/second)	Ultimate pressure (Torr)	Rotational speed (rev/min)	Bake-out temperature (°C)
Turbovac 350	330	3×10^{-3}	24 000	120

Starting-up time is 4 to 5 minutes, cooling water consumption is 30 litres per hour and the recommended backing pump is model D12A.

Sargent-Welch

'Turbotorr' models 3102, 3102D, 3103, 3103D, 3116, 3117 (to 10^{-10}) and 3120, 3120D (to 10^{-7}).

Veeco

3102 series providing a speed of 260 litre/second over a range of 1×10^{-2} to 1×10^{-9} Torr. 500 and 1000 litre/second pumps are also available.

2.1.12 Vapour pumps

(1) diffusion pumps, (2) ejector pumps. The gas to be pumped moves to the vapour stream under which molecular flow predominates in class (1) whereas viscous flow predominates in class (2). Multi-stage pumps are classed simply as diffusion pumps if they have diffusion-type first-stage jets. All pumps are metal unless otherwise indicated.

MANUFACTURER:	**Balzers**						
Manufacturer's model no.	Diff 10L	Diff 10	Diff 60L	Diff 60	Diff 260	Diff 900	Diff 1900
Diffusion or ejector type	Diffusion	Diffusion	Diffusion	Diffusion	Diffusion	Diffusion	Diffusion
No. of stages	3	3	3	3	3	3	3
Pump mouth internal diameter (mm)	25	25	46	20	158 (flange)*	220 (flange)	300 (flange)
Pumping speed for air (litre/second)	6·5 (with baffle)	6·5 (with baffle)	23 (with baffle)	23 (with baffle)	260 (no baffle)	950 (no baffle)	1800 (no baffle)
Cooling	Air	Water	Air	Water	Water	Water	Water

Remarks: * Diameter of flange to pump mouth.
A wide range of accessories is available for all these vapour pumps, including baffles, water flow control switch, temperature controller, five vacuum vessels with oil filters, diffusion pump oils and valves.

MANUFACTURER:	**Balzers** *cont'd*			**Brizio Basi**
Manufacturer's model no.	Diff 5000	Diff 500	Diff 630	BVH (Silica)
Diffusion or ejector type	Diffusion	Diffusion	Diffusion	Diffusion
No. of stages	3	3	3	3
Pump mouth internal diameter (mm)	420 (flange)	510	660	300
Pumping speed for air (litre/second) (Pressure at which measured in Torr)	5000 (no baffle	11 500 (no baffle)	20 000 (no baffle)	3000 (10^{-5})
Cooling	Water	Water	Water	Water

MANUFACTURER:	**Cenco**					
Manufacturer's model no.	93415*	93410†	93405‡	93403§	93423	93420-16
Diffusion or ejector type	Diffusion (oil)	Diffusion (oil)	Diffusion (oil)	Diffusion (oil)	Diffusion (oil)	Diffusion (oil)
No. of stages	3	3	3	3	3	3
Pump mouth internal diameter (mm)	250	150	100	75	50	38
Pumping speed for air (litre/second) (Pressure at which measured in Torr)	4500 (10^{-4})	2000 (10^{-4})	1000 (10^{-4})	550 (10^{-4})	275 (10^{-4})	80 (10^{-4})
Cooling	Water	Water	Water	Water	Air‖	Water¶

Remarks:

* 93415 has stainless steel body; 93416 has carbon steel body.
† 93410, 93440 and 93440-1 have stainless steel body; 93411, 93411-1, 93441 and 93441-1 have carbon steel body.
‡ 93405, 93405-1, 93435, 93435-1 have stainless steel body; 93406, 93406-1, 93436 and 93436-1 have carbon steel body.
§ 93403, 93426-16, 93426-17 have stainless steel body; 93404 has carbon steel body.
‖ 93423-16 and 93423-17 are air-cooled; 93422-16 and 93422-17 are water-cooled.
¶ 93420-16 and 93420-17 are water-cooled; 93421-16 and 93421-17 are air-cooled.

MANUFACTURER:	**Cenco** *cont'd*			
Manufacturer's model no.	Supervac OD-25	93201	MCF-300*	MCF-60*
Diffusion or ejector type	Diffusion (oil)	Diffusion (mercury)	Diffusion (oil)	Diffusion (oil)
No. of stages	3	3	3	3
Pump mouth internal diameter (mm)		24 (reducer in position) 63 (without reducer)	98	45
Pumping speed for air (litre/second) (Pressure at which measured in Torr)	28 (10^{-4})	7 (with reducer) (10^{-4})	300 (10^{-4})	60 (10^{-4})
Cooling	Water	Water	Water	Water

Remarks:

* All-metal fractionating diffusion pumps with enlarged boilers and capable of ultimate pressure of 5×10^{-7} Torr without cold trap or baffle.

MANUFACTURER:	Cenco *cont'd*				
Manufacturer's model no.	93236 (glass)	GF-25 (glass)	GF-26 (glass)	GF-20 (glass)	G-4 (glass)
Diffusion or ejector type	Diffusion*	Diffusion*	Diffusion*	Diffusion*	Diffusion†
No. of stages	3	3	3	2	1
Pump mouth internal diameter (mm)	48	30	30	30	30
Pumping speed for air (litre/second) (Pressure at which measured in Torr)	40 (10^{-5})	26 (10^{-5})	26 (10^{-5})	29 (10^{-5})	8 (10^{-5})
Cooling	Water	Water	Air	Water	Air
Remarks:	* A glass fractionating diffusion pump. † A glass semi-fractionating diffusion pump.				

MANUFACTURER:	Cooke			Ditric
Manufacturer's model no.	DPD4	DPD6	DPD10	8020*
Diffusion or ejector type	Diffusion	Diffusion	Diffusion	Diffusion
No. of stages	4	4	4	
Pump mouth internal diameter (mm)	143	193	289	100
Pumping speed for air litre/second. (Pressure at which measured in Torr)	1050 (10^{-3}–10^{-5})	2000 (10^{-3}–10^{-5})	4700 (10^{-3}–10^{-5})	385
Cooling	Water	Water	Water	Water
Remarks:	All pumps provided with thermostatic protection.			* Combines a valve-trap-baffle assembly.

MANUFACTURER:	Edwards High Vacuum								
Manufacturer's model no.	EMG (glass)	EM1	EM2	6M3A	9M3A	12M3	16M4	24M4	24MB*
Diffusion or ejector type	Diffusion	Diffusion	Diffusion	Diffusion	Diffusion	Diffusion	Diffusion	Diffusion	Diffusion
No. of stages	2	2	3	3	3	3	4	4	4
Pump mouth internal diameter (mm)	25	25	51	152	229	305	406	610	51
Pumping speed for air (litre/second). (Pressure at which measured in Torr)	10 (10^{-4})	10 (10^{-4})	150 (10^{-4})	650 (10^{-4})	1500 (10^{-4})	2700 (10^{-4})	5500 (10^{-4})	11 000 (10^{-4})	70 (10^{-4})
Cooling	Water	Water	Water	Water	Water	Water	Water	Water	Water

Remarks: Pump fluid is mercury; in all cases pumping speed quoted is in accordance with International Standards Organisation recommendations. Measurements based on American Vacuum Society practice give speed figures more than 10% higher.
* Special purpose pump with critical backing pressure of 35 Torr.

MANUFACTURER:	Edwards High Vacuum *cont'd*								
Manufacturer's model no.	E01	F203A*	E203D†	E02	403A	E04	E06	E09	E012
Diffusion or ejector type	Diffusion	Diffusion	Diffusion	Diffusion	Diffusion	Diffusion	Diffusion	Diffusion	Diffusion
No. of stages	2	3	3	4	3	4	4	4	4
Pump mouth internal diameter (mm)	25	51	51	51	102	102	152	229	305
Pumping speed for air (litre/second). (Pressure at which measured in Torr)	10 (10^{-4})	70 (10^{-4})	80 (10^{-4})	150 (10^{-4})	300 (10^{-4})	600 (10^{-4})	1350 (10^{-4})	2500 (10^{-4})	4200 (10^{-4})
Cooling	Water or air	Water	Water	Water or air	Water	Water or air	Water	Water	Water

Remarks: Pump fluid is oil; in all cases pumping speed quoted is in accordance with International Standards Organisation recommendations. Measurements based on American Vacuum Society practice give speed figures more than 10% higher.
* Fractionating pump.
† Special purpose pump: fast warm up with special heater.

MANUFACTURER:	Edwards High Vacuum *cont'd*							
Manufacturer's model no.	E020/4	F3605	9B3	9B4	18B4A	30B4A	30B5A	100B4
Diffusion or ejector type	Diffusion	Diffusion	Ejector	Ejector	Ejector	Ejector	Ejector	Ejector
No. of stages	4	5						
Pump mouth internal diameter (mm)	508	914	229	305	305	406	610	
Pumping speed for air (litre/second). (Pressure at which measured in Torr)	14 000 (10^{-4})	45 000 (10^{-4})	850 (10^{-4})	3800 (10^{-4})	4000 (10^{-4})	5000 (10^{-4})	12 500 (10^{-4})	22 000 (10^{-4})
Cooling	Water	Water	Water	Water	Water	Water	Water	Water
Remarks:	Ejector types are classified by the manufacturer as 'oil vapour booster pumps'.							

MANUFACTURER:	General Electric	General Engineering			Hick Hargreaves
Manufacturer's model no.	2/50/200 (Glass)	ODP150	ODP150A	ODP650	A wide range of air and gas operated ejector pumps is supplied.
Diffusion or ejector type	Diffusion				
No. of stages	2	4	4	4	
Pump mouth internal diameter (mm)	45	75	75	125	
Pumping speed for air (litre/second). (Pressure at which measured in Torr)	30 (10^{-3})	150 (10^{-5})	150 (10^{-5})	650 (10^{-5})	
Cooling	Water	Water	Air	Water	
Remarks:	A glass mercury diffusion pump.				

MANUFACTURER:	**Leybold-Heraeus**							
Manufacturer's model no.	Leybodiff 30	Leybodiff 170	Leybodiff 400	Leybodiff 1000	Leybodiff 3000	Leybodiff 6000	Leybodiff 12000	Leybodiff 30000
Diffusion or ejector type	Diffusion (oil)	Diffusion (oil)	Diffusion (oil)	Diffusion (oil)	Diffusion (oil)	Diffusion (oil)	Diffusion (oil)	Diffusion (oil)
No. of stages	3	4	4	3	3	3	3	3
Pump mouth internal diameter (mm)	40	65	100	150	250	350	500	800
Pumping speed for air (litre/second). (Pressure at which measured in Torr)	30 (10^{-5})	170 (10^{-5})	400 (10^{-5})	1000 (10^{-5})	3000 (10^{-5})	6000 (10^{-5})	12 000 (10^{-5})	30 000 (10^{-5})
Cooling	Water	Water	Water	Water	Water	Water	Water	Water

Remarks: A variety of baffles, fore-vacuum condensers and other accessories are available for each pump.

MANUFACTURER:	**Leybold-Heraeus** *cont'd*					
Manufacturer's model no.	DI 1000/2	DI 3000/6	DI 6000/10	DI 12000/12	DI 30000/24	DI 50000/30
Diffusion or ejector type	Diffusion (oil)	Diffusion (oil)	Diffusion (oil)	Diffusion (oil)	Diffusion (oil)	Diffusion (oil)
No. of stages	4	4	4	4	4	4
Pump mouth internal diameter (mm)	150	250	350	500	800	1000
Pumping speed for air (litre/second). (Pressure at which measured in Torr)	1000 ($<10^{-4}$)	3000 ($<10^{-4}$)	6000 ($<10^{-4}$)	12 000 ($<10^{-4}$)	30 000 ($<10^{-4}$)	50 000 ($<10^{-4}$)
Cooling	Water	Water	Water	Water	Water	Water

Remarks: In pump performance at low pressures and in height, DI pumps are comparable with Leybodiff pumps; in throughput and critical backing pressure they are more related to Leybojet pumps.
A wide range of accessories is available for each pump.

MANUFACTURER:	**Leybold-Heraeus** *cont'd*						
Manufacturer's model no.	Leybojet 170/1	Leybojet 1000/4	Leybojet 3000/7	Leybojet 6000/10	Leybojet 12000/12	Leybojet 30000/24	Leybojet 75000/50
Diffusion or ejector type	Diffusion (oil)	Diffusion (oil)	Diffusion (oil)	Diffusion (oil)	Diffusion (oil)	Diffusion (oil)	Diffusion (oil)
No. of stages	4	4	4	4	4	5	4
Pump mouth internal diameter (mm)	65	150	250	350	500	800	1250
Pumping speed for air (litre/second). (Pressure at which measured in Torr)	170 ($<10^{-4}$)	1000 ($<10^{-4}$)	3000 ($<10^{-4}$)	6000 ($<10^{-4}$)	12 000 ($<10^{-4}$)	30 000 ($<10^{-4}$)	75 000 ($<10^{-4}$)
Cooling	Water	Water	Water	Water	Water	Water	Water

Remarks: A wide range of accessories is available for each pump.

MANUFACTURER:	**Leybold-Heraeus** *cont'd*	
Manufacturer's model no.	Leybodiff 30L	Leybodiff 170L
Diffusion or ejector type	Diffusion (oil)	Diffusion (oil)
No. of stages	3	4
Pump mouth internal diameter (mm)	40	65
Pumping speed for air (litre/second). (Pressure at which measured in Torr)	30 (10^{-5})	170 (10^{-5})
Cooling	Air	Air

Remarks: A wide range of accessories is available for each pump.

MANUFACTURER:	Leybold-Heraeus *cont'd*			
Manufacturer's model no.	Quick 105	Q16*	Q20†	Q30/31†
Diffusion or ejector type	Diffusion (mercury)	Diffusion (mercury)	Diffusion (mercury)	Diffusion (mercury)
No. of stages	2	3	3	2
Pump mouth internal diameter (mm)	65	ground joint	ground joint	ground joint
Pumping speed for air (litre/minute). (Pressure at which measured in Torr)	100 (10^{-5})	5 (10^{-5})	8 (10^{-5})	40 (10^{-5})
Cooling	Water	Water	Water	Water

Remarks:

* Silica pump.
† Hard glass pumps.
The critical backing pressures of pumps Q16 and Q20 are so high that even water jet pumps may be used as backing pumps.
A wide range of accessories is available for each pump.

MANUFACTURER:	Leybold-Heraeus *cont'd*			N.G.N.
Manufacturer's model no.	Hg3	Hg12	Hg45	506A
Diffusion or ejector type	Ejector (mercury)	Ejector (mercury)	Ejector (mercury)	Diffusion
No. of stages	2	3	3	2
Pump mouth internal diameter (mm)	20	32	50	25
Pumping speed for air (litre/speed). (Pressure at which measured in Torr)	2–3 (10^{-1}–10^{-2})	12 (10^{-1}–10^{-2})	45 (10^{-1}–10^{-2})	10 (10^{-3}–10^{-5})
Cooling	Water	Water	Water	Air

Remarks:

Leybold-Heraeus: Unusually high critical backing pressure requiring only single-stage backing pumps which may even operate at full gas ballast.
A wide range of accessories is available for each pump.

N.G.N.: An educational pump used in conjunction with other eductional equipment for vacuum technology supplied.
A 515A butterfly valve may be used at intake to this pump.

MANUFACTURER:	Pope Scientific						
Manufacturer's model no.	20038 (glass)	20041* (glass)	20045* (glass)	20048 (glass)	20051 (glass)	20053 (glass)	20055 (glass)
Diffusion or ejector type	Diffusion (mercury)	Diffusion (mercury)	Diffusion (mercury)	Diffusion (mercury)	Diffusion† (oil)	Diffusion† (oil)	Diffusion† (oil)
No. of stages	3	2	2	2	3	2	1
Pump mouth internal diameter (mm)	56	37	37	21	28	28	28
Pumping speed for air (litre/second). (Pressure at which measured in Torr)	35 (10^{-4})	6·3 (10^{-4})	6·3 (10^{-4})	3·5 (10^{-4})	30 (10^{-4})	26 (10^{-4})	8 (10^{-4})
Cooling	Water	Water	Water	Water	Air‡	Air§	Air‖

Remarks:

* 20041 has cartridge heater type boiler; 20045 has jacket heater type boiler.
† Fractionating oil diffusion pump.
‡ Water-cooled version is no. 20052.
§ Water-cooled version is no. 20054.
‖ Water-cooled version is no. 20056.
Range of accessories including cold traps supplied for all these pumps.
Also available: Toepler pumps (nos. 20029, 20030 and 20031) for quantitative transfer of gases.

MANUFACTURER:	Precision Scientific		Teltron	Torr Vacuum Products	
Manufacturer's model no.	10444	10445	TEL506B	DHG800	DHG500
Diffusion or ejector type	Diffusion (oil)	Diffusion (oil)	Diffusion	Diffusion	Diffusion
No. of stages	2	2	2	4	4
Pump mouth internal diameter (mm)	22*	22*	25		
Pumping speed for air (litre/second). (Pressure at which measured in Torr)	10 (10^{-4})	10 (10^{-4})	10 (10^{-3})	760	4500
Cooling	Water	Air		Water	Water

Remarks: * Intake nipple outside diameter.

MANUFACTURER:	Vactronic Lab Equipment				Varian	
Manufacturer's model no.	HVP-450W	HVP-450A	HVP-150	HVP-150A	NHS4	NHS6
Diffusion or ejector type	Diffusion (oil)	Diffusion (oil)	Diffusion (oil)	Diffusion (oil)	Fractionating diffusion	Fractionating diffusion
No. of stages	3	3	3	3	4	4
Pump mouth internal diameter (mm)	160	125	65	65	Inlet flange: 23 mm o.d.	Inlet flange: 28 mm o.d.
Pumping speed for air (litre/second). (Pressure at which measured in Torr)	450* (5×10^{-5})	450* (5×10^{-5})	150† (5×10^{-5})	150† (5×10^{-5})	750 (10^{-6})	1500 (10^{-6})
Cooling	Water	Air	Water	Air	Water	Water

Remarks:

Vactronic Lab Equipment:

* Baffle supplied reduces speed, when in position, to 280 litre/second.

† Baffle supplied reduces speed, when in position, to 85 litre/second.

Varian:

Stainless steel bodies: operating pressure range from 10^{-3} to $<10^{-9}$ Torr.

NRC 0315-1 series circular chevron cryo-baffle: with NHS4, speed with baffle is 320 litre/second; with NHS6, speed with baffle is 660 litre/second. High vacuum slide gate valves (manually operated) available to fit inlet ports of these pumps.

MANUFACTURER:	Varian *cont'd*						
Manufacturer's model no.	SHS-2	HSA-2	HS-2	M4	VHS-4	M6	VHS-6
Diffusion or ejector type	Diffusion (oil)	Diffusion (oil)	Diffusion (oil)	Diffusion (oil)	Diffusion (oil)	Diffusion (oil)	Diffusion (oil)
No. of stages	3	3	3	4	4	4	4
Pump mouth internal diameter (mm)	86	76	86	153	146	195	198
Pumping speed for air (litre/second). (Pressure at which measured in Torr)	175 (10^{-5})	150 (10^{-5})	293 (10^{-5})	800 (10^{-5})	1200 (10^{-5})	1500 (10^{-5})	2400 (10^{-5})
Cooling	Water	Air	Water	Water	Water	Water	Water

MANUFACTURER:	**Varian** *cont'd*							
Manufacturer's model no.	VHS-10	HS-10	HS-16	HS-20	HS-32	HS-35	HS-48	H4X
Diffusion or ejector type	Diffusion (oil)	Diffusion (oil)	Diffusion (oil)	Diffusion (oil)	Diffusion (oil)	Diffusion (oil)	Diffusion (oil)	Diffusion (oil)
No. of stages	4	3	5	5	5	5	5	4
Pump mouth internal diameter (mm)	305	292	457	520	816	900	1320	146
Pumping speed for air (litre/second). (Pressure at which measured in Torr)	5160 (10^{-5})	4160 (10^{-5})	10 000 (10^{-5})	17 500 (10^{-5})	33 300 (10^{-5})	46 700 (10^{-5})	95 000 (10^{-5})	800 (10^{-5})
Cooling	Water	Water	Water	Water	Water	Water	Water	Water

MANUFACTURER:	**Veeco**			
Manufacturer's model no.	EP2A	EP2W	EP4W	EP776
Diffusion or ejector type	Diffusion	Diffusion	Diffusion	Diffusion
No. of stages	3	3	3	3
Pump mouth internal diameter (mm)				
Pumping speed for air (litre/second)	90	90	425	2000
Cooling	Air	Water	Water	Water

2.1.13 Water jet pumps

A type of pump in which water from a high pressure supply issues as a high velocity jet through a nozzle in a suction chamber where gas is entrained in the jet.

Edwards High Vacuum

Models 7 and 8 (stainless steel) and plastic model: water pressure 1·4 kgf/cm², pumping speed 3·9 litre/min, water consumption 10·5 litre/min, ultimate vacuum 12 to 14 Torr. Vacuum connection: riffled to suit 7 mm bore tubing: water connection ¼ inch BSP female or 12 mm tubing (8), ¼ inch BSP male or 12 mm tubing (7) 12 mm tubing (plastic model).

Kinney Vacuum

Model no.	Water pressure (N/m²)	Pumping speed (litre/min)	Vacuum connection i.d. (mm)	Water connection i.d. (mm)
KLJ-5	1	99	25·4	38·1
KLJ-6	1	127	25·4	38·1
KLJ-9	1	213	25·4	38·1
KLJ-15	1	410	50·8	50·8
KLJ-21/22	1	453/495	50·8	50·8
KLJ-32	1	991	50·8	50·8
KLJ-60	1	1290	63·5	63·5
KLJ-90	1	1833	63·5	63·5
KLJ-120	1	3400	76·2	76·2
KLJ-170	1	4816	76·2	76·2
KLJD-10	1	198	25·4	38·1
KLJD-12	1	255	25·4	38·1
KLJD-18	1	425	25·4	38·1
KLJD-30	1	766	50·8	50·8
KLJD-42/44	1	905/991	50·8	50·8
KLJD-64/66	1	1833	50·8	50·8
KLJD-120	1	2266	63·5	63·5
KLJD-180	1	3116	63·5	63·5
KLJD-240	1	6666	76·2	76·2
KLJD-340	1	9166	76·2	76·2

Pumps are constructed from cast iron and brass or stainless steel.

Leybold-Heraeus

Light alloy water jet pumps, also in red brass or polyethylene.

Model no.	152 10	152 11	152 12	152 24	152 27	152 28
Water pressure	3 kgf/cm²	3 kgf/cm²	3 kgf/cm²	3 kgf/cm²	3 kgf/cm²	3 kgf/cm²
Pumping speed	400	400	400	850	180	180
Vac. conn. o.d.	11	11	11	11	11·5	11·5
(mm)		Union nut i.d.				
Water conn. o.d.	17 mm	¾ inch	½ inch	¾ inch	½ inch	¾ inch
Construction	Anodized light alloy			Red brass	Polyethylene	

Ralet-Defay
Vacothene water-jet pump in $\frac{1}{2}$ inch and $\frac{3}{4}$ inch sizes, anti-return valve available as an extra.

Schutte-Koerting
Water jet laboratory vacuum pumps: type 474 with bronze body and nozzle, type 475 of chromium-plated bronze.

Type 474, operated at a water pressure of 30 lbf/sq. inch will produce a vacuum of approximately 125 Torr (25 inch mercury) in a five litre vessel in two minutes. Type 475 has approximately half the capacity of 474.

2.1.14 Valves

AEI Scientific Apparatus
A range of Viton and metal sealed valves of 1 to $1\frac{1}{2}$ inch bore is available.

Airco Temescal
Series 1000/Brass angle and in-line valves have replaceable bellows (stainless steel is optional), interchangeable hand-wheel, toggle or pneumatic actuators, Buna-N O-ring seals (Viton A is optional) for tubing o.d. in inches of $\frac{3}{8}$, $\frac{5}{8}$, $\frac{3}{4}$, $2\frac{3}{8}$, and $2\frac{9}{16}$.
Series 2000/Stainless steel valves have replaceable bellows, interchangeable hand-wheel, toggle, electropneumatic or remote pneumatic actuators, fluorocarbon O-rings, are bakeable to 255°C and available in $\frac{5}{8}$, $1\frac{1}{8}$ and $1\frac{1}{2}$ inch port diameters of which angle valves have molecular conductances in litre/second of 2, 14 and 40 respectively whereas inline types have molecular conductances in litre/second of 1, 10 and 30 respectively.
Series 2500/Aluminium angle valves have interchangeable toggle or electropneumatic actuators, Viton O-ring seals and port sizes of 2, 3, 4 and 6 inch.
Series F2500/Fabricated steel angle valves are available with electropneumatic actuation, are Viton O-ring sealed and in port diameters in inches of 6, 8, 10, 12, 16, 20 and 35.
Series 5000/Unit body gate valves available are actuated by side lever, toggle, hand-wheel, pneumatically or side-pneumatically, interchangeably, in port diameters in inches of 2, 3, 4 and 6.
Series ST and 5000/Split body aluminium gate valves (interchangeable actuated by hand-wheel, toggle, electropneumatic and as manifold valves either hand-wheel or electropneumatically operated) are in port diameters in inches for ST series: 2, 3, 4, 6 and 5000 series 8, 10 and 12, whereas the manifold valve (STM series) port diameters supplied are 2, 4 and 6 inch.
Series 6000/Fabricated steel gate valves are electropneumatically actuated where gate, carriage and actuator are removable without breaking the vacuum line. Port diameters in inches are 14, 16, 20, 24, 30, 32, 35, 36, 48 and 60.

Andonian Cryogenics

High vacuum bakeable, adjustable leak valve

Model no.	Flow range (std, cc/s of air with 760 Torr differential pressure)
11	5×10^{-2} to 5
20	6×10^{-3} to 5×10^{-1}
42	1×10^{-4} to 3×10^{-2}
73	10^{-7} to 10^{-3}

Adjacent to model number: M signifies for non-bakeable manifold whereas MB denotes bakeable manifold.

Also supply model no. A-10, a combination vacuum and relief valve.

Balzers

Gate valves: Common features to all types are that they have Viton seals, will withstand bakeout to 150°C, have a cast aluminium housing, a stainless steel valve plate and have a leak rate of below 10^{-9} Torr litre/second. Opening against atmospheric pressure is possible, but only if it is on the seat side of the valves.

Manually operated gate valves are of the following models; in each case, the number given in brackets is the molecular conductance in litre/second of the open valve:
SVV 32H (34); SVV 52H (240); SVV 76H (470); SVV 95H (1530); SVV 138H (2400); SVV 205H (8450); SVV 330H (20000).

The same models are also available with electropneumatic operation; the final letter in the model code is then P instead of H, i.e. the models are:
SVV 32P; SVV 52P; SVV 76P; SVV 95P; SVV 138P; SVV 205P and SVV 330P.

High vacuum air inlet and control valves: Needle valve RNV 10H, for fine control of gas flow. Exhaust valves VAH 005 and VAH 010, for air admittance and for draining out liquids. Air inlet angle valve VFH 005, for admitting air or gas and for fitting in vacuum lines, venting valve VFH 010 for venting small systems. Emergency venting valve VFS 010, which is closed when electrically energized; if the power fails the valve is opened by a spring. Solenoid valve VFM 010. Control valve RV 10H, a manually operated valve for controlling the inlet of large quantities of gas into vacuum chambers. Air inlet valve FV 10 ML, solenoid operated and particularly suitable for remote control.

Automatic regulating valve RME 010 (control unit: RVG 010; control unit: RVG 030) in an automatic and instantaneous regulating device for maintaining a constant pressure and admitting gas to vacuum plants. Gas flow range is from $<10^{-4}$ to 50 Torr litre/second.

Manual and electropneumatic shut-off valves: All models utilize Viton seals, have leak rate below 1×10^{-5} Torr litre/second, withstand temperatures up to 200°C and are of essentially cast-iron or steel construction. Manually operated angle valves are models VEH 010, VEH 025, VEH 040, VEH 065, VEH 100 and VEH 150. The same valves are available with electropneumatic operation in which case a P is used in the code number instead of an H. Manually operated straight through valves are models VEH 200, VDH 010, VDH 025, VEH 040, VDH 065, VDH 100, VDH 150, VDH 200, VDH 250 and VDH 350. Again, the straight through models are available with electro-pneumatic operation: model numbers

are VEP 200, VEP 500, VEP 800, VDP 025, VDP 040, VDP 065, VDP 100, VDP 150.
Vacuum sealed, pressure resistant valves, designed for pressure range from 10^{-3} to 2×10^{3} Torr. Three ball valves each with light alloy housing, Teflon valve ball and Viton housing seal and having a leak rate of below 1×10^{-3} Torr litre/second are models VKH 010, having a leak rate of below 1×10^{-3} Torr litre/second are models VKH 010, VKH 020 and VKH 032.
High vacuum flap valves for fore lines and high vacuum with a cast aluminium housing, Viton seals and leak rate below 5×10^{-8} Torr litre/second.
Manually operated valves are V52H and V76H which respectively have conductances at 10 Pa of 360 litre/second and 1000 litre/second. The corresponding solenoid valves are V52M and V76M whereas the corresponding pneumatic valves are V52P and V76P. Flap valves with small flanges are manually operated types, V20H, V32H and V50H with conductances at 10 Pa of respectively 20, 40, and 100 litre/second. The corresponding valves with pneumatic operation are V20P, V32P and V50P. Solenoid operated versions are V10ML, V20M, V32M and V50M with conductances at 10 Pa of respectively 2·2, 20, 40 and 100 litre/second.
High vacuum plate valves for fitting above diffusion pumps have nickel-plated steel housings, stainless steel internal mechanism and valve plate and with Viton seals. The leak rate is below 10^{-8} Torr litre/second. Straight-through valves are tabulated as follows:

Model no.	PV52H*	PV9SH	OVK38G	OV295G‡	OV339G‖	PV500P**	PV63OP††
Molecular conductance (litre/second)	80	300	1050	3000	11 000	20 000	38 000
Cooling water (litre/hour)	100	150	150	150	150	none	none
Operation	Hand	Hand	Hand†	Hand§	Hand¶	Electro-pneumatic	Electro-pneumatic

* This valve can be converted into the angle model by changing over a blanking plate on the straight-through type.
† Electropneumatically operated model is PV138P.
‡ PV205H-2 is for connection to horizontal chamber and has conductance of 2200 litre/second.
§ Electropneumatically operated model is PV205P. Model PV205P-2 is the electropneumatically operated version of PV205H-2.
‖ PV330H-2 is for connection to horizontal chamber and has conductance of 6700 litre/second.
¶ Electropneumatically operated model is PV330P.
** PV500P-2 is for connection to horizontal chamber and has conductance of 15 000 litre/second.
†† PV630P-2 is for connection to horizontal chamber and has conductance of 30 000 litre/second.

Customatic regulating valve REM 010 (with control unit RVG 030) forms an automatic regulating device for maintaining a constant pressure and for admitting gas in the rough, medium and high vacuum ranges. Gas flow range is from 10^{-4} to 50 Torr litre/second.
Ultra-high vacuum valves, manually operated in all-metal construction, bakeable to 450°C. UV-8H with glass/metal sealed connection tubes has connection diameter of 15 mm, an open conductance of 0·5 litre/second and a leak rate when closed of 10^{-10} Torr litre/second. Ultra-high vacuum valves with Balzers or CF-flanges have a conductance when closed of below 10^{-14} litre/second; models are UVH035 (CF-flange) and UVH 052 (Balzers flange) with

molecular conductances in the open condition of 25 litre/second whereas UVH152 (Balzers flange) and UVH 163 (CF-flange) have open conductances of 75 litre/second.
Ultra-high vacuum gas metering and inlet valves UDV 035 and UDV 052. The former is for a CF-flange and the latter for a Balzers flange. Their gas inlet orifice outside diameters are 14·3 mm; working pressure range is up to 10^{-9} Pa, conductance when closed in $<10^{-9}$ Torr litre/second.
Vacuum bleed valve V4-160 is used to bring the vacuum chamber up to ambient pressure. It has a Monel body and a phosphor-bronze bellows, with Ke2-F gasket sealing. A copper-sealed flange version (model V4C-160) is not bakeable.

Brizio Basi
A range of valves, bore diameter 10 to 800 mm, of the following types: isolating valves, regulating valves, bellows sealed valves, diaphragm valves, plate valves, butterfly valves, gate valves, plug valves, magnetic valves, pneumatic valves, air admittance valves.

Cenco
VRC high vacuum gate valves with rotary stem seal and cam actuation. Sizes 1 up to 12 inch i.d. are of dense, non-porous copper–silicon–aluminium alloy; sizes 16, 20 and 24 inch are constructed of carbon steel in rectangular parallelepiped design with cast aluminium gate assemblies. Manual (lever or throttle) or pneumatic operation. Buna-N (Viton optional) O-ring sealing between plate and seat. Leak rate against atmosphere when closed is 1×10^{-9} Torr litre/second. Port inside diameters in inch available are 1 (code nos. 1T1 and 1N5), 2 (code nos. 2T1, 2T2 and 2N5), 3 (code nos. 3T1, 3T2 and 3N5), 4 (code nos. 4T1, 4T2 and 4N5), 6 (code nos. 6T1, 6T2 and 6N5), 8 (code no. 8T1), 10 (code no. 10T1), and 12 (code no. 12T1), 16 (code no. 16T1), 20 (code no. 20T1) and 24 (code no. 24T1).
VRC hyvac valves: LP (large port) series; cast aluminium; actuation by 3 interchangeable methods, manual (200 series), throttle (300 series) or electropneumatic (100 series). Wide range of compatible flanges. Sizes are 2, 3, 4, 6, 8 and 10 inch (standard).
LP valves have a conductance in litre/second of 816 for the 2 inch valve ($3\frac{3}{8}$ inch port), 3035 for the 4 inch valve ($5\frac{3}{8}$ port), 6668 for the 6 inch valve ($7\frac{1}{8}$ port) and 21800 for the 10 inch valve ($11\frac{5}{8}$ inch port).
Angle valves: Cenco 90° angle valves made with forged brass bodies with machined seats. Valve plate discs of brass utilize a replaceable O-ring. Bellows-sealed. Manual actuation or pneumatic-solenoid control available.
Models are:
With female pipe thread ports

Catalogue numbers		Pipe thread
94608-1	94604-1*	$\frac{1}{2}$ NPTF
94608-2	94604-2*	$\frac{3}{4}$ NPTF
94608-3	94604-3*	1 NPTF
94608-4	94604-4*	$1\frac{1}{4}$ NPTF
94608-5	94604-5*	$1\frac{1}{2}$ NPTF

* Have rising stem actuators.

Catalogue numbers		For tube o.d. (in inch)
94609-1	94605-1*	$\frac{5}{8}$
94609-2	94605-2*	$\frac{3}{4}$
94609-3	94605-3*	$\frac{7}{8}$
94609-4	94605-4*	$1\frac{1}{8}$
94609-5	94605-5*	$1\frac{3}{8}$
94609-6	94605-6*	$1\frac{5}{8}$

* Having rising stem actuators.

Some models as above are also available with lever actuators; those with female pipe thread ports have catalogue numbers 94606-1, 94606-2, 94606-3, 94606-4, and 94606-5; those with female solder joint ports have catalogue numbers 94607-1, 94607-2, 94607-3, 94607-5, and 94607-6.

No. 94552 three-way valve: Low pressure solenoid valves for use on intake port of Cenco vacuum pumps to close off automatically when motor drive is stopped.

Cenco no.	IPS thread (inch)
94280-1	$\frac{3}{8}$
94280-2	$\frac{1}{2}$
94280-3	$\frac{1}{2}$
94280-4	$\frac{3}{4}$
94280-5	$\frac{3}{4}$
94280-6	1
94280-7	1
94280-10	$1\frac{1}{2}$
94280-11	$1\frac{1}{2}$

Universal vacuum release valve: An air admittance valve.

Sylphon packless high vacuum valves: A Teflon disc valve with stem enclosed in metallic Sylphon-type bellows. Two sizes: Cenco 94551-1 for connection size of 1 inch and Cenco 94551-2 for connection size of $1\frac{1}{2}$ inch.

Toggle release valve: For admitting gas or air to a vacuum chamber without disturbing pumping operation. No. P35428.

Vacuum release valve: No. 94271.

Virtis quick seal valves: Made of pure gum rubber and polypropylene, vacuum tight to 10^{-1} Pa. For quick connection between manifold ports and freeze-drying flashes and other bottles or ampoules and to permit samples to be removed without breaking the vacuum. No. 45407-1 for $\frac{1}{2}$ inch o.d. ports; No. 45407-2 for $\frac{3}{4}$ inch o.d. ports.

Vacuum bleed valve: No. 94275-4.

Cooke Vacuum Products

Manufacture SS valves from 3 mm to 250 mm bore in most shapes with optional flanging. Sealing is either by O-ring or metal with bellows in bakeable models. Flat seals supplied in Teflon, Neoprene, Viton or Copper.

Duriron
A range of valves for generally engineering purposes. Full details in Bulletin V-21D of the variety of 'Sleeveline' valves, types G and CP and Bulletin V-22C for the range of Durco butterfly valves.

Edwards High Vacuum
Air admittance valves: RS series: chromium plated brass, sealed by a non-rotating nitrile rubber washer on valve seat. Leak rate across seat and leak rate through body = 10^{-7} Torr litre/second. Models: RSIA (panel mounted, 6·4 mm vacuum union), RS2 (pipeline tee boss or elbow or direct to chamber, body tapped 6·4 BSP), RS3 (pipeline supported, 12·8 mm pipeline coupling systems).
Solenoid operated: supplied with 1·59 mm or 3·2 mm air admission orifices; nitrile rubber washer seals and dc operated. Maximum body and seat leak rate 10^{-6} Torr litre/second.
Baffle and isolation valves: Water cooled and available for hand, pneumatic or electrical operation, in straight through or right-angled versions. Models: 2L1B (manual, 30 litre/second conductance), H5L2 (manual), H52LB (manual), S5LB (solenoid), E5L2B (motor) (all 180 litre/second conductance).
Bakeable isolation valves (UHV): Suitable for clean, mercury-contaminated or chemical systems, available with flangeless (tubular), 150 or copper compression gasket type flange connections. Models: BV4 (bore size 12·7 mm, conductance open 5 litre/second, BV8 (bore size 25·4 mm, conductance open 15 litre/second, BV16 (bore size 50·8 mm conductance open 70 litre/second. Leak rate across seat 10^{-9} Torr litre/second for all; baking range 350°C (open) 300°C (closed).
Butterfly valves: QSB series, quarter swing, hand or pneumatic operation. Models: QSB1B (45 litre/second conductance, manual), QSB2 (220 litre/second conductance, manual or pneumatic), QSB2P (220 litre/second conductance, pneumatic).
Foreline trap changeover valve: Automatic in operation, permits the major portion of the atmospheric air/gas to be roughed out of a system, bypassing the foreline trap. Valve actuated by the diaphragm, changeover completed when system pressure reaches approximately 30 Torr.
Needle valves: Series OS1D (fine control) O-ring sealed SS spindle, replacement seat and 100 division 360° dial and pointer; mountable on panels up to 4·8 mm thick. Chromium plated brass with two 1·6 mm right angled vacuum unions or one union and a B14 cone (also in stainless steel with two vacuum unions). Body and seat leak rates 10^{-7} Torr litre/second, maximum flow rate 0·03 litre/second.
Series LB (extra fine control) 18 turn 100 division dial and lever movement, bellows mounted stainless steel needle with strong return spring to eliminate backlash, and renewable brass seat. Models: LB1B (in-line mounting, pipeline supported), LB2B (panel mounted, enclosed lever movement, revolution counter with positive stops). Body and seat leak rates 10^{-7} Torr litre/second, maximum flow rate 0·09 litre/second.

Pipeline valves: 'Speedivalves' manually operated, in line mounted. Valve body mercury resistant silicon aluminium alloy. Diaphragms of nitrile rubber (Grades C and CV) or Viton. Sizes $\frac{1}{4}$, $\frac{1}{2}$, $\frac{3}{4}$, and 1 inch. (Body and seat leak rates 10^{-5} Torr litre/second.)
DS1A right angle, manually operated pipeline isolation valve. Nitrile rubber diaphragm. Size $\frac{1}{2}$ inch. (Body and seat leak rates 10^{-7} Torr litre/second.)
Right angle manually operated (lever) models PLV10 ($\frac{1}{2}$ inch) PLV25 (1 inch) and PLV50 (2 inch). Vacuum range 760 to 10^{-9} Torr. Viton valve seat. Bakeable to 70°C (120°C with glass fibre pistons). Leak rate better than 10^{-9} Torr litre/second.
Right angle pneumatically operated models PLV10P, PLV25P, PLV50P. Vacuum range 760 to 10^{-9} Torr. Viton valve seat. Bakeable to 70°C (120°C with microswitches removed). Leak rate better than 10^{-9} Torr litre/second.
Solenoid operated pipeline valves. Three models for isolation only, one combines isolation with air admittance to one side and one combines isolation when shut with a water cooled plate baffle when open.
Diaphragm valves: A range of corrosion resistant diaphragm type valves is available in single or two way versions for bench or vertical mounting or for T piece or elbow attachment. Handwheel operated.

GEC Elliott Control Valves

Regulators: Series Y600. Body: cast iron, ductile iron or steel. Sizes $\frac{3}{4}$, $\frac{3}{4}$ x 1, 1, 1 x $1\frac{1}{4}$, $1\frac{1}{4}$ inch.
Type Y610: vacuum breakers; type Y611: vacuum breakers or relief valves; type Y612: vacuum regulators.
Series 66 regulators: types 66-111 to 66-114 (vacuum regulators) types 66-125 to 66-128 (vacuum breakers).
Control valve bodies: designs A (globe body, direct acting) AR (globe body, reverse acting) ES (single seated globe body).

General Engineering

Valves of the following types are available: butterfly valves, isolating valves, packless diaphragm valves, regulating valves.

Granville-Phillips

Type C $\frac{1}{2}$ inch bakeable ultra-high vacuum valves, series 202: Variable conductance from 1 to 10^{-14} litre/second, bakeable to 450°C, all-welded construction in stainless-steel. Connections available are glass or metal tubulation or CuSeal flanges.
Variable leak valve, series 203: Gas flow continuously variable from 300 to below 10^{-10} Torr litre/second. Bakeable to 450°C. Constructed entirely of low vapour pressure metals. A variety of connections is available in glass, metal or flanges.
1 and 2 inch gold seal ultra-high vacuum valves: series 20A is 1 inch and series 205 is 2 inch. Bakeable to 450°C with valves open and to 250°C with valve closed. Open conductance in litre/second: 10 for 1 inch valves and 60 for 2 inch valve.

6 inch gate valve, gold seal, series 205: Seals to conductances less than 10^{-12} litre/second. Bakeable to 300°C. Clear opening of 5·88 inch diameter with flange-to-flange flow path of 3 inch. Open conductance is 1340 litre/second. Gold seal. Pneumatically operated.

Gresham and Craven
Manufacture valves for vacuum brake systems: Graham 3 inch strainer and check valve unit and Gresham diaphragm operated vacuum relief valve.

High Voltage Engineering
Model no. HVA6025 gate valve of straight-through type with welded bellows seal, OFHC copper disc plate-to-seat seal of outside flange diameter 130 mm; body of 304 stainless steel, pneumatically operated with leak rate against atmosphere when closed of $<10^{-9}$ Torr litre/second.
Model no. ED-1100, a 6 inch all-metal gate valve: Bellows sealed stem, pneumatic operation, bakeable to 450°C in the open position. Offers a 6 inch diameter unobstructed orifice when open. Fabricated from 300 series stainless steel with OFHC copper plate, easily replaceable.
B-VAC-120 and B-VAC-97: 2 and 4 inch gate valves with stainless steel bellows stem seal, available with either manual or pneumatic actuators. Leakage rate against atmosphere when closed is $<5 \times 10^{-9}$ Torr litre/second.

Hoke
Packless valves

Body material	Pressure range	Temperature range (°F)	Bellows types	Orifice sizes (inch)
Monel	Vacuum to 100 psig	–65 to 240 –65 to 350 –65 to 600	Aluminium or Kel-F gasket or silver solder diaphragm	$\frac{1}{8}$
316 SS or Monel	Vacuum to 2000 psig at 600° F or 500 psig at 350°F	–320 to 1200	316 SS or Monel	$\frac{5}{32}$
Brass or Monel	Vacuum to 200 psig	–20 to 250	Phosphor bronze or Monel	$\frac{5}{32}$, $\frac{9}{32}$
Brass	Vacuum to 600 psig	–40 to 300	Blunt Vee or Kel-F	$\frac{11}{64}$, $\frac{1}{16}$
316 SS	Vacuum to 1000 psig	650	Blunt or Vee	0·170 0·059
316 SS	Vacuum to 2000 psig at 600°F	–320 to 1500	316 SS	$\frac{5}{8}$
Brass or 303 SS	Vacuum to 3000 psig	–	–	–
304 SS or 316 SS	Vacuum to 1000 psig	–320 to 1500	316 SS	$\frac{5}{8}$

Note: SS = stainless steel; 1 psi = 1 lbf/in^2 = 0·689 N/cm^2.

Remotely actuated valves: Electrically operated ball valves, electrically operated metering valves, pneumatically operated ball valves, air operated bellows valves, air operated packless valves, pneumatic-hydraulic operated packless valves, pneumatically operated packless valves are available each with a range of inlet and outlet connections.
Miniature valves: 'Ultra-mite', 1200 series, forged needle valves in 316 SS or Monel with NPT and 'Gyrolok' connections.
Miniature forged needle valves, 3100, 3200 and 3300 series in aluminium, brass 316 SS or 303 SS with 'Gyrolok' tube fitting connections, pipe ended connections or NPT connections.
Bar stock needle valves, 2200 series; designed specifically for corrosive fluids, in 316 SS or Monel with NPT and 'Gyrolok' tube fitting connections.
Forged needle valves, 3400 series, for general hydraulic and pneumatic utilities, in weldable carbon steel with male and female NPT connections.
Bar stock needle valves, 2100 series, in brass, 430 F SS, 316 SS or carbon steel, with NPT and 'Gyrolok' connections.
Check valves (ball-poppet types), 6100–6200 series, designed to prevent flow and pressure reversal, in brass or 303 SS with NPT and 'Gyrolok' tube fitting connections.
Bar stock needle valves, 2300 series, designed to give fine metering performance in gas and liquid flow measurement, in brass or 316 SS with female NPT connections.
'Mill-mite' forged needle valves, 1300 series, designed for use in medical or biochemical gas or vapour analysis work, in brass or 316 SS with male NPT and 'Gyrolok' tube fitting connections.
'Flo-mite' miniature forged ball valves, 7100 series, in brass, 303 SS or 316 SS with 'Gyrolok' tube fitting and female NPT connections.
'Selecto-mite' 3-way ball valves, 7165 series, designed for both single inlet-double outlet or double inlet-single outlet services, enabling selection (or diversion) of two fluid streams into a single line, in brass or 303 SS with 'Gyrolok' tube fitting and female NPT connections.
Toggle valves, spring closing 1500 series, in brass or 303 SS with NPT and 'Gyrolok' tube fitting connections.

Ion Equipment

UHV bakeable valves: Stainless steel, copper alloy gasket, bakeable in open or closed position to 450°C, right-angle or Tee patterns. Bellows-sealed drive. BVV-150 series: 1½ inch size; open conductance 32 litre/second. BVV-250 series: 2½ inch size; open conductance 118 litre/second.
Ultra-high vacuum valves: Bakeable to 150°C in open or closed position, all metal parts exposed to high vacuum are 300-series stainless steel, stainless steel bellows-sealed; valve seal is Viton O-ring. Sizes in inch are ½, 1, 1½, 2, 2½, 3, 4, and 6 having respectively open molecular conductances in litre/second of 1, 11, 40, 85, 135, 200, 490 and 1410. Available with pneumatic actuators.
2, 4, and 6 inch gate valves supplied in both manual and pneumatically operated version.

Kinney Vacuum Company

Kinney 3-way ball valves: Aluminium valve block, bronze end caps, chrome-plated bronze ball and brass stem guide. The seat rings are Buna-N and O-rings are Viton. Available in ¾

and 1 inch nominal sizes. Manually operated. Pneumatically operated valves are available on request. Serve as roughing and backing valves.
BB series of globe type valves: Bronze body with flanged cover, brass bellows and non-rising stem. Designed for applications employing soldered or brazed manifolding. Positive isolation of rotating parts is insured by sealing the brass bellows to the disc holder and cover flange by O-rings.
Disc seals and O-rings are of Buna-N and operating temperatures to 135°C are permissible. Available in 1 to 3 inch sizes. Solenoid, air-operated actuation.
Butterfly valves: Manual (KBV), pneumatic (KBVP) and electric (KBVE) in sizes of 2, 4 and 6 inch.
DC series of diaphragm-sealed vacuum valves for systems operating at pressures above 10^{-1} Pa at temperatures up to 83°C. Cast iron Y-body and neoprene diaphragm and valve disc. Available with screwed ($\frac{1}{2}$ to 2 inch) and flanged (1 to 6 inch) connections and for manual or pneumatic operation.

Leybold-Heraeus

Air inlet valves: Manually operated: No. 173 24 screw cap valve, steel, wing nut operated (leak rate less than 10^{-5} Torr litre/second). No. 273 51 lever operated valve, steel and aluminium (leak rate less than 10^{-7} Torr litre/second, bakeable to 200°C). No. 173 27 screw cap corrosion resistant, stainless steel valve (leak rate less than 10^{-7} Torr litre/second).
Solenoid operated: air inlet No. 173 26 brass body, stainless steel connecting flange 10 seconds venting time for 10 litre vessel (leak rate less than 1×10^{-7} Torr litre/second). No. 173 25, same design and construction but larger air flow path (10 mm diam). 14 seconds venting time for 50 litre vessel (leak rate less than 1×10^{-7} Torr litre/second).
Solenoid operated: power failure airing valves. No. 174 26, steel body, 4 mm diam. air inlet; venting time for 10 litre vessel: 8·2 seconds with filter in place, 7·8 seconds without filter (leak rate less than 10^{-7} Torr litre/second). No. 174 28, steel body, 10 mm diam. air inlet, venting time for 50 litre vessel: 40 seconds with filter, 10 seconds without filter (leak rate less than 1×10^{-7} Torr litre/second).
Ball valves: Simple straight through valves operated by lever. Sizes NW 10 and NW 20 are made of brass, NW 32 valves is of steel with brass inner parts these valves must be protected against mercury and should not be heated above 80°C (leak rate less than 1×10^{-5} Torr litre/second).
Bellows sealed right angle valves: With handwheel: sizes NW 10 to NW 32 with nickel plated cast-iron bodies, sizes NW 50 to NW 150 with light alloy bodies, sizes NW 10 to NW 50 also in stainless steel (leak rate less than 1×10^{-9} Torr litre/second).
Lever operated: sizes NW 10 to NW 32 with nickel plated cast iron bodies, sizes NW 50 to NW 150 with light alloy bodies (leak rate less than 1×10^{-9} Torr litre/second).
Electropneumatically operated: sizes NW 50 to NW 150 with light cast alloy bodies, sizes NW 250 to NW 800 with bodies of nickel plated steel and Viton gaskets (bakeable to 200°C). Closing time from 0·1 to 7 seconds depending on size (leak rate 1×10^{-9} Torr litre/seconds).
Bellows sealed straight through valves: Manually operated (lever): sizes NW 32, NW 65 to

NW 250 with steel bodies nickel-plated inside fitted with Viton gaskets (bakeable to 200°C) (leak rate less than 1 x 10^{-9} Torr litre/second).
Electropneumatically operated: sizes NW 65 to NW 350, construction as manually operated models. Supply of dry, lightly oiled compressed air at a pressure between 42 and 85 psig is required. Flameproof version available.
Gate valves: Manually operated (lever): sizes NW 65, 100 and 150 with bodies of corrosion resisting light alloys (leak rate less than 10^{-7} Torr litre/second).
Electropneumatically operated: sizes NW 100 to NW 800. NW 100, 150 and 250 with bodies of corrosion resisting light alloy; NW 350, 500 and 800 with steel bodies. In sizes 250 and upwards two side flanges are provided for connecting bypass lines (leak rate less than 10^{-7} Torr litre/second).
Plate valves: Manually operated (handwheel): straight through valves, nickel plated steel body, designed for connection to the intake side of small vapour pumps. No. 174 13: conductance 50 litre/second; No. 174 16 conductance 100 litre/second.
Right angle valves: Manually operated (handle): sizes NW 10, 20 and 32 with silumin bodies (leak rate less than 1 x 10^{-7} Torr litre/second).
Electromagnetically operated: sizes NW 10 to NW 32 with nickel plated cast iron bodies, NW 50 with light alloy body. NW 10 to NW 32 are tight against atmospheric pressure in both directions, NW 50 is gas-right against atmospheric pressure in one direction only (body leak rate less than 10^{-9} Torr litre/second, seat leak rate less than 1 x 10^{-7} Torr litre/second.
Electropneumatically operated: sizes NW 250 to NW 1000. (70–100 psig compressed air needed). NW 250 and NW 350 are light against atmospheric pressure in both directions, NW 500, 800 and 1000 only in one direction. Conductances NW 250: 2900 litre/second, NW 350: 6000 litre/second, NW 500: 12 000 litre/second, NW 800: 32 000 litre/second, NW 1000: 56 000 litre/second (leak rate less than 1 x 10^{-7} Torr litre/second).
Safety valves: 'Secuvac' valves combine a high vacuum isolation valve with an air inlet valve. Electromagnetically opened, they close due to the pressure difference between atmospheric pressure and vacuum. The valves are right-angled and made of steel and light alloy (leak rate less than 10^{-7} Torr litre/second).
Two-way valves: Cross-over valves for two pipelines, lever operated. Sizes NW 10, 20, 32. Body and flanges of corrosion resisting light alloy. (Leak rate less than 10^{-7} Torr litre/second.)
UHV valves: Bellows sealed right angle, bakeable to 200°C, leak rate less than 10^{-9} Torr litre/second. Sizes NW 16 and NW 35. Also electromagnetically operated right angle: NW 16 and NW 35.
Vacuum locks and seal-off fittings: These fittings are useful for evacuated vessels which are seldom re-pumped. The lock allows the seal-off fitting to be inserted under vacuum. The seal off fitting is of stainless steel with viton gaskets, the lock of cast, mercury-proofed, silumin (leak rate less than 10^{-7} Torr litre/second).
Variable leak valves: No. 173 18: needle valve, coarse adjustment range up to 400 Torr litre/second, steel (chromium plated inside).

No. 173 19: fine adjustment range 1–1250 Torr litre/second. Body and operating mechanism are of nickel plated steel, valve plug guide of stainless chromemolybdenum steel, plug of polished hare chromium steel, interior sealed by Teflon bellows. Bakeable when open to 150°C.
No. 173 20: coarse adjustment range 1×10^{-4} to 100 Torr litre/second. Construction and materials as 173 19.
No. 173 17: fine adjustment range 1×10^{-4} to 50 Torr litre/second. Valve is of chromium plated steel, needle and flow diaphragm of corrosion resisting material, sealed by an annular greased gasket.
Model SM 21: operated by an electric servo-motor. Gas throughput from 10^{-5} to 50 Torr litre/second. Supplied with either remote control unit SM 22 or pressure controller SM 20.
Valve accessories: Valve position indicator (shows position of valve plate in remotely actuated valves) for connection to bellows sealed right angle valves (manual and electro-pneumatic), bellows sealed straight-through valves and electromagnetic right angle valves (No. 172 99).
Switching unit (No. 173 95). Small compressor (No. 172 98) produces 13·8 litre/minute of air at 100 psi for the operation of compressed air driven valves.

Lucifer Ltd
Solenoid operated valves: pressure range 10^{-6} to 1520 Torr; orifice size 3 to 400 mm; tightness 10^{-9} Torr litre/second.

Ch. Lumpp
Designed for the various needs of the chemical industry, membrane and plug type valves with orifice diameters from 20 mm to 200 mm.

Materials Research
Vacuum bleed valve: Model V4-160: Monel body and phosphor-bronze bellows. Model V4C-160 is a copper-gasketed version, not bakeable.

Mill Lane Engineering
Model nos. 5042R and 4042: 5042R is right-angle pattern, 4042 is straight-through type of gate valve. Seal between stem and bonnet is O-ring or bellows (optional). Seal between plate and seat is O-ring. Bore diameters are from 50 to 1500 mm in range of sizes. Leak rate against atmosphere when closed is below 5×10^{-12} litre/second. Main body constructional material: carbon steel or stainless steel. Pneumatic operation available.

Nanotech
NV01, NV02 and NV03: plate valves, magnetically operated, right-angle pattern with O-ring seal between stem and bonnet and Viton O-ring seal between plate and seat. Leak rate against atmosphere when closed is 10^{-7} Torr litre/second. Main body is of aluminium alloy. Bore diameters in mm are NV01: 6; NV02: 12 and NV03: 25.

NBV3, NBV5 and NBV7: regulating, bellows-sealed straight-through plate valves, magnetically operated, with Viton O-ring seal between plate and seat. Leak rate against atmosphere when closed is 10^{-8} Torr litre/second. Main body is of aluminium alloy. Bore diameters in mm are NBV3: 73; NBV5: 127; NBV7: 179 with molecular conductances when open in litre/second of respectively 190, 510 and 800. These are baffle/isolation valves for mounting above diffusion pumps.

Neyrpic

Butterfly vacuum valves with O-ring or lip seal between stem and bonnet, O-ring seal between plate and seat, bore diameters 100–1200 mm, leak rate against atmosphere when closed: 10^{-3} Torr litre/second. Air admittance valves with bore diameters of 80, 150 and 250 mm.

N.G.N.

Model no. PV1/NO: A right-angle disc valve used for air admittance to vacuum system (usually to mechanical rotary pump). Bore diameter 2 mm. Magnetically operated.
Model no. 515A: A packed vacuum valve of quarter-swing butterfly type, straight through with rubber O-ring seal between stem and bonnet and rubber O-ring plate seal. Bore diameter 25 mm. Lever operated.

Nupro

H. series bellows valves in brass and stainless steel.
B. series bellows valves in brass or 316 stainless steel.
BKT series toggle operated bellows valves in brass and stainless steel, designed for quick operation.
BK series air operated bellows valves, designed for remote control applications.
U series bellows valves.
T series bellows valves, similar to U series except that the secondary packing system is eliminated.
BM series bellows metering valves for very fine flow control, in stainless steel primarily.
M and S series fine metering valves for use in liquid control on gravity feed laboratory systems, also gas analysis and sampling instruments; also supplied with vernier handles.
Cross pattern and double pattern fine metering valves for series and parallel installations of metering valves.
2C-4C-6C-8C series check valves for unidirectional flow and as in-line or pop-off relief valves: $\frac{1}{8}$, $\frac{1}{4}$, $\frac{3}{8}$, $\frac{1}{2}$, and $\frac{5}{8}$ inch connections.
12C-14C-16C series check valves have larger soft seat with $\frac{3}{4}$, $\frac{7}{8}$, and 1 inch 'Swagelok' connections.
Safety relief valves for the automatic venting of gas or liquid systems from over pressure.
Also supply F series in-line filters, FR series in-line removable filters, V and VD series plastic valves, purge valves, heat exchange and condenser tube plugs and the SNOOP liquid leak detector for locating leaks on air or gas lines.

Phönix

Supply a range of valves for general engineering and chemical applications, some of which can be used (or modified for use) in vacuum work. Models are:
Globe valves, NP25 up to NP320, of straight-through and angle types.
Regulating valves, NP25 up to NP160, with parabolic disc and lift indication; fine adjustment valves NP25/40 with narrowed seat for extremely fine regulation; ND15, ND25 and ND50 each with free sections and ND65 onwards upon request; also available as bellows sealed valves.
Check valves, NP25 up to NP40, of straight-through and angle types.
Globe valves, NP25/40, for chlorine service, straight-through and angle types with stainless steel interior.
Bellows sealed globe valves, NP25 from ND10 up to ND500 and NP40 from ND10 up to ND300, of straight-through type, angle type and with inclined seat, particularly suitable for aggressive and volatile fluids.
Globe and regulating valves, NP25/40, for refrigerating agents.
Special valves for cryogenic work, NP10 up to NP320.
Special valves for thermofluid work, NP25: ND10 up to ND500, of straight-through and angle types.
Pneumatically operated safety valves, NP25, ND40, particularly suitable for chlorine, with interior ball-check valve.
Quick closing valves and quick opening valves, NP25 up to NP40: (a) operated by cable, (b) operated by hydraulic or pneumatic piston, (c) operated by magnetic release, also available with electric command.
Three-way changeover valves, NP25 up to NP160.
Combination safety valve with change-over valves for oxygen, NP40, ND15 and ND25.
Proportional safety valves, NP25/40 from ND15 up to ND50.
Globe valve with electric command, bellows-sealed with safety gland.
Globe valves for pressure gauges, also available as bellows-sealed valves up to NP160.
Small sized valves, NP100 up to NP400, available as globe and regulating valves, straight-through and angle types, also as bellows-sealed valves up to NP160.
Bellows-sealed valves, NP10, with Fluoroflex-T lining of straight-through and single types available as globe and regulating valves with manual operation or operated by piston or diaphragm, safety valves, check valves, suitable for temperatures from – 70°C to 250°C.
Globe valves, for measuring circuits NP100 up to NP640, also available as bellows-sealed valves up to NP160.
Manifold, NP100 up to NP400, also available with bellows-sealed valves up to NP160.

RIBER

Bellows-sealed valves, made of stainless steel: VME and VV ranges of both right-angle and straight-through types with copper (bakeable) or Viton sealing; bore diameters of 14 to 63·5 mm with open conductances of 5 to 100 litre/second. Also VT gate valve with bore of 196 mm and open conductance of 3000 litre/second.

Schuf

Flush-off valves of all types, other special application valves.

Teltron

Butterfly valve: TEL 715, straight-through, aluminium alloy O-ring sealed 25 mm bore diameter. Designed to plug into backing and diffusion pumps.

Torr Vacuum Products

Butterfly valves: BV series available in sizes from 2 to 24 inch inside diameter with manual or pneumatic operators. These valves can operate in the 10^{-6} Torr range and can be baked to 100°C.
Gate valves: SMT 2 series, stainless steel, rotary stem seal, standard sizes 2, 3, 4 and 6 inch, capable of operation in 10^{-9} Torr range, bakeable to 200°C. SVB 2 series, stainless steel, bellows shaft seal, standard sizes from 2 to 12 inch, capable of operation in 10^{-9} Torr range, bakeable to 200°C. SVB 3 series, stainless steel, bellows stem seal, standard sizes from 2 to 16 inch, capable of operation in 10^{-10} Torr range, bakeable to 300°C. SVL series, available in stainless steel, carbon steel and carbon steel electrolen nickel plated, sizes from 10 to 52 inch inside diameter, pneumatically operated. SVS 2 series, stainless steel, rotary shaft actuated, standard sizes from 2 to 48 inch, capable of operation in 10^{-9} Torr range, bakeable to 200°C. SVS 3 series, stainless steel, bellows stem seal, standard sizes from 2 to 12 inch, capable of operation in 10^{-10} Torr range, bakeable to 300°C. SVS 4 series, austinitic stainless steel, standard sizes 4, 6, and 12 inch, bakeable to 450°C (open).
General purpose vacuum valves: 2100 series bronze ball valve, sizes $\frac{1}{4}$ to 2 inch, manual operation. 2300 series stainless steel solenoid valve, sizes $\frac{1}{8}$ to $\frac{1}{4}$ inch. 2500 series brass solenoid valve, sizes $\frac{1}{8}$, $\frac{1}{4}$ and $\frac{3}{8}$ inch.
Right angle valves: RBS and RVS series, sizes from $1\frac{1}{2}$ to 35 inch, stainless steel or carbon steel construction, manual or air operation, metal or elastomer body seals, bakeable to 200°C.

Ultek

$1\frac{1}{2}$ inch in-line valve: All stainless steel with welded bellows and Viton seals; bakeable to 200°C; model nos. 252-1512 has $1\frac{1}{2}$ inch ($2\frac{3}{4}$ inch o.d.) 'curvac' flanges and 252-1520 has 2 inch ($3\frac{3}{8}$ inch o.d.) 'curvac' flanges.

$\frac{3}{8}$ inch high vacuum stainless steel valve has replaceable Viton bellows seal, 150°C max. bake out temperature, all stainless steel body with aluminium knob $\frac{3}{8}$ inch o.d. tubulations. Mode no. 251-1000.

Right-angle, in-line and gate high vacuum valves, stainless steel, bakeable to 200°C. Valve sizes in inches are 1, 1½, 2 and 3 with molecular conductance in open position in litre/second of respectively 11, 40, 85, and 200.

UHV

Isolating/air admittance diaphragm valves: models DV6, QAV and SV, HV (magnetically operated), straight-through type; QAV and HV sealed by Viton 'O' ring, SV by Viton gasket and DV6 by copper pad and knife-edge.

Vactronic Lab. Equipment

Vari-vac adjustable leak valve, model VV-50: Bellows-sealed, micrometer drive; adjustable flow-rate from 10^{-5} to 5×10^{-3} litre/second against air at stp. Made of plated brass or stainless steel. Model VVB-50Q has 'quick connect' fittings; model VV-50S has brazed fittings. Model VVB-505M is motorized. Model CVB-50TH throttling valve is for larger flow rate.
Forged high vacuum valves in brass, bellows-sealed. A range is available under model nos. CVB-37, CVB-50, CVB-62, CVB-75, CVB-87, CVB-100, CVB-112, CVB-150, CVB-162, CVB-200, CVB-212 which, successively, are for tubes of o.d. in inches of ⅜, ½, ⅝, ¾, ⅞, 1, 1⅛, 1½, 1⅝, 2 and 2⅛. In-line valves of type L have brazed connections and are denoted by CVB-37L-S, CVB-50L-S etc.; whereas those with quick connections are denoted by CVB-37L-Q, CVB-50L-Q etc. Angle valves of type R with quick connections are denoted by CVB-37R-Q, CVB-50R-Q etc. Valves with pipe thread (n.p.t.) connections are available as models CVB-12, CVB-25, CVB-37, CVB-50, CVB-75, CVB-100, CVB-150 for pipe sizes (n.p.t.) in inches of successively ⅛, ¼, ⅜, ½, ¾, 1 and 1½. In-line valves of type L are denoted by CVB-12L-P, CVB-25L-P etc. Angle valves of type R are denoted by CVB-12R-P, CVB-25R-P etc. Buna N sealing is commonly used in most vacuum systems; Viton A sealing is formulated especially for vacuum systems.
Stainless steel high vacuum angle valves: Bellows-sealed, special conical stainless steel seat, stainless steel 304 used. Includes manually operated high vacuum angle valves: CVS-225R and CVS-450 and pneumatically operated (compressed air with solenoid actuation) high vacuum angle valves: AVS-225R and AVS-450.
High vacuum butterfly valves made from stainless steel, conductance <125 litre/second. Bakeable to 200°C. Viton sealing. Model HSV-150 with mating flanges to connect to ½, ¾, 1 and 1½ inch tube sizes.

Vacuum Accessories Corp. of America

MV-25-XL-series of precision gas metering valves: Meters helium and other gases from 1×10^{-9} std cc/second. Leak-tight in closed position to helium leak detector having sensitivity to 5×10^{-13} std cc/second. Precision micrometer actuator; bakeable to 225°C; easily disassembled for cleaning. Motorized drive available.
Gold seal ultra-high vacuum valves, bakeable to 450°C, closed conductance less than 10^{-14} litre/second. Bore sizes: ¼, ⅜, ½, ¾, 1, 1½, and 2½ inch.

Stainless steel high conductance bellows-sealed angle valves for ultra-high vacuum. Bakeable to 225°C. Viton O-ring sealing. Bore sizes, $\frac{1}{2}$, $\frac{3}{4}$, 1, $1\frac{1}{2}$, 2, and $2\frac{1}{2}$ inch.
AOV-ATSS series pneumatic stainless steel type 304 high vacuum bellows-sealed angle valves: Viton O-ring sealing. Bore sizes: 1, 2 and $2\frac{1}{2}$ inch; flanged, socket-weld and butt-weld models.
High vacuum straight through solenoid valves for fail-safe automation. Viton O-ring sealing. Bore sizes, $\frac{1}{2}$ and 1 inch. Magnetically operated.
High vacuum butterfly valves in aluminium and type 304 stainless steel. Viton sealing. Bore sizes: $\frac{1}{2}$, $\frac{3}{4}$, 1, $1\frac{1}{2}$, 2, $2\frac{1}{2}$, 4, 6, 8, 10 inch. Motorized versions of high vacuum butterfly valves are available in 2, 4, 6, 8, and 10 inch sizes.
Forged brass angle and inline bellows-sealed high vacuum valves: Sealing choice of O-ring seat seal or flat disc seal with Viton optional. Flat disc valves supplied with teflon or KEL-F seals.
Pneumatic brass angle and inline high vacuum bellows valves: (AOV-AT series) Viton sealing. Solenoid (magnetically) or compressed air operated. Sizes: $\frac{3}{4}$, 1, $1\frac{1}{2}$ and 2 inch.

Vacuum Generators
Specialize in stainless steel, manually operated, bellows-sealed isolation valves, bakeable and most useful in the ultra-high vacuum range.

Type	*Features*
CRD:	right angle, demountable, all-metal, bakeable.
CRPD:	right angle, partially demountable, all metal, bakeable.
CR:	right angle, all-welded, all-metal, bakeable.
CSD:	straight through.
VRD:	right angle, demountable, Viton-sealed.
GH:	right angle, demountable, Viton sealed.

CRD, CRPD, CR and CSD have seal consisting of stainless steel knife edge and copper pad. VRD and GH have Viton O-ring seal.
Special types supplied have stainless steel knife-edge and gold-plated copper pad, also stellite knife edge and stainless steel pad.

Closed conductance: $<10^{-14}$ litre/second for all-metal valves.
$<10^{-12}$ litre/second for Viton-sealed valves.

Open conductances (molecular flow)	Conductance in litre/second
CR14, CSD14, GH14, VRD14	1·0
CR25, CRD25, CRPD25, CSD25, VRD25	18
CR38, CRD38, CRPD38, CSD38, VRD38	25
CR50, CRD50, CRPD50, CSD50, VRD50	75
CR64, CRD64, CRPD64, VRD64	110
CRD100, VRD100	400
CRD150, VRD150	1000

In above the numbers 14, 25, 38, 50 etc. are approximately the bore diameters in mm.
MD6 is an all-metal bakeable leak valve for gas-metering; also available as the MO142 servo-motor operated valve.
Pneumatic operation can be supplied on VRD series valves.

Varian

Bakeable valves: For use in UHV systems with temperatures of up to 450°C, giving sealing to below 10^{-11} Torr. All metal parts exposed to vacuum are stainless steel except for the copper alloy sealing gasket. Right angle and Tee configurations are available in three sizes: $1\frac{1}{2}$ (32 litre/second conductance), $2\frac{1}{2}$ (118 litre/second) and 4 (440 litre/second) inch.
Bellows-sealed forged brass body valves: Manufactured in both in-line and right angle bodies, hand-operated or air-operated (three-way solenoid valve). Available in sizes from $\frac{3}{8}$ to $1\frac{5}{8}$ inch.
Gate valves: Stainless steel, Viton-sealed, manually or pneumatically operated. Available in 2, 4 or 6 inch sizes (3 and 8 inch on special order). Bakeable up to 125°C (closed) or 200°C (open).
Mini valves: Stainless steel with copper alloy gasket, bakeable to 400°C (open or closed), $\frac{3}{4}$ inch size. Pressure range down to 10^{-11} Torr.
Polyimide and Viton valves: Stainless steel with interchangeable Polyimide and Viton-A bellows and main seals. Operate at pressures down to 1×10^{-9} Torr. Sizes $\frac{1}{2}$ or $1\frac{1}{2}$ inch, manually operated. Viton bakeable to 150°C (open) or 125°C (closed), Polyimide bakeable to 300°C (open) or 300°C (closed).
Right angle valves (*Viton sealed*): Stainless steel body, welded stainless steel bellows and Viton-A seals. Sizes: $2\frac{1}{2}$ (135 litre/second conductance), 4 (490 litre/second) and 6 (1410 litre/second) inch. Hand operated.
Slide valves: HC series, hand or air operated, sizes in inch: 4 (2450 litre/second conductance), 6 (6500 litre/second), 10 (25 200 litre/second); pressures down to 10^{-8} Torr and lower. Body of cast aluminium.
Variable leak valves: Stainless steel body, flange and fitting, gasket of copper alloy and stainless steel, Inconel and sapphire piston. Range to below 10^{-11} Torr, minimum controlled leak rate 1×10^{-10} bakeable to 450°C (open or closed).

Veeco

Forged brass angle valves: types SR and SL (O-ring seal), FR and FL (disc seal). Bellows sealed stem. SL and FL are in-line valves; FR and FL can be converted to pneumatic operation.
Throttle valves: Models R150ST and R62ST for admitting gas to mass spectrometer leak detector (flow rate variable from 10^{-2} to 40 Torr litre/second).
Brass angle valve: Type R300S, 3 inch bellows sealed valve.
Adjustable leak valves: Models VVB50S and VVB50Q (chrome plated brass), VVS50W and VVS50Q (stainless steel). Flow rates adjustable from 0·01 cc/second to 5 cc/second.
Solenoid valves: Models SV-62S-AC, SV-62P-AC, SV-62-DC.

Pneumatic valves: Models FPR (right angle) and FPL (in line) leak free at sensitivity of 1×10^{-10} std cc/second. Power: 115 V r.m.s.; Air: 50–90 psi.
Stainless steel butterfly valves: Models BV-776M (manual) and BV776P (pneumatic). Port diameter 7¾ inch, flange thickness 1⅝ inch. Conductance (fully open) 18 000 litre/second. Vacuum rating 10^{-9} Torr range.
Stainless steel miniature butterfly valves: Models BV 1600 and BU-1600-U (uhv valve). Conductance 125 litre/second; bakeable to 200°C; applications down to 10^{-9} Torr.
Stainless steel gate valve: Model VA600. Conductance for air at 25°C is 4860 litre/second. Gate size 6 inch; bakeable to 200°C; vacuum rating 10^{-10} Torr.
Stainless steel right angle valves (uhv): Models VA110 to 113 (1 inch), VA160 to 163 (1½ inch) VA210 to 213 (2 inch) VA 250 to 253 (2½ inch) VA300 to 303 (3 inch). Conductances: 11 litre/second (1 inch), 40 litre/second (1½ inch), 88 litre/second (2 inch), 130 litre/second (2½ inch) 210 litre/second (3 inch). Vacuum rating 10^{-10} Torr. Bakeable to 200°C.
Economy stainless steel right angle valves: Models R25 RSS, R38 RSS, R50 RSS.
Stainless steel piston valves: Models PR 400P (pneumatic), PR 400M (manual). Port 3·75 inch.
Gold plated copper seal uhv valves (ambient to 450°C): CR series, stainless steel, all welded construction, right angle (T configuration available). Conductances from 5 to 400 litre/second. CRD series, stainless steel, demountable valves. Conductances from 18 to 110 litre/second. CRD150 stainless steel, demountable, bakeable 6 inch valve. CRPD series stainless steel, partially demountable. Conductances from 18 to 75 litre/second. CSD series stainless steel, demountable straight-through valves.
Stainless steel plate valve: P series; open conductance 350/400 litre/second, closed conductance 10^{-13} litre/second. Temperature range: open 350°C (closed: baking not recommended).
Stainless steel precision metering valve: Model MD6. Micrometer screw with numbered indicator, closed conductance of less than 10^{-14} litre/second, bakeable to 450°C.

Wallace and Tiernan

Ball valves: McCannaseal top entry ball valves for temperatures from –200°C to 540°C for pressures up to 1400 lb/sq. inch; McCannaflo ball valves for pressures up to 1000 lb/sq. inch in sizes from 15 mm to 350 mm; Trunnion ball valves for pressures up to 720 lb/sq. inch in sizes from 250 mm to 400 mm; McCanna 800 ball valves with 1850 lb ASA rating at 65°C in sizes from 5 mm to 50 mm; McCanna 500 ball valves for pressures up to 600 lb/sq. inch and temperatures up to 260°C in sizes from 15 mm to 50 mm; McCanna plast ball valves with bodies made from PVC or KYNAR in sizes from 15 mm to 50 mm.

Whitey

Ball valves with top-loaded Teflon packing seals with orifice sizes 0·187 inch (catalogue nos. 43S4, 43M4-S4, 43F2 and 43F4), 0·281 inch (4456, 44F4 and 44F6), and 0·406 inch (45S8, 45S12 and 45F8).

These catalogue nos. are for those with brass body and stem; 316 is added for 316 stainless steel types.
3-way ball valves: Nos. 43 x S4, 43 x F4 and 43 x S4-S4-M4 with brass or 316 stainless steel body and stem.
Union bonnet valves in orifice sizes 0·062, 0·156, 0·250 and 0·312 inch. NB union bonnet shut-off valves in orifice sizes 0·156, 0·250 and 0·437 inch.
Screwed bonnet valves in orifice sizes 0·093, 0·200 and 0·312 inch.
Micro-metering valves: For fine flow control of orifice size 0·020 inch. Models are 22RS4 with micrometer vernier handle and Viton A O-ring stem seal and 21RS4 with phenolic knob handle and Teflon packing; other varieties also supplied. For all types body is brass and 316 stainless steel; stem, spring and retainer is 316 stainless steel and needle is 303 stainless steel.
Forged body valves: 13 main types each in three models according to stem type (i.e. regulating, Kel-F tip or Vee) offering a complete range of sizes (orifice diameters: 0·08, 0·172 and 0·250 inch) with integral bonnet construction, Teflon cylinder packing, 316 stainless steel or Monel stems in straight-through, angle and cross patterns. Normally have phenolic handles. Also large orifice (0·375 inch) types 18VS8-CS with Vee stem and 18KM8-CS with Kel-F tip stem, in straight-through angle and cross patterns.
Toggle operated forged body valves, quick opening or closing types in 11 different models with orifices of 0·125 or 0·250 inch.
DK forged body shut-off valves: 16 DK valves with rupture discs, providing protection against overpressure.
Three valve manifold model SS-300-S4 and five valve manifold SS-500-S4: Air operated forged body valves for remote or automatic operation. Lift check valves in which forward flow lifts the metal plug, thus opening the valve whereas reverse flow is checked by the fluid pressure sealing the plug against the orifice. Orifice sizes are 0·156, 0·250 and 0·437. Also supply feed pumps and compressors and safety clamps.

2.1.15 Vacuum accessories

AEI Scientific Apparatus

A range of couplings and leadthroughs is available.

Aeroquip

Conoseal tube joints, pipe joints and fittings can be supplied to withstand temperatures ranging from –270°C to 1100°C and pressures up to 20 000 lb/sq. inch, helium mass spectrometer leak tested. Wide range of sizes. V-band couplings and flanges and other couplings for joining tubing. Gimbal joints, bellows, expansion joints and duct assemblies. Flexmaster tube and pipe joints: straight, elbow, TEE, right-angle, and mitred types in sizes from $\frac{3}{8}$ to $4\frac{1}{2}$ inch.

Airco Temescal

R.L. fittings: nipples and other couplings. Feed throughs for wide range of currents and voltages. 2, 4 and 8 current feedthroughs. Octal connector. Dual water/electrical feed-through. Water feedthroughs. Liquid nitrogen feedthrough (two pass). Blank plugs and flanges. Bellows-sealed rotary feedthroughs.

Alcatel

Baffles and cold traps: Chevron, water-cooled baffles in sizes 80 to 300 mm outside diameter. Liquid nitrogen cold traps models PEA 80, 150, 200, 300 (model numbers indicate diameter in mm).

Anaconda American Brass

Manufacture wide range of sizes of flexible corrugated hose in various alloys: useful for cryogenic and vacuum service. Appropriate hose fittings also supplied. Also hoses for a variety of other purposes, including flexible metal connectors designed to absorb vibration.

Andonian Associates

Modular cryogenic instrumentation and accessories: temperature controls and sensors; modular liquid helium and nitrogen Dewar bodies; storage Dewar flasks and transfer lines. Cryogenic containers for liquid oxygen, nitrogen, hydrogen and helium. Special cryogenic assemblies for studies in optics, superconductivity, nmr and epr, X-ray and neutron diffraction, Mössbauer effect, gamma detection, thermal and mechanical testing, sensing and control, thin film evaporation. Electrical feedthroughs including multi-pin. Level gauges and controls, particularly for liquefied gases. Transfer lines for liquefied gases. Cryopump baffles B-1-300 (bottom-opening) and B-2-300 (side opening).

Apiezon Products

Greases: AP 100: anti-sieze grease suitable for use down to 5×10^{-11} Torr at 15°C, with a softening point of 30°C.

AP 101: general purpose stop-cock grease suitable for use down to 10^{-5} Torr at 20°C, has a softening point of 185°C and can be used over a wide range of temperatures (–40°C to 180°C).

Grease L: general purpose high vacuum lubricant, softening at 30°C and melting at approximately 47°C.

Grease M: similar to grease L but contains more wax and has a somewhat higher vapour pressure.

Grease N: similar to Grease L but contains a high molecular weight polyteric hydrocarbon additive and can be used as a combined high vacuum lubricant and general purpose laboratory lubricant.

Grease T: similar to grease N but can be used over a wider temperature range: contains a gelling agent.

Grease H: similar to grease T but can be used over a wider temperature range (–10°C to 24°C); at temperatures above 40°C it stiffens rather than melts.

Oils: Oil A: for use in small vapour diffusion pumps, ultimate pressure of 10^{-4} to 10^{-5} Torr; permits the pump to operate at a higher backing pressure than with oils B or C (maximum backing pressure with a typical 3 inch pump is 0·4 Torr).
Oil B: a general purpose fluid for large and small vapour diffusion pumps required to obtain baffled ultimate pressures of 10^{-6} Torr.
Oil BW: similar to oil B though it enables slightly higher pumping speeds to be achieved; its high purity makes it suitable for use in the evacuation of mass spectrometers.
Oil C: the lowest vapour pressure diffusion pump oil in the range can be used to produce baffled ultimate pressures of 10^{-7} Torr without a cold trap.
Oil G: for large vapour diffusion pumps required to reach ultimate baffled pressures of 10^{-5} Torr.
Oil AP 201: vapour diffusion booster pump fluid, based on a hydrocarbon oil incorporating a non-toxic, thermally stable anti-oxident.
Lubricating and sealing oils (for use in rotating seals etc.): Oil J: moderately viscous, with a low vapour pressure.
Oil K: very viscous, lower vapour pressure.
Sealing compounds: Sealing compound Q: a putty-like substance suitable for sealing at not too low a pressure (vapour pressure: 10^{-5} Torr at 20°C).
Waxes: Wax W: softening between 80–90°C, suitable for joints which may become warm in operation.
Wax W100: softening between 50–60°C, used when soft wax is needed to reduce danger of joints cracking because of vibration.
Wax W40: designed for use when the sealing medium must be flowed into or around a joint without heating the apparatus to any great extent. Flows at temperatures of 40–50°C.

Balzers
Supply a very wide range of all kinds of vacuum accessories, including vacuum connections (pipes of various geometries, flanges both in a variety of metals and sizes), electron-beam welded uhv-components with Conflat flanges or knife edge flanges, tube fittings, bellows, adaptors, glass joints and metal-to-glass joints, O-ring and other gaskets in a variety of elastomers, also aluminium and gold. High vacuum and ultra-high vacuum lead-ins in a range of configurations and sizes, rotary feedthroughs. Metal-ceramic inserts for uhv feed-throughs. Uhv sight glasses. Baffles. Devices for manual and automatic supply of liquid nitrogen to cooled baffles. Filling device for liquid nitrogen.

Cajon
Ultra-torr fittings designed to provide a vacuum-tight seal with quick, finger-tight assembly useable on glass, metal or plastic tubing and in a variety of sizes and configurations. VCO O-ring vacuum couplings (wide range). VCR vacuum couplings for a variety of functions. 321 stainless steel flexible hose with accessories in the form of adapters for brazed, 'Swagelok', O-ring and welded connections. Flexible glass-end tubing designed to isolate vibration from glass systems. Glass-to-metal transition tubes. Vacuum butt weld fittings. Vacuum flanges. Copper, Viton and 316 stainless steel O-rings.

Cenco

Base-plates (e.g. for bell-jars) in a range of diameters. Gauge ports. Glass bell jars of various forms and sizes. Aluminium bell jars. Vacuum feedthrough collars. Blank-off vacuum flanges. Sight glass port. Liquid nitrogen feedthrough. Water connector feedthrough. Rotary feedthroughs. Push-pull rotary feedthrough. 60-degree cone movement feedthrough. Universal manipulator feedthrough. Adjustable leak assembly. Electrical feedthroughs, including multi-pin assemblies and an r.f. feedthrough. Connectors. Cenco vacuum couplings. High vacuum and ultra-high vacuum flanges. Cap screws. Gold O-ring seals. Mandrels for stretching gaskets. Cenco vacuum couplings in a variety of sizes and configurations in series 100, 200, 300 and 500. Cenco vacuum coupling caps for series 100 and series 200 couplings. Nipple adapters. Vacuum system baffles: air and water-cooled chevron types. Stainless steel 'Fluo-war' flasks for cryogenics. Liquid nitrogen traps in stainless steel and in glass. Glass Dewar flasks. Double Dewar flasks for liquid helium/nitrogen. Vacuum system traps. High frequency (Tesla) coil. Buna-N and Viton O-rings. Liquid nitrogen level controller. Vacuum ovens. Vacuum control apparatus. Rubber vacuum tubing, with clamps. Glass tubing. Metal stopcocks. Glass stopcocks. Tee and elbow fittings. Hand-operated vacuum and air pressure pumps. Seals, waxes and cements. Vacuum greases. Diffusion pump oils.

Ceramaseal

Vacuum flanges and fittings: ultra-high vacuum bakeable. Range of sizes for tubes of o.d. from $1\frac{1}{2}$ to 8 inch. Miniflanges for $\frac{3}{4}$ inch o.d. tubes of General Electric beaded seal and Varian conical seal designs. Vacuum fittings: Tee, cross, elbow, nipple (straight-through) and flexible nipple tubulations with flanges for uhv connections.
Connectors acceptable for uhv and bakeable to 450°C: grounded shield and floating shield electrical connectors for various voltages and currents. In-vacuum cable assemblies with six different kinds of termination. In-vacuum connectors: high current conductor clamps, lap type connector clamps, in-line connectors and push-on conductor sockets.
High-vacuum electrical feedthroughs: tube adapter mounted, connector fitted, connector fitted-octal MS, flange and adapter mounted with 1 conductor, flange and adapter mounted with 4 conductors, 8 conductors and 12 conductors, high voltage, for sputtering, special designs.

Cooke Vacuum Products

DAB cold cap baffles of 143, 183, and 289 mm o.d. in aluminium or stainless steel, 4300 series of sandwich type baffles with water-cooling of 143, 183 and 289 mm o.d. in stainless steel. Cold traps: 4000 series of thimble type, right-angle in stainless steel for liquid nitrogen; DMC series of chevron type in aluminium or stainless steel, multi-coolant; Model 4100 coaxial stainless steel anti-migration trap for liquid nitrogen. A wide range of couplings and other unions in stainless steel. Feedthroughs for current, voltage, cryogenic liquids, motion and instrumentation connections.

Crawford Fitting

'Swagelok' tube fittings in any machineable metal or plastic in a wide variety of sizes and configurations, with appropriate nuts and ferrules and also flexible metal hose connectors. 'Strip teeze' tape for joint wrapping 'Goop' sealing/lubricating materials: 'silver goop' for antisieze on threaded ports of stainless steel and super alloys; 'blue goop' for antisieze on titanium, stainless steel, steel, aluminium and high-temperature alloys; 'high purity goop' for antisieze and sealing on stainless steel, steel, aluminium and high-temperature alloys.

Denton

Bell jars: Pyrex (sizes 12 x 12, 12 x 18, 14 x 18, 18 x 18, 18 x 24, 18 x 30 inch D x H); metal (sizes 18 x 30, 22 x 30, 24 x 24, 24 x 30, 26 x 30, 36 x 30 inch).
Collars: Available for four bell jar sizes in stainless steel with a finished surface on top and an O-ring groove in the bottom. Supplied with Buna-N O-rings (Viton-A available).
Feedthroughs: Feedthroughs for vacuum collars (blank flange, 4 header individual, Octal, $\frac{3}{4}$ inch seal, lucite window, dual water, 10 kV hermetic, liquid nitrogen, ion tube).

Feedthrough for base plates (high voltage, rotary motion, low voltage, plug, safety switch).

Gauge mounted feedthroughs (low voltage O-ring, ball-bearing rotary motion, low voltage (metal), high voltage, 3 inch window).
Power supplies: DCG2; glow discharge power supply, LV 2 (2 kVA), LV 7·5 (7·5 kVA) low voltage power supplies.

Diversey Corporation

Brighteners and scale conditioners: DS-9-33 brighteners for Kovar and similar alloys, brightener and descaler for stainless steel. Diversey 299 scale conditioner for stainless steels and allied materials.

Edwards High Vacuum

Baffles: CB series: water cooled (refrigerated units available), cool to below −25°C.
DCB series: thermo-electrically cooled chevron baffles, water cooled to −25°C, air cooled to −5°C.
Slip-in oil baffles: available for 25 and 51 mm oil vapour diffusion pumps.
Couplings: Demountable unions with screwed body (hand tightened), sizes 1·6 to 50·8 mm in brass with screwed or plain ends; sizes 1·6 to 12·7 mm also available in stainless steel with plain ends.

Plain tailpieces, blind tailpieces and solid nuts (all brass) available in sizes covering majority of union and couplings.

Metal bellows, 12·7 to 50·8 mm. Also wide range of coupling nuts, clamping clips, inserts, fixed face of screwed tail pieces, reducers, elbows, tees, crosses, reducing crosses, etc.

Feedthroughs:
Rotary shaft vacuum seals (rotational or longitudinal movement)

Shaft size mm	3·2	6·4	9·5	12·7
Maximum speed of rotation rev/min	200	120	100	75
Maximum longitudinal movement mm	65	178	183	173

High voltage leadthroughs: Type 6D (150 A, 10 000 V d.c. continuous), 7D (15 A, 5000 V d.c. continuous).

High current leadthroughs: Type 8A (200 A, 60 V d.c. continuous), 9 A (400 A, 60 V d.c. continuous).

Multipin leadthrough: Type TL4 (2 A at 250 V r.m.s.).

UHV leadthrough: Type TL8 (8 way leadthrough), Type TL1/12 kV (high voltage leadthrough), Type TL1/200 (high current leadthrough), Type TL1/500 (high current leadthrough).

UHV motion feedthroughs: RD1 (rotary motion drive), LMD1 (linear motion drive).

Flanges: UHV pipeline couplings with 150 or copper compression gasket type flanges. Demountable flanged joints (used in pairs on a pipeline or singly on a flat machined surface). Sizes 38·1, 50·8, 76·2 mm.

Fluids, oils and greases
Rotary pump oils: 8A for non-gas ballast pumps; 15 medium viscosity, for pump with displacement of 2000 litre/minute and above and for EDM2; 16 high vacuum oil, alternative to 16; 29 for mechanical booster pumps; Fomblin rotary pump fluid type YR.

Silicone fluids: type 702 (10^{-6} Torr ultimate pressure); 704 (10^{-6}–10^{-8}); 705 (10^{-8}–10^{-10}).

Greases: Edwards greases: No. 3 soft (stopcock lubrication, flows freely at 40°C), No. 3 hard (sealing flat or conical joints, flows freely at 50°C).

Fomblin* greases: for anti-sieze in glass system connections, operating range 10°C to 23°C.

Silicone greases: normal grade stopcock grease (stable from 40°C to 200°C), HV grade for system pressures below 10^{-6} Torr.

Sealing compounds: MS Silicone leak sealant (aerosol) (MS Silicone is a product of Dow Corning Ltd).

Seals: VOR series of O-rings (nitrile rubber). VIT series of O-rings (Viton). Sizes from 4·75 to 380·37 mm i.d. IE metallic seal (aluminium gasket) bakeable: bore sizes from 10 to 250 mm.

Spacers: provide clearance for valve seal plate swing on quarter swing valves when baffles and other components are too close. Available for valve sizes 51–305 mm.

Traps: Cold traps: NTM series liquified gas cooled traps.

Vapour traps: VTG series (for use with EMG mercury diffusion pumps).

Foreline traps: 01–100 (activated alumina charge) UHV models 25 mm and 50 mm sizes.

Viewing ports: UHV: 38 mm diameter (glass 38 mm diameter (quartz), 51 mm diameter (glass) 102 mm diameter (glass), bakeable to 400°C (1000°C for quartz).

* Fomblin is a registered trade mark of Montecatini Edison SpA.

Ferrofluidics

'Ferrometic' rotary feedthroughs for vacuum use, high torque transmission and speeds of up to 10 000 rev/minute. Exposed surfaces of stainless steel; no rubbing parts; permanently lubricated. Utilize 'magnetic fluid' technology. Variety of sizes under model numbers SB253, SB500, SB750 and SB1000.

Fisons Scientific Apparatus

Joint seals: FIVAC p.t.f.e. sleeves, reduction adaptors and bellows adaptors, leakage rate of less than 1 x 10^{-4} Torr litre second^{-1}.

Flextube

Piping and ducting systems and components: Bellows expansion joints and assemblies, V flange duct joints, thin-wall tubing, flexible metallic hose.

General Engineering

Coupling systems: The VacLok system is designed to provide all the necessary combinations of joints and fittings; it is used with $\frac{1}{4}$, $\frac{1}{2}$, $\frac{3}{4}$, and 1 inch nominal bore copper tubing. Types J1 and J2 (slip and swaged joints), J5 and J6 (soldered joints), J8 and J9 (component to component joints). Flexible connections: VFC flexible rubber coupling, VFMC flexible metal coupling, stainless steel flange bellows.

Oils and greases: Rotary pump oil: GST 31, GFQ 37, GO 10, Kinney super X, Mobile DTE heavy medium.

Mechanical booster pump oil: Mobile DTE extra heavy, Mobile DTE BB.

Diffusion pump oil: Silicone 702 (10^{-1} Torr at 150°C: vapour pressure)
Silicone 704 (10^{-2} Torr at 150°C: vapour pressure)

O-rings: Nitrile and Viton: sizes from 0·114 inch internal diameter to 15·475 inch.

Granville Phillips

Stainless steel, bakeable, 'cryosorb' cold traps for liquid nitrogen: series 224, 2 inch; series 250, 4 inch and series 251, 6 inch of conductances in litre/second respectively, 30, 275, and 475. Variety of type 304 stainless steel connections for 'cryosorb' cold traps. 'Cuseal' flanges of type 304 stainless steel with OFHC copper sealing: range of sizes of rotatable and non-rotatable types. 'Cuseal' tapped flanges for permitting the close coupling of pumps and fittings. ASA style flanges of 304 stainless steel: two basic types, smooth face and O-ring grooved. Wide view windows: pyrex-Kovar and Kovar-stainless steel sealing for tubes of o.d. $1\frac{1}{2}$, $2\frac{1}{2}$ and 4 inch. Fittings: Tees, crosses, straight nipples and reducing nipples. Pyrex bellows for tube diameters of 8 and 14 mm. Vacuum stainless steel hose. Work chambers. Pyrex-Kovar seals. Rotary-linear motion feedthroughs. Bolt and stud sets. Gaskets and O-rings.

High Voltage Engineering

Type 304 stainless steel components: coupling, bolted; metal seals, indium coated, Inconel X-750; quick-opening coupling; insulating coupling; nipple; Tee; Tee reducing; cross; cross reducing; elbow, 90°C; weld adapter; reducer, bell; reducer, plate; short bellows; long bellows; bellows support assembly; blank-off. Window, regular glass; window, optical glass; feedthrough assembly; 7-pin connector; adapter, ASA to 'dependex', Conflat to 'dependex' and Pyrex to 'dependex'.

Couplings, standard 'dependex': A-XK-72-000 series, aluminium to Buna-N; A-XK-241-000 series, mild steel to Viton; A-XK-281-000 series, stainless steel to Viton. The standard coupling is used to join various vacuum components via the 'dependex' system. The equivalent pipe length (or insertion length) of a standard coupling is $\frac{3}{32}$ inch. The aluminium, mild steel and stainless steel flanges are available in 1, 3, 6 and 8 inch sizes.
6 inch stainless steel vacuum components: multi-ported vacuum chamber.

Hoke International

Union swivel adapters, hex nipples, hex couplings, hose connectcrs, reducing bushings, solder/weld tube fittings, pipe plugs.

Huntington

Flanges: Vac-U-Flat flanges (manufactured under licence from Varian) interchangeable with Conflat and compatible with other metal gasket sealed flanges. Stainless steel, rotatable or non-rotatable sizes from 1·33 to 13·25 inch outside diameter.
Traps: Coaxial foreline traps: stainless steel, sizes mini (0·75 inch inlet internal diameter) to 3 inch, hose or flange connections.

Ion Equipment

'Optic-dense' fore-line traps: trapping effect relies on a large area of metallic material with high adsorption characteristics: types 100, 150, 200, 250 and 300 of respective sizes, 1, $1\frac{1}{2}$, 2, $2\frac{1}{2}$ and 3 inch. Molecular sieve fore-line traps: 1, $1\frac{1}{2}$, 2 and 3 inch sizes.
IEC 'Lo-torr' mini high vacuum components: mini-flanges; fittings with compact mini-flanges; electrical and liquid feedthroughs; glass and copper adapters; sapphire and glass view ports; metal-to-glass seals.
Ultra-high vacuum flanges of sexless design, bakeable to 500°C. Rotatable and non-rotatable flanges are available for a range of tube sizes from o.d. 1 to 8 inch. Larger sizes to special order. All these IEC flanges are made from type 300 stainless steel and utilize 0·080 inch OFHC copper gaskets. Double faced flanges. OFHC copper gaskets.

Kinney Vacuum

Kinseal: a cement for metal-to-metal ground joints with a vapour pressure of $<10^{-1}$ Pa at 20°C, a softening temperature of 100°C after curing and a maximum working temperature of 150°C after curing.

Larson Electronic Glass

Glass to metal seals: Glass to copper, glass to stainless steel and glass to Kovar.
Viewing ports: Sealed to copper, stainless steel or Kovar, diameters from 1 to 6 inch.
Feedthroughs: Range of feedthroughs from single lead to 54 lead most of which are glassed into tubular metal. The leads can be of solid or tubular Kovar or of tungsten and can be mounted in tubular copper, stainless steel or Kovar.

Leybold-Heraeus

Baffles: Cold cap baffles: for oil diffusion pumps up to NW350 the cold cap baffles are designed to be mounted inside the pump body. The cap designed for a particular nominal width up to NW500 can be used for all water and air cooled diffusion pumps of the same nominal width.
Shell and astrotorus baffles: chevron, stainless steel; shell baffles available in sizes NW32 to NW350, astrotorus baffles in sizes NW100 to NW1000.
Air cooled baffles: available for air cooled oil diffusion pumps.
Chevron baffles: for large oil diffusion pumps from NW500.
Refrigeration units are available for baffles from NW65 to NW1250. Separate anti-creep barriers are also available.
Condensers: Models K05 and K1 with a simple drain and shut off valve fitted to the receiver.
Models K05S, K2S, KGS with quick drainage manual or solenoid valves. Area of condensing surface m^2: K05, 0·2; K1, 1; KO5S, 0·5, K2.5, 2·0; and K6S; 6·0.
Filters: Dust filters: Models FS 2, 3, 5, 6, and 10 (dry filters for intake line of a mechanical vacuum pump).
Dust separators: Models AS 3, 5, 6, and 10 (two stage: first stage is a cyclone, second stage is a wet filter).
Exhaust filters: Models AF 201, 2-2, 3-1, 3-2 (ceramic porous filters), AF 5-0, 5-1, 5-3, 5-2, 6 (impregnated glass wool filters).
Oil filters: Oil filter attachments (driven by pump drive shaft) and oil filter units (driven by own motor) are available for Simplex and Duplex pumps. Mechanical and chemical oil filter units and an automatic oil level regulator for Monoblock pumps E and DK models.
Flanges: Small flange fittings KF: down to pressures of 10^{-8} Torr, bakeable in 200°C, self centering. Sizes NW 10, 20, 32, 50 (with long or short tubulation). Flanges of stainless steel or ordinary steel, centering rings of nickel plated steel, clamping rings of light alloy.
Complete range of components: bellows, crosses, tees, elbows etc.
Clamp flange fittings LF: down to pressures of 10^{-8} Torr, bakeable to 200°C, self centering. Sizes NW 65, 100, 150, 250, 350, 500. Clamp flanges of standard or stainless steel, collar ring flanges of standard steel, clamps of cadmium plated steel. Centering rings of nickel plated steel (sizes 350 and 500 of aluminium). Range of components.
Special seals: elastomer seals and vacuum sealing discs for systems in which gaskets or components are often changed or where excessive pressure as well as vacuum may occur. Sizes NW 10, 20, 32, 50, 25/20, 25/32. Metal gaskets: ultra-high vacuum discs (sizes NW

65, 100, 150, 250); aluminium foil seals (sizes NW 10, 20, 32, 50 in KF design, sizes NW 65, 100, 150, 250 in LF design). Note: flange sizes 10, 20, 32, 50 are now replaced by Pneurop standards 16, 25 and 40.

UHV flange connections, CF flanges: bakeable to 450°C. Sizes NW 35, 63, 100, 150, 200. Flange of specially hardened stainless steel, flat OHFC copper gasket (silver coated on request) can also be sealed with a Vitilan O-ring. Fixed and rotary flanges in all sizes. Components: elbows, tees, crosses, diaphragm bellows, flexible metal tubing, intermediate piece.

UHV flange connections, mini CF flanges: construction as CF flanges. Size NW 16 (33·8 mm o.d.). Components: elbows, tees, crosses, bellows, flexible metal tubing, current lead-throughs, flange with glass tubing. Connection of different-type flanges: Adaptors available for: KF small flanges to LF clamp flanges, small flanges to EF unit flanges and Herseus flanges, LF clamp flanges to EF unit flanges and other bolted flanges.

Glass connections: Ground joint connections, 90°: NS 19/38 ground steel cone, NS 19/38 ground steel socket, NS 19/38 ground soft glass socket. Standard ground glass joints: hard or soft glass NS 14·5/35, 19/38, 29/42, 49/50, 75/52.

NS figures = ratio of largest diameter to the length of the joint.

Glass fittings and systems for UHV: Mercury diffusion pumps made of hard glass are available along with graded glass seals (join glass gauge tubes to glass systems), glass bellows and flanges with glass tubulation.

Oils, fluids and greases:

Oils for mechanical vacuum pumps: Special oil N62 (viscosity class SAE 30) has a vapour pressure of 1×10^{-3} Torr at 20°C. Backing pump oil TO20 (viscosity class SAE 20) vapour pressure as for N62. Corrosion protective oil, Protelen, suitable for pumping acidic vapours, very hygroscopic and not for use where water vapour is evolved, vapour pressure slightly higher than for N62. Anti corrosion oil KS20, mineral oil with aldaline corrosion inhibitor additives.

Pump fluids for vapour pumps: Diffelen mineral oil fluids (light for pressures to 10^{-5} Torr, normal to 10^{-7}, ultra to 10^{-8} with water cooled baffle or 10^{-10} with cold tap). Silicone oils: SO66 (similar to Diffelen normal), DC705 (for UHV range with extremely low vapour pressure). Also Convalex 10, a polyphenyl ether used where contamination by silicone oils has a disturbing effect.

Greases: Ramsay grease (soft for taps, viscous for ground joints) for applications down to 10^{-2} Torr, vapour pressure at 20°C, 10^{-4} Torr. LH high vacuum grease P and vacuum grease R, P has slightly lower vapour pressure, R is more viscous and hydrocarbon free, both are for ground joints and taps. Lithelen (for ground joints and taps) has a viscosity which varies very little and so can be used in a wide temperature range (below 0°C to above 150°C). Silicone high vacuum grease (polymerization over 200°C). DD joint grease, used to lubricate and seal rotary transmissions and certain valves.

Sealing compounds: Pizein, made of bituminous substances, in thermoplastic and adheres to metal and glass. Sealing lacquer.

Leadthroughs:

Current leadthroughs: With plastic insulation or glass insulation, multi-way leadthroughs with glass insulation (up to 10 conductors). Leadthroughs with ceramic insulation, water-

cooled leadthroughs with plastic insulation. Current leadthroughs for UHV (bakeable to 450°C).
Rotary transmissions: Maximum axial movement is 150 mm, sealed by oil-sealed or grease packed lip washer seals, 3 models with threaded connection, 1 model with flange connection.
Mechanical motion leadthroughs for UHV: Manipulator (rotary and linear motion leadthrough) maximum rpm 150, 10 mm readable on linear scale. Magnetic rotary motion leadthrough, maximum rpm 1000.
Tubular feedthroughs: Two-way feedthrough (8 mm bore copper tubes, 1 mm wall thickness). UHV Simplex feedthrough (one 4 mm i.d. stainless steel tube, drillable up to 10 mm bore). UHV Duplex feedthroughs (two 8 mm o.d. copper feed tubes, wall thickness 1 mm).

Observation windows:
Window flanges (with synthetic acrylglas window): Sizes NW 10, 20, 32, 50, 65, 100, 150.
Sight glasses (with hard glass window): Sizes NW 50, 65.
UHV observation windows (tube shaped or wide-view): Sizes NW 35, 63 (tube), NW 100 (wide-view).

Traps:
Condensate traps: Sizes NW 19, 20, 32, 50, 65, 100. Fitted with inspection window, enamelled exterior and interior.
Foreline sorption traps: Zeolite filled, pressures of 1×10^{-5} Torr attainable with two-stage mechanical vacuum pumps. Sizes NW 10, 20, 32 (zeolite filling 300, 700, 1250 g).
Cold traps: Screen cold traps (originally designed for Quick mercury diffusion pumps, also used with oil diffusion pumps). Stainless steel coolant chamber, LN_2 filled, copper screen-like deflectors. Spherical cold traps.

Materials Research

V-4 series: vacuum collars (stackable), flanges, adaptors, feedthroughs. Pyrex bell jars. Bell jar guards. Spacer rings. Stainless steel chambers. Neoprene and Viton O-rings. Boot type gaskets in neoprene and in Viton. Rotary seals. Miniature rotaries. Magnetic and bellows-sealed rotaries. Push-pull rotaries. 3-mode feedthrough. Manipulator – a 'hand in vacuum'. Thermocouple lead-ins. Octal feedthroughs. Instrument feedthroughs. Electrical feedthroughs: single, two-stud, high current, heavy duty, high voltage, double triple and quadruple, power, high power. Pyrex windows. Light assembly for providing a light source for a stainless steel or other metal chamber. Fibre optic window. LN_2 feedthrough. Liquid N_2 traps. Universal vac seals. Water feedthrough. Low volume water feedthrough. R.F. feedthrough. Inert gas feedthrough and purifier. Adjustable leak assembly.

Monsanto

Pump fluid: Santovac 5 polyphenyl ether vacuum pump fluid. Vapour pressure at 25°C is 4×10^{-10} Torr.

N.G.N.

Silicone high vacuum grease.
Small vacuum systems for educational purposes.

Nanotech

Leadthroughs: A range of rotary, water, and liquid nitrogen leadthroughs is available. Current leadthroughs: Models 50E, 50ME and 30E (bakeable to 130°C).

Model	50E	50ME	30E
Leadthrough metal	Aluminium alloy	Gold plated copper	Aluminium alloy
Insulation	Glass	Glass	Glass
Number of conductors	1	9	1
Load capacity	10 000 V, 100 A.	1000 V, 5 A	5000 V, 30 A
Conductor diameter	9·5 mm	1 mm	4·5 mm
Bore diameter	26·98 mm	26·98 mm	–

Neslab Instruments

Coolers for vapour/freezing traps: Models CC60, CC60f achieve temperatures of down to –60°C. Model CC100f: temperatures down to –100°C.

Nordiko

High vacuum feedthroughs for liquid nitrogen, twin-tube liquid transfer, quick coupling, rotary motion, right angle rotary motion, stainless steel bellows-sealed rotary motion, mechanical wire feed.

Pope Scientific

Glassware: Vacuum manifolds, gas bulbs, gas collecting bulbs and tubes, mercury float valves, check valves, break seals, separable cold traps, plain cold traps, spherical cold traps (for horizontal and for vertical flow) cylindrical cold traps, U-tube traps, sorption traps, high conductivity traps, copper foil traps, dewar flasks, glass to metal seals and viewing ports, stop cocks.

Heat guns: Flameless guns delivering up to 1000°F for outgassing vacuum systems.

RIBER

Electrical leadthroughs

Model no.	TH1	T9C	T9P	TM1	TM1D
No. of conductors	1	9	9	1	2
Load capacity per conductor					
Volt		1000	1000	12 000	12 000
Amp	500	5	5	150	150
Diameter of single conductor (mm)	18	2	2	9	9
Bore diameter for fitting in container wall (mm)	38	38	38	38	38

Sadlavac

Coupling systems: 'Quik-vac' coupling system (elbows, tees, crosses, couplers and reducer sets). Conversion couplers for Edwards, GEC-AEI, Genevac, Leybold, NGN and Sogev units, Glass cone adaptors.

Sigma Laboratories

Viewing ports

Model	Type	Sighting opening	Visual angle
8030-372	Long tube and intermediate flange	20 mm	30°
8030-375	Mirror head	80 mm	30°
8040-374	Conical tube	80 mm	40°
4055-369	Tube	40 mm	55°
2030-368	Tube	20 mm	30°

Sloan

Feedthroughs:

Feedthrough collar no 113-700: Stainless steel, Viton O-ring, eight 1 inch diameter ports, 5 inch high, 8 inch diameter.

Electron beam service feedthrough nos. 113-400, 113-551: Through-hole mounted, fits standard 1 inch port, two $\frac{3}{8}$ o.d. stainless steel tubes.

High voltage feedthrough nos. 113-425, 113-451: Will accommodate maximum of 15 000 V d.c. at 50 amp.

Dual HV feedthrough no. 113-475: Maximum rating of 15 000 V d.c. at 50 amp with a maximum rating of 500 V d.c. between the two terminals.

Octal feedthrough no. 113-050: Maximum rating of 300 V at 5 amp per pin.

Instrumentation feedthroughs nos. 113-100, 113-200, 113-300: 100 has two water cooling lines and one coaxial feedthrough, 200 and 300 have two water cooling lines and two coaxial feedthroughs.

Spirax-Sarco

Spirax 'Ogden' automatic pump: Lifts condensate and, when suction is not required, can handle many other hot or cold liquids. Operated by steam, compressed air or gas. Four sizes, from 1 to 3 inch, with capacities up to 1260 gallons/hour. Also in packed pump units comprising the pump, vented receiver and all necessary ancillary tubulation, valves, etc.

Springham

High vacuum stop cocks (figures in model numbers indicate bore in mm):

Models STV/1/B to STV/25/B: Boro-silicate, plain arms, hollow key, straight-through.

Models STS/2/B to STS/6/B: Boro-silicate, solid key, plain arms, straight-through.

Models STVC/1/B to STVC/3/B: Boro-silicate, hollow key, capillary arms, straight-through.

Models STSC/2/B and STSC/3/B: Boro-silicate, solid key, capillary arms, straight-through.

Model MSTV: micro pattern, Boro-silicate, straight-through, key diameter 10 mm, bore diameter 1–2 mm.
Models SOV/1/B to SOV/10/B: Boro-silicate, plain arms, hollow key, single oblique pattern.
Models SOVC/1/B to SOVC/3/B: Boro-silicate, capillary arms hollow key, single oblique pattern.
Models DOV/1/B to DOV/10/B: Boro-silicate, plain arms, hollow key, double oblique pattern.
Models DOVC/1/B to DOVC/3/B: Boro-silicate, capillary arms, hollow key, double oblique pattern.
Models TBV/1/B to TBV/10/B: Boro-silicate, plain arms, hollow key, T-bore three-way pattern.
Models TBV/1/B to TBV/3/B: Boro-silicate, capillary arms, hollow key, T-bore three-way pattern.
Models YV/1/B to YV/10/B: Boro-silicate, plain arms, hollow key, Y shape with V bore pattern.
Models YVC/2/B and YVC/3/B: Boro-silicate, capillary arms, hollow key, Y shape with V bore pattern.
Models SP/4/B to SP/15/B: Boro-silicate, plain arms, hollow key, right angle single pump pattern.
Models SPH/4/B to SPH/15/B: As SP/4/B but with bottom tube horizontal.
Models DP/4/B to DP/15/B: Boro-silicate, plain arms, hollow key, tee shape double pump pattern.
Models SSP/2/B to SSP/15/B: Boro-silicate, plain arms, hollow key, Springham safety pattern.
Spring loading keys are available for all the above stop cocks.
High vacuum greaseless stopcocks: These stopcocks can be used at pressures down to 10^{-6} Torr and within the temperature range $-30°C$ to $+300°C$. It employs either a Neoprene or Fluorocarbon (Viton A) diaphragm. Bores from 1 to 10 mm.

Torr Vacuum Products

Baffles: Series MCB1 liquid nitrogen cooled chevron baffles, stainless steel, bakeable to 200°C, nominal sizes 4, 6, 10 inch.
Bell jar hoists: Model LPH 150 (150 lb on 12 inch arm); model LPH 500 (500 lb on 18 inch arm). Adjustable for 24 or 30 inch travel.
Cryotraps: Series MCB3 liquid nitrogen traps with ‘two bounce blocking-plate design’, stainless steel, available for 4, 6 and 10 inch size diffusion pumps.
Feedthroughs: Stainless steel, mounted through a 1 inch diameter hole, bakeable to 150°C, capable of operation in the 10^{-9} Torr range. Models: HCF 50 (50 amp capacity electrode), HCF 400 (400 amp capacity electrode), HVF 10 (10 kV electrode), 1FT8 (8 pin instrumentation feedthrough, 5 amp capacity), TFT8 (8 pin thermocouple feedthrough), RMF1R (rotary, push-pull, mechanical motion feedthrough giving 360° rotation or 4 inch linear travel), RFF1 (RF feedthrough) LNF1 (cryogenic re-entrant feedthrough, $\frac{1}{2}$ inch diameter tube), WGF1 (water or gas feedthrough, $\frac{1}{2}$ inch diameter tube).

Flanges: Foil-seal: bakeable (450°C) metal gasket seal wing aluminium foil. V-seal bakeable (450°C) copper gasket flanges.

Torvac

Feedthroughs: Wobble drive shaft Mark II: bakeable, all stainless, transmits up to 5 kg cm, maximum speed 50 rev/min.

UHV

Couplings: A range of flanges, observation windows, Pyrex adaptors, crosses, tees, elbows and flexible couplings is available.
Current leadthroughs: A range of various types is available.

Ultek

Vac-Seal epoxy resin, model 288-6000: Useable to 10^{-7} Pa, low outgassing rate, curing at 100°C for 2 hours, withstands bakeout at 150°C. Valuable for temporary leak repair.
'Curvac' high vacuum flanges: Sexless 1 to 8 inch in nine different sizes, made from 300 series stainless steel, utilize OFHC copper gaskets, 0·080 inch thick and are bakeable to 500°C. Rotatable and integral flanges available.
'Curvac' flanges with valves and thermocouple gauge tubes: Eight different models.
'Curvac' flange accessories and supplies: Thread lubricant; Viton gaskets L-shaped for $1\frac{1}{2}$, 2 and 4 inch flanges.
Wire seal flanges 10 to 48 inch diameter: Bakeable to 450°C; made from OFHC copper wire of 0·080 inch diameter. Flanges for tubes of o.d. in inch: 10, 12, 14, 16, 18, 24, 36, 48. Assembly bolts needed supplied in boxes of 25 sets, each set consisting of one hex head bolt, hex nut and washer.
'Curvac' convenience kit: Plate nuts with tab gaskets.
Fore-line traps of sizes 1, $1\frac{1}{2}$ and 2 inch: Utilize molecular sieve; large sieve capacity, replaceable sieve and bakeout heaters; 9 different models, 3 of each size.
Versa-trap for mechanical pumps: Interchangeable molecular sieve, cryotrap or surface area modules. Fabricated from 304 stainless steel. Available in 6 models and 18 combinations.
26 pin instrumentation feedthrough: Bakeable to 400°C, mounted on $1\frac{1}{2}$ inch ($2\frac{3}{4}$ inch o.d.), 'Curvac' flange, ceramic/metal construction.
Linear motion feedthroughs: Model 282-6200, micrometer spindle; model 282-6100, push-pull. Bakeable to 450°C.
Low profile viewports: 7 sizes (1, $1\frac{1}{2}$, 2, 3, 4, 6 and 8 inch), glass protected by high vacuum flange, mounted on 'Curvac' non-rotatable flanges, bakeable to 400°C, 7056 glass.
Custom electrical feedthroughs: Wide variety of custom power and medium current types for vacuum applications. Mounted on 'Curvac' high vacuum flanges, bakeable to 450°C, ceramic-sealed, multi-conductor types to suit many needs.
High vacuum electrical feedthroughs: Bakeable to 450°C, designed for pressures from $10^{-2} \rightarrow 10^{-9}$ Pa; utilize 'Curvac' all-metal rotatable flange connectors with copper gaskets and inserts of 304 stainless steel. Both ceramic-to-metal and glass-to-metal seals are utilized.

Models are: high current, r.f., high voltage, instrumentation (multi-conductor), high current water-cooled, r.f. instrumentation, dual medium current, single medium current.
Motion feedthroughs: Direct drive, 300 rev/min, welded stainless steel bellows, bakeable to 300°C; micrometer spindle, bakeable to 450°C; magnetic coupled, 50 rev/min, bakeable to 400°C; push-pull, bakeable to 450°C.
High vacuum rotary motion feedthroughs: 300 rev/min; bakeable to 300°C; bellows-sealed direct drive mechanism; 1½ inch i.d., 2¾ inch o.d. flange mounted.
Magnetically-coupled rotary motion feedthrough: Similar except bakeable to 400°C.
Glass adapters: Metal-to-glass seals consisting of 7052 glass to Kovar or 7740 Pyrex to 7052 glass to Kovar; Kovar then joined to 'Curvac' stainless steel flange; five sizes, nominally 1, 1½, 2, 2½ and 4 inch; bakeable to 400°C.
Pyrex bell jars of i.d. 11½, 13½ or 16⅞ inch; bell jar guards; L-shaped Viton gaskets.

Vactronic Lab. Equipment
Quick-connectors, Vac-Caps and quick-connect fittings of in-line, elbow and Tee forms; brass but available also in 303 or 304 stainless steel or special order; Buna-N O-rings are standard but Viton or silicone rubber O-rings are available.
Circular base-plates of aluminium or 304 stainless steel of o.d. in inch 9, 11, 13, 15, 16, 20 or 25 with 8 to 24 holes. Feedthrough collars of 304 stainless steel of o.d. in inch 9⅜ (8 holes), 10¾ (10 holes), 13⅛ (12 holes), 15⅛ (12 holes), 18¾ (18 holes) and 24¾ (24 holes).
Feedthroughs of following types: cryogenic, high current, low current, high voltage, liquid, rotary push-pull, thermocouple, thin film shutter; variable leak; blank-off plugs.
Bell jars in Pyrex-type BJ: four sizes with diameters of 8 to 10 inch. Bell jar guards.
Cylinders in Pyrex-type CY: five sizes with diameters of 8, 10 or 12 inch. Boot gaskets of Buna-N and of Viton.

Vacuum Accessories
'Vacoa' high vacuum 'quick-seal' connectors made of stainless steel or brass, for rapid and positive sealing of leak-tight joints in metal-to-metal or metal-to-glass; utilize Viton O-rings.
13 models of each of 'quick-seal' connectors, 'quick-seal' cap connectors and 'quick-seal' straight-through connectors.
Vacoa brass and stainless steel flange sets: SFB series for tubing of 18 different sizes ranging from an o.d. of ¼ inch to an o.d. of 4⅛ inch.

Vacuum Barrier
'Semiflex' liquid nitrogen transfer systems: semiflex line of corrugated copper tubing with Teflon spacer, coated with PVC: four sizes. 'Semiflex' bayonet for quick connection of sub-assemblies. 'Semiflex' Tees, crosses and elbows. Gas elimination system: assures a constant and instant supply of gas-free and sub-cooled low pressure liquid at the cold trap valve in cryogenic engineering. End terminations.

Vacuum Generators

Cold traps: Models CCT 50, CCT 100, CCT 150: right angle configuration, top filling, nickel chevron with integral water cooled baffle plate, stainless steel body.

Model no.	CCT 50	CCT 100	CCT 150
Pump size	2 inch	4 inch	6 inch
LN_2 capacity cm^3	4100	5000	11 000
LN_2 consumption $cm^3\ hr^{-1}$	170	275	500
Conductance	$50\ l\ s^{-1}$	$375\ l\ s^{-1}$	$750\ l\ s^{-1}$

Couplings etc.:

Flanges: All fittings and feedthroughs are fitted with FC series copper gasket flanges as standard. The standard fixed and rotatable flanges are

1·33 inch o.d. x ¾ inch maximum i.d.
2¾ inch o.d. x 1½ inch maximum i.d.
4½ inch o.d. x 2½ inch maximum i.d.
6 inch o.d. x 4 inch maximum i.d.
8 inch o.d. x 6 inch maximum i.d.
10 inch o.d. x 8 inch maximum i.d.

Fittings: Stainless steel fittings (tees, crosses, elbows, bellows couplings, flanged glass-to-metal seals, etc.) provided with FC series flanges standard.
Special tubulation: Stainless steel flexible tubing 18 inch long, 1 inch diameter (FLX25) and 1½ inch diameter (FLX38).
Viewports: Viewports from 1 inch to 6 inch diameter. The glass is sealed to Kovar tubing and the Kovar tubing welded to the flange.
Some sizes of sapphire and quartz viewports are available.
Feedthroughs: Liquid feedthroughs for water and liquid nitrogen; Mechanical feedthroughs:

RD1 mechanically coupled bellows sealed rotary motion drive
MRD1 magnetically coupled rotary motion drive
RD2 mechanically coupled bellows sealed rotary motion drive providing two rotations of a specimen (consider a specimen mounted on the face of a pocket watch, the two rotations are the twisting of the watch on its suspension chain and the rotation of the hands).
UMD1 Universal motion drives for positioning specimens
UMD2 with respect to an incident (typically electron) beam.

Electrical feedthroughs:

Model	EFT930	EFT91	EFT1	EFT9A	EFT1A	EFT3	EFT3A
Conductors	3	1	1	1	1	3	3
A	5	40	40	40	40	40	40
kV	1	6·5	6·5	15	15	6·5	15

Model	EFT911	EFT11	EFT13	EFT94	EFT4	EFT9	EFT6
Conductors	1	1	3	1	1	1	1
A	100	100	100	200	200	600	1000
kV	3	3	3	15	15	15	1

Model	EFT10	EFT15	EFT920	EFT20	EFT5	EFT19	EFT55
Conductors	1	1	1	1	8	19	55
A	30	30	3	3	5	7	7
kV	60	150	0·5	0·5	2	0·7	0·7

Conductor material: stainless steel for EFT930 to EFT3A; OFHC copper for EFT911 to EFT15; bakeable BNC coaxial plug and socket for EFT920 and EFT20; copper-nickel for EFT5, internal stainless steel barrel connectors for EFT5; gold-plated nickel iron for EFT19 and EFT55. Internal crimp connectors for EFT19/55.

Vacuum Research Manufacturing

High vacuum gaskets: mates 2 to 10 inch ASA flanges, in Buna-N or Viton. Movable shaft seals: Wilson type double seal shaft seals providing a feedthrough for rotational or push-pull motion; 3 sizes.

Vac Torr

Smoke eliminators for mechanical vacuum pumps; 5 models for the range of Vac Torr pumps. Vacuum tubing of red gum rubber: 7 sizes of i.d. from $\frac{1}{4}$ to $1\frac{1}{2}$ inch. Adjustable tubing clamps. Vac Torr in-line molecular sieve (sorption) traps for mechanical rotary pumps: 6 sizes for tubulation of i.d. in inch: $\frac{1}{2}$, $\frac{3}{4}$, 1, $1\frac{1}{4}$, $1\frac{1}{2}$ and 2. Molecular sieve within stainless steel cartridge with cast-iron trap. If the principal function of the trap is to prevent back-streaming of pump oil, an optical glass bead cartridge is available. The 4 mm diameter glass beads within the cartridge are easily cleaned by washing.
Vac Torr vacuum fittings: couplings, rubber tubing to rubber tubing, five sizes for tubing i.d. in inch of $\frac{1}{4}$, $\frac{3}{8}$, $\frac{1}{2}$, $\frac{3}{4}$ and 1; rubber tubing to rubber tubing reducing, 10 sizes from $\frac{3}{8}$ to $\frac{1}{4}$ inch up to 1 to $\frac{3}{4}$ inch; male NPT to rubber tubing, five sizes from i.d. in inch of $\frac{1}{4}$ to 1; male NPT to tubing specified by i.d., five sizes from $\frac{1}{4}$ to 1 inch. Connectors: male NPT to tubing specified by o.d. (five sizes, $\frac{1}{4}$ to 1 inch); female NPT to female NPT (five sizes, nominally $\frac{1}{4}$ to 1 inch); tubing to tubing specified by i.d. (five sizes, $\frac{1}{4}$ to 1 inch); tubing to tubing specified by o.d. (five sizes, $\frac{1}{4}$ to 1 inch); male NPT to female NPT, reducing (five sizes); NPT to tubing specified by o.d. (five sizes, $\frac{1}{4}$ to 1 inch); tubing specified by i.d. to tubing specified by o.d. (five sizes, $\frac{1}{4}$ to 1 inch); tubing to tubing specified by o.d. (one half is sealed by compression, five sizes, $\frac{1}{4}$ to 1 inch); tubing to tubing specified by o.d. (both halves sealed by compression, five sizes, $\frac{1}{4}$ to 1 inch); tubing to tubing specified by

o.d., reducing (ten sizes). Nipples: male NPT to rubber tubing (five sizes); male NPT to rubber tubing, reducing (ten sizes). O-rings.
Vacuum coupling adapters: caps; nipple, rubber tubing; nipple, reducing rubber tubing; female NPT; for i.d. tubing; for o.d. tubing; of nominal sizes in inch of $\frac{1}{4}$, $\frac{3}{8}$, $\frac{1}{2}$, $\frac{3}{4}$ and 1 to fit any of these.
Vacuum coupling bodies: 90° elbow; 45° elbow; Tee, Y, cross, bellows, 1-inch ASA flange.

Varian
Couplings:
Stainless steel fittings: Elbow, cross, flexible, nipple, tee; vacuum range to below 10^{-11} Torr, temperature range from –196°C to +500°C. Sizes $1\frac{1}{2}$ to 6 inch and 'mini' series.
Glass adaptors: Corning 7052 glass or Pyrex sealed to Kovar tubing and mounted on a non-rotatable Conflat flange. Sizes $\frac{3}{4}$ to 4 inch o.d. and 'mini' series.
UHV compression ports: Demountable, leak free, bakeable connections between copper tubing and stainless steel vacuum systems.

Baffles:
Cold cap: 'Mexican Hat' optional fitting to VHS-6 diffusion pump, stops backstreaming as effectively as an optically dense baffle but retains 66% of the pump's intrinsic speed.
Water cooled baffles: Models NRC 334, 336 (low profile).

Feedthroughs:
Ceramic/metal header assemblies: Under continuous operating temperature of 400°C these assemblies maintain at least 10^{-9} atmosphere/cc/second.
Feedthrough spools: 8 ports, 12 or 18 inch diameter, stainless steel.
Electrical feedthroughs: Model 954-5012 (8 pin instrumentation), 954-5013 (20 pin instrumentation), 954-5030 (4 pin power), 954-5007 (high voltage), 954-5019 (medium current), 954-5008 (dual medium current), 954-5009 (high current), 954-5014 (8 wire instrumentation), 954-5015 (8 tube thermocouple), NRC 1332 (400 amp), NRC 1373 (1000 amp, water cooled), NRC 1349 (100 amp), NRC 1554 (10 000 volt), NRC 1348 (8 tube instrumentation), 'mini' series 954-5142 (high voltage/medium current), 954-5143 (high voltage), 954-5144 (coaxial instrumentation).
Liquid feedthroughs: Models 954-5022 (water, two tubes), 954-5023 (LN_2 one tube).
Mechanical motion feedthroughs: Models 954-5120 (positive drive rotary), 954-5026 (standard magnetic rotary), 954-5039 (large magnetic rotary), 954-5049 (linear motion), NRC 1301 (rotary bellows), NRC 1371 (rotary push-pull), NRC 1372 (rotary ball-bearing $\frac{1}{4}$ inch), NRC 1316 (rotary ball-bearing $\frac{1}{2}$, $\frac{3}{4}$, 1 inch).

Flanges:
ConFlat flanges: Mating flanges are identical, temperature range –196°C to +500°C, pressure range down to below 10^{-13} Torr, rotatable and non-rotatable, nominal o.d. sizes $2\frac{3}{4}$ to 10 inch. Also 'Mini-ConFlat' 1·33 inch o.d.

Wheeler flanges: Temperature range −196°C to + 450°C, pressure range to below 10^{-11} Torr. Accept tubing from 10¾ to 24 inch o.d.

Fluids, sealants:
Torr seal: Epoxy resin, bakeable to 120°C, for use in systems at 10^{-9} Torr and below.

Traps:

Type	Model	Conductance (below 10^{-4} Torr) in litre/second	Volume of LN_2 required in litre	Size
Circular chevroncryobaffle	0315-1-4	550 litre/second	500 cc	4 inch
Circular chevroncryobaffle	0315-1-6	1200 litre/second	1000 cc	6 inch
Circular chevroncryobaffle	0315-1-10	3000 litre/second	1750 cc	10 inch
Circular chevroncryobaffle	0316-0-6	1800 litre/second	1750 cc	6 inch
Cryobaffle	0325	460 litre/second	3500 cc	2 inch
Circular chevron cryotrap	315-20	12 000 litre/second	17 litre	20 inch
Cryotrap	316-4	950 litre/second	3300 cc	
Cryotrap	316-6	1800 litre/second	1750 cc	
Long life cryotrap	326-4	1000 litre/second	14 litre	4 inch
	326-6	1700 litre/second	15 litre	6 inch
	326-10	2700 litre/second	16 litre	10 inch

Also molecular sieve trap-valve and cryotrap-valve (traps with built-in valves).

Viewing ports:
Sapphire optical windows: Series CS 5000 (sizes 0·25–1·5 inch diameter) CS 6000 (0·25–2·0 inch) CS 7000 (2·0–4·0 inch) CS 8000 (0·25–2·0 inch). Also model 954-5140 'mini' series (⅝ inch).
Viewing ports: Zero-length glass/Kovar (1½, 4, 6 inch) non-magnetic glass/stainless steel (1½ inch) sapphire/Kovar (1 inch).

Veeco

Baffles and cold traps:
Water cooled baffles: BAF 200 (2 inch), BAF 400 (4 inch). Brass body, copper disc and cooling coils (BAF 400 nickel plated body).
Cold traps: CT200 (2 inch) CT 400 (4 inch) 8 to 10 hours operation per filling of 0·8 litre LN_2 (ADP 200 adapter joins CT 200 to BAF 200).
Integrally housed water baffle and cold traps: LL 776 (7 inch), LL 1000 (10 inch) LL 1200 (12 inch).
Foreline traps: Molecular sieve (zeolite) foreline traps: TR 101, 102, 103 (tubing o.d. 1 inch, flange o.d. 2⅛ inch), TR 151, 152, 153 (tubing o.d. 1½ inch, flange o.d. 2¾ inch), TR 201, 202, 203 (tubing o.d. 2 inch, flange o.d. 3⅜ inch), TR 301, 302, 303 (tubing o.d. 3 inch, flange o.d. 4⅝ inch). Last figure refers to configuration, 1: hose/hose, 2: flange/hose, 3: flange/flange.

Coaxial foreline traps: Use a high surface area metallic adsorbent, removing need for bake-out etc. VS 120, 121, 122 (1, $2\frac{1}{8}$ inch), VS 160, 161, 162 ($1\frac{1}{2}$, $2\frac{3}{4}$ inch), VS 220, 221, 222 (2, $3\frac{3}{8}$ inch), VS 260, 261, 262 ($2\frac{1}{2}$, $4\frac{1}{2}$ inch).

Couplings:
Quick vacuum couplings (brass): C 06 ($\frac{1}{16}$ inch), C 09 ($\frac{3}{32}$ inch), C 12 ($\frac{1}{8}$ inch), C 18 ($\frac{3}{16}$ inch), C 25 ($\frac{1}{4}$ inch), C 31 ($\frac{5}{16}$ inch), C 38 ($\frac{3}{8}$ inch), C 50 ($\frac{1}{2}$ inch), C 62 ($\frac{5}{8}$ inch), C 75 ($\frac{3}{4}$ inch), C 87 ($\frac{7}{8}$ inch), C 100 (1 inch), C 112 ($1\frac{1}{8}$ inch).
Quick coupling reducers: (models CR 03 to CR 100).
Flanged quick couplings: Sizes from $\frac{1}{32}$ inch to $1\frac{1}{8}$ inch (models FC 03 to FA 112).
Stainless steel uhv fittings (sizes $\frac{3}{4}$ to 6 inch): Elbows (models FI 072 to FI 062), crosses (models FI 071 to FI 601), tees (FI 074 to FI 604), nipples (FI 073 to FI 603). Flexible couplings models FI 070 to FI 600 (bellows i.d. $\frac{3}{4}$ to $5\frac{1}{2}$ inch).
Glass adaptors: Sizes $\frac{3}{4}$ to 4 inch (glass o.d.) Pyrex model nos. GL 070 to GL 400; glass model nos. GL 071 to GL 401.

Hose and hose clamps:
Hose: Gum-rubber, sizes: $\frac{7}{16}$ inch i.d. x $\frac{5}{16}$ inch wall, $\frac{1}{2}$ i.d. x $\frac{3}{8}$ inch wall, $\frac{5}{8}$ inch i.d. x $\frac{3}{8}$ inch wall, $\frac{5}{8}$ i.d. x $\frac{1}{2}$ inch wall.
Clamps: Type S1; available for hose sizes $\frac{1}{2}$ inch o.d. to $2\frac{1}{4}$ o.d.

Feedthroughs:
Current feedthroughs: Ambient to 200°C: Type CFT (water cooled high current), AFT (ambiently cooled, high current), HVF (high voltage), GDP (glow discharge port), HFT (resistance heater), DFT (dual high voltage), HFF (high frequency induction heating), OCTR (octal ring pin header), OCTO (octal hollow pin header).
Bakeable (to 450°C): TF910 (high current), TF908 (water cooled, r.f., coaxial), TF912 (high voltage, single), TF915 (high voltage, dual), TF911 (medium current), TF901 (8 tube instrumentation), TF902 (8 pin instrumentation).
Motion feedthroughs: Linear: TF 925 (bakeable).
Rotary: BMT (bellows sealed), DMT (direct drive), WCH (water cooled rotary hearth), PPR (push pull rotary), RSF (2 position, motor driven, rotary shutter), DDR (direct drive).
Bakeable: TF904 (magnetically coupled) RD1, RD2, RD3 (bellows sealed).
Motion drives: Series LMD (linear), Series UMD (universal).
Other feedthroughs: WFM1 (wire feed mechanism), QCF (quick-coupling), LNF (liquid nitrogen), WFT (water), BPT (blank fixturing support), BPF (blank plug feedthrough), TF 903 (liquid nitrogen, single tube), TF 913 (liquid nitrogen, dual tube).
Feedthrough collars: FTC18 (20 inch, 14 holes maximum), FTC24 (26 inch, 18 holes maximum).

Viewing ports:
Flush mount, tube and non-magnetic types available (sizes $1\frac{1}{2}$ to 4 inch).
Bakeable to 400°C.

Base plates:
Stainless steel, four sizes (18, 12, 10 and $8\frac{3}{4}$ inch jars and chambers), positions

for up to 49 optional feedthrough holes. Elliptical port stainless steel baseplates, two sizes (20 and 26 inch).

Bell jars:

Stainless steel (18 x 30 inch, 24 x 30 inch). Cylindrical bell jar chambers 18 x 10 inch (Pyrex), 24 x 10 inch (stainless steel and water cooled stainless steel). Pyrex bell jars (18 x 30 inch, 18 x 18 inch, 12 x 18 inch, 10 x 12 inch). Also guards and manual counter-balanced hoists. 'Vakit': bell jar fixturing kit allows flexibility of design and construction.

Flanges:

Hall uhv flanges (edge sealed): Rotatable flanges, models FL 125, 126 (¾ inch), FL 135, 136 (1 inch), FL 175, 176 (1½ inch), FL 225, 226 (2 inch), FL 275, 276 (2½ inch), FL 425, 426 (4 inch), FL 625, 626 (6 inch), Integral, non-rotatable flanges; Models FL 121, 122 (¾ inch), FL 151, 152 (1 inch), FL 191, 192 (1½ inch), FL 231, 232 (2 inch), FL 281, 282 (2½ inch), FL 431, 432 (4 inch), FL 631, 632 (6 inch), FL 831, 832 (8 inch).
Doubleface flanges: Models FL 940, 942, 943, 944.
Blank flanges: Models FL 945, 946.
Hall uhv 'Kwik klamp' metal sealed flanges (stainless steel, bakeable to 500°C, rotatable, two bolts only): Models FL 127 (flange), 128 (blank flange), 129 (clamp), (nominal tube size ¾ inch), FL 137, 138, 139 (1 inch), FL 177, 178, 179 (1½ inch), FL 227, 228, 229 (2 inch), FL 277, 278, 279 (2½ inch), FL 377, 378, 379 (3 inch), FL 427, 428, 429 (4 inch), FL 627, 628, 629 (6 inch).
Economy uhv flanges: Rotatable; models SF 101 (flange), 103 (insert), 104 (blank insert) (nominal tube size 1 inch), SF 151, 153, 154 (1½ inch), SF 201, 203, 204 (2 inch), SF 251, 253, 254 (2½ inch), SF 301, 303, 304 (3 inch). Integral non-rotatable; models SF 401 (flange), 402 (blank flange) (4 inch), SF 501, 502, (5 inch), SF 601, 602, (6 inch), SF 801 802 (8 inch).
Large uhv flanges: Temperature range −196°C to + 450°C, vacuum range to 10^{-12} Torr; non-rotatable, available as open flanges, blank flanges or cover plates. Sizes 10, 12, 14, 16 and 18 inch (standard), custom built up to 36 inch.
Brass flanges: Viton O-ring gasketed flanges: models BF25 (tubing o.d. size ¼ inch), BF38 (⅜ inch), BF50 (½ inch), BF62 (⅝ inch), BF 75 (¾ inch), BF87 (⅞ inch), BF100 (1 inch), BF112 (1⅛ inch), BF150 (1½ inch), BF162 (1⅝ inch), BF 200 (2 inch), BF212 (2⅛ inch), BF 250 (2½ inch), BF262 (2⅝ inch), BF300 (3 inch), BF312 (3⅛ inch), BF400 (4 inch), BF412 (4⅛ inch).

Pump fluids:

DC704 and 705 silicone diffusion pump fluids. Santovac 5 polyphenyl ether fluid for use in uhv systems in the 10^{-9} to 10^{-10} Torr range. Octoil general purpose diffusion pump fluid (ester).

Resin:

Ve-seal: epoxy resin, seals systems at 10^{-9} Torr and below and withstands bakeout temperatures up to 120°C. Solvent free, cures at room temperature.

2.2 Vacuum instrumentation

2.2.1 Thermal conductivity gauges

A thermal conductivity gauge depends on the change with pressure, of the rate of loss of heat from a heated element in the gas (BS 2951, Part 1: 1969). These gauges include: Pirani gauges in which the temperature of the element is measured in terms of its electrical resistance; thermistor gauges in which the heat element is a thermistor; thermocouple gauges in which the temperature of the heated element is measured by a thermocouple.

The gauge heads containing the pressure sensitive elements are listed separately from their corresponding gauge control units, containing the power supplies and pressure indicators.

Pirani gauge heads

MANUFACTURER:	Balzers			Brizio Basi	Cooke
Model no.	TRP 010 (constant temperature)	NV4 (constant temperature)	NVR4 (constant temperature)	VPN	PG1, PG2
Total pressure range (torr)	$100–1 \times 10^{-3}$	$50–10^{-3}$	$50–10^{-3}$	$1–1 \times 10^{-3}$	$0{\cdot}1–10^{-4}$
Interchangeability				Yes	Yes
Construction	Aluminium alloy housing	Steel	Stainless steel	Metal	Metal
Temperature range (°C)					
Operating	0–50	0–50	0–50		−40–+50
Maximum					
Response time (s) (10^{-3} to 10^{-1} Torr pressure change)	$< 10^{-2}$				1

MANUFACTURER:	**Edwards High Vacuum**						
Model no.	G6B	M6B	M7B	G6A	M6A	G5C1	M5C1
Total pressure range (torr)	$3–10^{-3}$	$3–10^{-3}$	$3–10^{-3}$	$5 \times 10^{-1}–10^{-3}$	$5 \times 10^{-1}–10^{-3}$	$10–10^{-3}$	$10–10^{-3}$
Interchangeability	Yes	Yes	Yes	Yes	Yes	Yes	Yes
Construction	Glass	Glass, metal	Stainless steel	Glass	Glass, metal	Glass	Glass, metal
Temperature range (°C)							
Operating	0–60	0–60	0–60	0–60	0–60	0–60	0–60
Maximum	70	70	70	70	70	70	70
Response time (s) (10^{-3} to 10^{-1} Torr pressure change)							

MANUFACTURER:	**Edwards High Vacuum** *cont'd*				
Model no.	G5C2	M5C2	G9 (constant temperature)	M9 (contant temperature)	M9B (constant temperature)
Total pressure range (torr)	$1–10^{-4}$	$1–10^{-4}$	$500–10^{-3}$	$500–10^{-3}$	$500–10^{-3}$
Interchangeability	Yes	Yes	Yes	Yes	Yes
Construction	Glass	Glass, metal	Glass	Stainless steel	Stainless steel
Temperature range (°C)					
Operating	0–60	0–60	0–60	0–60	0–60
Maximum	70	70	70	70	300
Response time (s) (10^{-3} to 10^{-1} Torr pressure change)					

MANUFACTURER:	General Engineering	Granville-Phillips	LKB
Model no.	PGH3	Series 260	Autovac (constant temperature)
Total pressure range (torr)	$1–10^{-3}$	$1000–1 \times 10^{-3}$	$100–10^{-3}$
Interchangeability	Yes	Yes	
Construction	Metal	Metal	Steel and sodium or pyrex glass
Temperature range (°C)			
Operating	0–35		
Maximum	35		450
Response time (s) (10^{-3} to 10^{-1} Torr pressure change)			

MANUFACTURER:	Leybold-Heraeus		
Model no.	TR111	TR112	TR201
Total pressure range (torr)	$10–10^{-2}$	$1–10^{-3}$	$760–10^{-3}$
Interchangeability			
Construction	Metal and glass (hard or soft)	Metal and glass (hard or soft)	Metal or hard glass
Temperature range (°C)			
Operating			
Maximum	Hard glass bakeable	Hard glass bakeable	Hard glass bakeable
Response time (s) (10^{-3} to 10^{-1} Torr pressure change)			2×10^{-2}

MANUFACTURER:	NGN	Nanotech	Sadla-Vac	Vacuum Generators	
Model no.	PRH3	NPI 2	PIH 1	PVG 1	PVG 2
Total pressure range (torr)	$760–10^{-3}$	0–1000	$1–10^{-3}$	$0{\cdot}5–10^{-3}$	$0{\cdot}5–10^{-3}$
Interchangeability	Yes				
Construction	Metal	Metal	Metal	Metal	Metal
Temperature range (°C)					
Operating		25			
Maximum		35			400 (bakeout)
Response time (s) (10^{-3} to 10^{-1} Torr pressure change)					

Thermistor gauge heads

MANUFACTURER:	Brizio Basi	Kinney
Model no.	VPMT	Thermistor gauge
Total pressure range (torr)	$1–1 \times 10^{-3}$	$760–10^{-3}$
Interchangeability	Yes	Yes
Construction	Metal	Metal
Temperature range (°C)		
Operating		15–35
Maximum		100
Response time (s) (10^{-3} to 10^{-1} Torr pressure change)		$<5 \times 10^{-2}$

Thermocouple gauge heads

MANUFACTURER:	AEI	Alcatel	Balzers	Brizio Basi
Model no.	DV23	Thermocouple gauge	NV3	VPRN
Total pressure range (torr)	5–10^{-3}	3–1 x 10^{-3}	1–10^{-3}	1 x 10^{-3}–1 x 10^{-5}
Interchangeability	Yes			Yes
Construction	Metal		Metal	Metal
Temperature range (°C)				
Operating			0–50 (ambient)	
Maximum				
Response time (s) (10^{-3} to 10^{-1} Torr pressure change)				

MANUFACTURER:	Cenco	Cooke	Edwards	General Engineering
Model no.	94180	HTT	TC1	TCH1
Total pressure range (torr)	0·2–10^{-3}	2–10^{-4}	3–10^{-3}	1–10^{-3}
Interchangeability				Yes
Construction	Glass	Metal	Metal	Metal
Temperature range (°C)				
Operating		–40–+50	0–60	0–35
Maximum		80	70	35
Response time (s) (10^{-3} to 10^{-1} Torr pressure change)				

MANUFACTURER:	Hastings Raydist				Kinney
Model no.	DV 4D	DV 5M	DV 6M	DV 8M	Thermocouple gauge
Total pressure range (torr)	20–0·1 (last scale marking = 0·1)	0·1–2 x 10^{-4}	1–1 x 10^{-3}	10^{-2}–10^{-4}	3–1 x 10^{-3}
Interchangeability	Yes	Yes			Yes
Construction	Metal, glass, stainless steel	Metal, glass, stainless steel			Metal
Temperature range (°C)					
Operating	30	1·5			15–35
Maximum	250	48			100
Response time (s) (10^{-3} to 10^{-1} Torr pressure change)	0·3	25			<5 x 10^{-2}

MANUFACTURER:	Ormandy and Stollery			Sadla-Vac	Varian	Veeco	
Model no.	0122	0123	0124	TCH 1	NRC 531	DV 1M	DV 4AM
Total pressure range (torr)	3–10^{-3}	2–10^{-4}	1–10^{-4}		2·5–10^{-3}	1–10^{-3}	0–20
Interchangeability	Yes	Yes	Yes				
Construction	Metal or glass	Metal or glass	Metal or glass	Metal	Stainless steel	Metal	Metal
Temperature range (°C)							
Operating	−20–+100 (metal)	−20–+100 (metal)	−20–+100 (metal)				
Maximum	450 (pyrex)	450 (pyrex)	450 (pyrex)				
Response time (s) (10^{-3} to 10^{-1} Torr pressure change)	1	15	25		3		

Thermal conductivity gauge control units

MANUFACTURER:	AEI	Balzers			
Type of gauge	Thermocouple	Pirani	Pirani	Pirani	Pirani and cold cathode
Model no.	EVH3	TPG022	TPG030	TPG040	PKG010
Model number of associated gauge head	DV23	TPR010 or NV4, NVR4	TPR010 or NV4, NVR4	TPR010 and NVR4	TPR010, IKR010
Number of gauge heads	1, 2 or 5	3	1	1–6	1–3 Pirani 1 cold cathode
Total pressure range (torr)	$5–10^{-3}$	$100–1 \times 10^{-3}$	$2–1 \times 10^{-3}$	$100–10^{-3}$	$100–5 \times 10^{-8}$
Number of pressure scales	1	1 (+ auxiliary)	1	1	2
Pressure range of each scale (torr)					
Scale readings at FSD (torr)					
10%	2×10^{-2}				
50%	0·2				
90%	1·5				
Mains and battery supply stabilization					
Power input (V)	240	230 or 115	230 or 115	230 or 115	230 or 115
Dimensions (mm)	101·6 x 88·9	214 x 144 x 184	105 x 90 x 113	214 x 144 x 264	214 x 144 x 264
Ancillaries	Switching units	Vacuum relay Pressure time recorder	Angled holder	Vacuum relay VWS 101 Automatic gas inlet and pressure control valve RME010 Recorders	Vacuum relay Pressure/time recorder Selector switch Poral filter

MANUFACTURER:	**Balzers** *cont'd*			
Type of gauge	Thermocouple	Pirani component assembly kit	Thermocouple component assembly kit	Thermocouple and cold cathode component assembly kit
Model no.	KV301B	KV403E	KV503E	KV613E
Model number of associated gauge head	NV3B	NV4, NVR4, TPR010	NV3	NV3, HV3
Number of gauge heads	1	3	3	4 (3 thermocouple, 1 cold cathode)
Total pressure range (torr)	$1–10^{-3}$	$50–1 \times 10^{-3}$	$1–1 \times 10^{-3}$	$1–2 \times 10^{-6}$
Number of pressure scales	1	1	1	1
Pressure range of each scale (torr)				
Scale readings at FSD (torr)				
10%				
50%				
90%				
Mains and battery supply stabilization				
Power input (V)	3	115 or 230	115, 220 or 240	115 or 230
Dimensions (mm)	178 x 133 x 80			
Ancillaries		Power supply unit NE4 indicating instrument. Selector switch	Power supply unit NE3 indicating instrument. Selector switch	Fore vacuum power supply NE3. High vacuum power supply HE2. Indicating instrument. Selector switch

MANUFACTURER:	**Balzers** *cont'd*	**Brizio Basi**			**Cenco**
Type of gauge	Thermocouple and cold cathode component assembly kit	Pirani	Thermistor	Thermocouple	Pirani
Model no.	KV713E	APN	APMT	APRN	94190-1
Model number of associated gauge head	NV3, HV5				
Number of gauge heads	3 thermocouple, 1 cold cathode				
Total pressure range (torr)	1 x 5 x 10^{-8}				2–1 x 10^{-3}
Number of pressure scales	1	3	1	1	2
Pressure range of each scale (torr)					0–50 x 10^{-3} 50 x 10^{-3}–2
Scale readings at FSD (torr)					
10%					
50%					
90%					
Mains and battery supply stabilization					
Power input (V)	115 or 230				115
Dimensions (mm)					260 x 165 x 139
Ancillaries	Fore vacuum power supply NE3 High vacuum power supply HE5 Indicating instrument Selector switch				

MANUFACTURER:	Cenco			Cooke	
Type of gauge	Pirani	Thermocouple	Thermocouple	Pirani	Thermocouple
Model no.	94190-8	94178-1	94178-8	CP26, CP27	TG42
Model number of associated gauge head				PG1, PG2	HTT
Number of gauge heads				1–10	1–11
Total pressure range (torr)	$2–1 \times 10^{-3}$	$1–10^{-3}$	$1–10^{-3}$	$1–10^{3}$	$2–10^{-3}$
Number of pressure scales	2	1	1	1	1
Pressure range of each scale (torr)	$0–50 \times 10^{-3}$ $50 \times 10^{-3}–2$				
Scale readings at FSD (torr)					
10%					
50%					
90%					
Mains and battery supply stabilization					
Power input (V)	230	115	230	Battery or 115	Battery or 115
Dimensions (mm)	260 x 165 x 139	260 x 165 x 139	260 x 165 x 139		
Ancillaries				Control units OA42, MA65	Control units OA42, MA65

MANUFACTURER:	**Edwards High Vacuum**				
Type of gauge	Pirani	Pirani	Pirani	Pirani	Pirani
Model no.	Series 70, Pirani 10	Series 70, Pirani 11	Monitorr 101	Model 8-1	Model 8-2
Model number of associated gauge head	M6B, M7B, G6B	M6B, M7B, G6B	M6B, M7B, G6B	M5C1, G5C1	M5C2, G5C2
Number of gauge heads	1	2	2	2	2, 6 or 10
Total pressure range (torr)	$3–10^{-3}$	$3–10^{-3}$	$3–10^{-3}$	$10–10^{-3}$	$1–10^{-4}$
Number of pressure scales	1	1	1	2	2
Pressure range of each scale (torr)	$760–10^{-3}$	$760–10^{-3}$	$760–10^{-3}$	$10–5 \times 10^{-2}$ $5 \times 10^{-1}–10^{-3}$	$1–5 \times 10^{-3}$ $5 \times 10^{-3}–10^{-4}$
Scale readings at FSD (torr)					
10%					
50%					
90%					
Mains and battery supply stabilization					
Power input (V)	100/115 or 200/250	100/115 or 200/250	100/115 or 200/250	100/130 or 200/250	100/130 or 200/250
Dimensions (mm)	110 x 133 x 130	110 x 133 x 130	110 x 133 x 130	165 x 216 x 187	165 x 216 x 187
Ancillaries					

MANUFACTURER:	Edwards High Vacuum *cont'd*		General Engineering	
Type of gauge	Pirani constant temperature	Thermocouple	Pirani	Thermocouple
Model no.	Model 9-A	Series 70 thermocouple 1, 2	PIR1	TCM1
Model number of associated gauge head	G9, M9, M9B	TC1	PGH3	TCH1
Number of gauge heads	1–4	1	2	1
Total pressure range (torr)	$500–10^{-3}$	$3–10^{-3}$	$1–10^{-3}$	$1–10^{-3}$
Number of pressure scales	3	1	2	1
Pressure range of each scale (torr)	$500–5$, $5–5 \times 10^{-2}$ $10^{-1}–10^{-3}$	$760–10^{-3}$	$1–5 \times 10^{-3}$ $25 \times 10^{-3}–10^{-3}$	$1–10^{-3}$
Scale readings at FSD (torr)				
10%			0·35, 0·022	0·4
50%			0·08, 0·012	0·1
90%			0·015, 0·003	0·02
Mains and battery supply stabilization			Transistorized circuit	Transistorized circuit
Power input (V)	100/130 or 200/250	100/125 or 200/250 (model 2 battery)		
Dimensions (mm)	213 x 235 x 241	110 x 133 x 130	225 x 106 x 165	225 x 106 x 165
Ancillaries				

MANUFACTURER:	Granville-Phillips		Hastings Raydist		
Type of gauge	Pirani	Pirani	Thermocouple	Thermocouple	Thermocouple
Model no.	224	260	VT4	VT5	VT6
Model number of associated gauge head	Series 260	Series 260	DV4D	DV5M	DV6M
Number of gauge heads			1	1	1
Total pressure range (torr)	$1000–10^{-3}$	$1000–10^{-3}$	20–0·1	$0·1–2 \times 10^{-4}$	$1–10^{-3}$
Number of pressure scales	1	1	1	1	1
Pressure range of each scale (torr)			20–0·1	$0·1–2 \times 10^{-4}$	$1–10^{-3}$
Scale readings at FSD (torr)					
10%			0·2	0·002	0·01
50%			1	0·01	0·07
90%			10	0·05	0·5
Mains and battery supply stabilization					
Power input (V)	105/125 or 210/250	105/125 210/250	115 or 220	115 or 220	115 or 220
Dimensions (mm)	483 x 89 x 292	437 x 89 x 292			
Ancillaries			Two and five position switching units for multi head use, single and two point controllers, recorders		

MANUFACTURER:	Kinney		LKB
Type of gauge	Thermistor	Thermocouple	Pirani (constant resistance)
Model no.	KTPG	KTG	Autovac
Model number of associated gauge head			Autovac
Number of gauge heads	1 or 2	1–6	4
Total pressure range (torr)	$760–1 \times 10^{-3}$	$3–1 \times 10^{-3}$	$100–10^{-3}$
Number of pressure scales	2	1	3
Pressure range of each scale (torr)	$760–10^{-1}$, $10^{-1}–10^{-3}$	$3–1 \times 10^{-3}$	$10^{-3}–10^{-1}$, $10^{-1}–100$, 0–100 (linear)
Scale readings at FSD (torr)	High / Low		
10%	5×10^{-3} / 1×10^{-2}	6×10^{-3}	
50%	65 / 5×10^{-2}	5×10^{-2}	
90%	500 / 9×10^{-2}	1	
Mains and battery supply stabilization	Solid state regulator	Solid state regulator	
Power input (V)	115 or 230	100/130	110, 130, 150, 220, 240 (selectable)
Dimensions (mm)	216 x 165 x 168	216 x 165 x 168	310 x 170 x 220
Ancillaries			

MANUFACTURER:	Leybold-Heraeus				
Type of gauge	Pirani	Pirani	Pirani	Pirani (constant resistance)	
Model no.	TM12/1	TM13/1	TM11/2	TM201	TM202
Model number of associated gauge head	TR111 series	TR112 series	TR111 or TR112	TR201	TR201
Number of gauge heads	1	1	2	1	2
Total pressure range (torr)	$10–10^{-2}$	$1–10^{-3}$	$10–10^{-3}$	$760–10^{-3}$	$760–10^{-3}$
Number of pressure scales	1	1	2	1	1
Pressure range of each scale (torr)	$10–10^{-2}$	$10^{0}–10^{-3}$	$10–10^{-2}$ $10^{-1}–10^{-3}$	$760–10^{-3}$	$760–10^{-3}$
Scale readings at FSD (torr)					
10%	10^{-2}	4×10^{-3}	Depends on range	7×10^{-3}	7×10^{-3}
50%	0·4	$2{\cdot}5 \times 10^{-2}$		0·3	0·3
90%	3	0·18		15	15
Mains and battery supply stabilization					
Power input (V)	220/240	220/240	220/240	220/240	220/240
Dimensions (mm)	100 x 80 x 160	100 x 80 x 160	200 x 140 x 140	100 x 80 x 160	100 x 80 x 160
Ancillaries					

MANUFACTURER:	Leybold-Heraeus *cont'd*				NGN
Type of gauge	Pirani (constant resistance)			Combination of TM202 and Penningvac 30	Pirani
Model no.	TM201S	TM201S2	TM203	Combitron CM 30	PRU3
Model number of associated gauge head	TR201	TR201	TR201	TR 201/PR 30	PRH3
Number of gauge heads	1	1	1		2
Total pressure range (torr)	$760–10^{-3}$	$760–10^{-3}$	$760–10^{-3}$		$760–10^{-3}$
Number of pressure scales	1	1	2		1
Pressure range of each scale (torr)	$760–10^{-3}$	$760–10^{-3}$	$760–10^{-1}$ $10^{-1}–10^{-3}$		$760–10^{-3}$
Scale readings at FSD (torr)			Depends on range	Depends on range	
10%					5×10^{-1}
50%					5×10^{-2}
90%					5×10^{-3}
Mains and battery supply stabilization					Mains Yes
Power input (V)	220/240	220/240	100, 110, 120, 200, 220 or 240	110, 220, 240	
Dimensions (mm)	100 x 80 x 160	200 x 140 x 140	200 x 140 x 270	265 x 135 x 270	178 x 130 x 152
Ancillaries					

MANUFACTURER:	Nanotech	Nuclide		Ormandy and Stollery	
Type of gauge	Pirani	Thermocouple	Thermocouple	Thermocouple	Thermocouple
Model no.	NPI1	TC6	TC6A	0120/B	0120/C
Model number of associated gauge head	NPI2	HVEC G-72	RCA1946	3	3
Number of gauge heads	2	1	1	1	1
Total pressure range (torr)	0–1000	$1–10^{-3}$	$1–10^{-3}$	$3–10^{-3}$	$3–10^{-5}$
Number of pressure scales	1			1	2
Pressure range of each scale (torr)				$3–10^{-3}$	$1–8 \times 10^{-3}$, $10^{-2}–10^{-5}$
Scale readings at FSD (torr)					
10%	15			7×10^{-1}	1
50%	80			10^{-1}	8×10^{-3}
90%	500			$1{\cdot}5 \times 10^{-2}$	5×10^{-4}
Mains and battery supply stabilization	Electronic, using zenier diodes			Voltage stabilized transformer (a.c.) or solid state regulator and dry cells (d.c.)	
Power input (V)	210–250			115 or 250 or 6 U_2 cells	
Dimensions (mm)	152 x 127 x 129			200 x 120 x 160	200 x 120 x 160
Ancillaries	Built-in variable setting electronic switch unit			Reference heads automatic current setting	Reference heads automatic current setting

MANUFACTURER:	Sadla-Vac				
Type of gauge	Pirani	Pirani	Thermocouple	Thermocouple	Thermocouple
Model no.	PIU-1	PIU-2	TCB-1	TCU-1	TCU-2
Model number of associated gauge head	PIH-1	PIH-1	TCH1 or DS2000	TCH-1 or DS2000	TCH-1 or DS2000
Number of gauge heads	1	2	1	1	2
Total pressure range (torr)	0·001–1	0·001–1	0·001–10	0·001–10	0·001–10
Number of pressure scales	2	2	1	1	1
Pressure range of each scale (torr)	0·001–0·025 0·010–1·0	0·001–0·025 0·010–1·0	0·001–10	0·001–10	0·001–10
Scale readings at FSD (torr)					
10%					
50%					
90%					
Mains and battery supply stabilization					
Power input (V)	220/250	220/250	1·5 (battery)	200/250	200/250
Dimensions (mm)					
Ancillaries					

MANUFACTURER:	Torr Vacuum Products		UHV	Vactorr	
Type of gauge	Thermocouple	Thermocouple	Pirani (constant resistance)	Thermocouple	Thermocouple
Model no.	TGC	GMT	PRG-1	10477	10479
Model number of associated gauge head	6343	6343	PIH-1		
Number of gauge heads	1	1	4	2	
Total pressure range (torr)	$1-10^{-4}$	$1-10^{-4}$	$500-10^{-3}$	$1-10^{-3}$	$1-10^{-3}$
Number of pressure scales	1	1	3		1
Pressure range of each scale (torr)			$500-4$, $7-10^{-1}$ $2 \times 10^{-1}-10^{-3}$		
Scale readings at FSD (torr)					
10%			$4{\cdot}5$, $0{\cdot}15$, 3×10^{-3}		
50%			12, 1, 4×10^{-2}		
90%			80, 4, 5×10^{-2}		
Mains and battery supply stabilization			Bridge rectifier and smoothing		
Power input (V)	105–125	105–125	110/130 or 200/250	120	120
Dimensions (mm)				170 x 180 x 220	280 x 90 x 240
Ancillaries			Vacuum switch automatic range changing		Includes ionization gauge

MANUFACTURER:	Vactronic			Vacuum Generators	Varian
Type of gauge	Thermocouple	Thermocouple	Thermocouple	Pirani	Thermocouple and thermistor
Model no.	HVI-1000	HVI-100	HVI-20	PIR1	NRC801
Model number of associated gauge head	HV6M	HV5M	MV4DM	PVG1, PVG2	NRC531
Number of gauge heads				2	1
Total pressure range (torr)	$1–10^{-3}$	$1–10^{-3}$	20–0·1	$0·5–10^{-3}$	$1–10^{-3}$ (additional readings to 2)
Number of pressure scales	1	1	1		1
Pressure range of each scale (torr)	$1–10^{-3}$	$1–10^{-3}$	20–0·1		$1–10^{-3}$
Scale readings at FSD (torr)					
10%					
50%					
90%					
Mains and battery supply stabilization					
Power input (V)	90–140 a.c. 110–150 d.c.	90–140 a.c. 110–150 d.c.	90–140 a.c. 110–150 d.c.	220/240	
Dimensions (mm)				140 x 70 x 100	
Ancillaries					

MANUFACTURER:	**Varian** *cont'd*				
Type of gauge	Thermocouple	Thermocouple	Thermocouple and ionization	Thermocouple and ionization	Thermocouple and ionization
Model number	NRC802-A	NRC804-A	NRC831	NRC853	NRC836
Model number of associated gauge head	NRC531	NRC531	NRC531 + NRC 563 573 or 507	NRC531 + NRC524-1	NRC531 + NRC 507 518, 527 or 538
Number of gauge heads	1	1–5	3 (2 thermocouples)	3 (2 thermocouples)	3 (2 thermocouples)
Total pressure range (torr)	$1–10^{-3}$	$2–10^{-3}$	$1–2 \times 10^{-9}$	$2–1 \times 10^{-9}$	$1–2 \times 10^{-9}$
Number of pressure scales	1	1	3	3	3
Pressure range of each scale (torr)	$1–10^{-3}$	$2–10^{-3}$	$1–10^{-3}$ $1–10^{-3}$ $1 \times 10^{-3}–2 \times 10^{-9}$	$2–1 \times 10^{-9}$ $2–1 \times 10^{-9}$ $1 \times 10^{-9}–5 \times 10^{-3}$	
Scale readings at FSD (torr)					
10%					
50%					
90%					
Mains and battery supply stabilization					
Power input (V)	105/130 or 210/260	105/130 or 210/260	105/130 or 210/260	115 or 230	
Dimensions (mm)	209 x 138 x 171	209 x 138 x 209	313 x 138 x 186	209 x 138 x 170	
Ancillaries					

MANUFACTURER:	Veeco				
Type of gauge	Thermocouple	Thermocouple	Thermocouple	Thermocouple and cold cathode	Thermocouple and ion
Model number	TG-7	TG-27	TC-77	DG2-2T	RG-8/TG-7
Model number of associated gauge head	DV1M	DV4M	DV1M	DV1M, DG211	RG81, TG7 or TG27
Number of gauge heads	1–3	1–3	1	3 (2 thermocouple)	3 (2 thermocouple)
Total pressure range (torr)	$1–10^{-3}$	0–20	$1–10^{-3}$	$1–5 \times 10^{-6}$	$1–2 \times 10^{-11}$
Number of pressure scales				3	
Pressure range of each scale (torr)				$1–10^{-3}$ $10^{-2}–10^{-5}$ $10^{-4}–5 \times 10^{-6}$	
Scale readings at FSD (torr)					
10%					
50%					
90%					
Mains and battery supply stabilization	Reliable from 100–130V	Reliable from 100–130V	±15%		
Power input (V)	115	115	115 or 220/240	115	
Dimensions (mm)	104 x 132 x 300	104 x 132 x 300	104 x 132 x 316	482 x 133 x 95	
Ancillaries					

2.2.2 McLeod gauges and compression manometers

MANUFACTURER:	Brizio Basi			Cenco		
Model no.	MLL50	MLL100	MLL300	94156	94160	94151
Pressure range (torr)	$2 \times 10^{-0}–1 \times 10^{-4}$	$2 \times 10^{-1}–1 \times 10^{-5}$	$5 \times 10^{-2}–1 \times 10^{-6}$	$1–5 \times 10^{-3}$ (two scales)	15–0·1	$1·6–5 \times 10^{-4}$
Quantity of mercury (kg)	2·5	3·3	6·8	13·6	4·5	4·5
Method of moving mercury				Plunger	Plunger	Plunger
Readout						
Dimensions (mm)						
Height				1900		1650
Base						

MANUFACTURER:	Cooke			Edwards High Vacuum		
Model no.	DMG101	DMG102	DMG123	1B2	2B2	1E2
Pressure range (torr)	$20–2{\cdot}5 \times 10^{-5}$	$20–2 \times 10^{-5}$	$1–2{\cdot}5 \times 10^{-7}$	$10–10^{-2}$	$1–10^{-3}$	$10–10^{-2}$
Quantity of mercury (kg)	0·15	0·15	1·25	0·10	0·15	0·10
Method of moving mercury				Rotary	Rotary	Rotary
Readout						
Dimensions (mm)						
Height				229	229	218
Base				216 diameter	216 diameter	83 diameter

MANUFACTURER:	Edwards High Vacuum *cont'd*			GEC Hirst Research	General Engineering
Model no.	2E2	2G	HS1A	Made to order	TD2
Pressure range (torr)	$1–10^{-3}$	$1–10^{-3}$	$4 \times 10^{-2}–10^{-6}$		$150–10^{-3}$
Quantity of mercury (kg)	0·15	0·105	9		0·26
Method of moving mercury	Rotary	Rotary	Airlift		Diaphragm
Readout					Scale
Dimensions (mm)					
Height	218	159	892		340
Base	83 diameter	87 diameter	302 x 200		Tripod 185 centres

MANUFACTURER:	Hastings Raydist	Kinney	Leybold – Heraeus				
Model no.	HM 504 100	TD	Vakuscop	Moser type	McLeod version 3	Kammerer 1	Kammerer 2
Pressure range (torr)	$0{\cdot}2–1 \times 10^{-5}$	$150–10^{-3}$	0·01–35	$760–10^{-4}$	$10^{-6}–10^{-1}$	$80–10^{-3}$	$80–10^{-4}$
Quantity of mercury (kg)	8·16	0·26	0·11	0·14	12	0·2	0·22
Method of moving mercury	Air lift	Diaphragm	Rotating	Rotary	Air lift	Diaphragm	Diaphragm
Readout	Linear mirror scale and computer chart readout	Direct			Direct		
Dimensions (mm)							
Height	863	342	220	130	890		
Base	330 x 330	215 x 215		120 x 120	165 x 130		

MANUFACTURER:	Pennwalt-Stokes			Pfeiffer	Pope Scientific		Virtis
Model no.	Kammerer 1	Kammerer 2	Moser type	Wohl type	2058	2059	
Pressure range (torr)	90–10^{-3}	90–10^{-4}	760–10^{-4}	10^{-1}–4 x 10^{-6}	Up to 10^{-6}	Up to 5 x 10^{-5}	5-5 x 10^{-3}
Quantity of mercury (kg)	0·2	0·22	0·14	10·9			0·14
Method of moving mercury	Diaphragm	Diaphragm	Rotary	Air lift			Rotary
Readout					Linear scale	Linear scale	
Dimensions (mm)							
Height	380	380	130	400			
Base	220 x 200	220 x 220	120 x 120	400 x 102			

2.2.3 U tube manometers

MANUFACTURER:	Cenco					Leybold	Pfeiffer	Pope Scientific
Model no.	94130	94106	94125	94162-2 (oil)	94162-3 (oil)	16017	Bennert type	20060
Pressure range (torr)	0·1–200	1–160	0–1500	0–0·15	0–2	1–200	1–200	0–1, 0–5, 0–10, 0–15
Quantity of mercury (kg)	0·45	0·16	0·35			0·15	0·08	
Method of moving mercury								
Readout						Linear scale		Linear scale
Dimensions (mm)								
Height	368	235				294	320	
Base						182 x 100	182 wide	

2.2.4 Mechanical gauges

Capsules, diaphragms and Bourbon gauges. The simpler ones are direct indicating, involving purely mechanical transmission to a pointer. The more complex types involve transducers. Specially sensitive transducer types are also known as micromanometers.

MANUFACTURER:	Brizio Basi		Datametrics	
Model no.	VQ	VQA	Sensor 511	Sensor 521
Pressure range (torr)	0–760	0–160	5000–10^{-5}	5000–10^{-5}
Pressure scales				
Sensitivity			Up to 5×10^{-5} %	Up to 5×10^{-5} %
Minimum readable pressure change (torr)				
Differential pressure of atmospheric pressure compensation			Differential	Differential
Construction			Viton O-ring	No elastomers or plastics
Temperature range (°C)				
Accuracy				
Percentage of full scale deflection				
Percentage of reading			0·2–0·1	0·2–0·1
Transducer method / Readout / Temperature range (Transducer types)			0 to ± 10 V d.c. (analogue), 4 and 5 digit plus 1 digit over-range, incline Nixies and parallel BCD output (digital)	

MANUFACTURER:	Datametrics *cont'd*				
Model no.	Sensor 538	Sensor 501	Sensor 531	Sensor 550	Sensor 523
Pressure range (torr)	2500–2·5 x 10^{-5}	5000–5 x 10^{-6}	1000–10^{-5}	5000–5 x 10^{-4}	5000–10^{-5}
Pressure scales					
Sensitivity	Up to 5 x 10^{-5}%	Up to 5 x 10^{-5}%	Up to 5 x 10^{-5}%	Up to 5 x 10^{-5}%	Up to 5 x 10^{-5}%
Minimum readable pressure change (torr)					
Differential pressure or atmospheric pressure compensation	Differential	Differential	Differential	Differential	Differential
Construction	Viton O-rings		UHV construction		Viton O-rings
Temperature range (°C)			Bakeable to 450		
Accuracy					
Percentage of full scale deflection					
Percentage of reading	±0·03	±0·5–0·25	±0·2–0·1	±0·25–0·1	±0·5–0·2
Transducer method, Readout, Temperature range } Transducer types	0 to ±10 V d.c. (analogue), 4 and 5 digit plus 1 digit over-range, in-line Nixies and parallel BCD output (digital)				

MANUFACTURER:	**Datametrics** *cont'd*				
Model no.	Sensor 523H	Transducer 1052E	Transducer 1014	Transducer 1018	Transducer 1085
Pressure range (torr)	$5000–10^{-5}$	0–5000	$1–1 \times 10^{-4}$ of sensor full range	$1–1 \times 10^{-3}$ of sensor full range	
Pressure scales		0–5, 0–5000	9 ranges	4 scales	7 ranges
Sensitivity	Up to 5×10^{-5}%	Up to 5×10^{-5}%			
Minimum readable pressure change (torr)					
Differential pressure of atmospheric pressure compensation	Differential	Differential	Differential	Differential	Differential
Construction					
Temperature range (°C)					
Accuracy					
Percentage of full scale deflection		0·1			
Percentage of reading	+ 0·5 – 0·2		±0·03		±0·003 of reading plus one count
Transducer method / Readout / Temperature range } Transducer types	0 to ±10 V d.c. (analogue), 4 and 5 digit plus 1 digit over-range, in-line Nixies and parallel BCD output (digital)		±10 V d.c.	4 digits plus 1 digit over-range (10 V d.c. available)	5 digits plus 1 digit over-range

MANUFACTURER:	Datametrics *cont'd*			Dunellen	Edwards
Model no.	Transducer 1015	Transducer 1023	Transducer 1083	AAA11	Series CG3
Pressure range (torr)	x1–x0·001 of sensor full range	x1–10^{-4} of sensor full range	x1–10^{-4} of sensor full range	11–760	0–760
Pressure scales	7 ranges	8 ranges	8 ranges	0–11, 0–19 0–55, 0–110 0–760	0–20, 0–40 0–100, 0–760
Sensitivity					
Minimum readable pressure change (torr)					
Differential pressure of atmospheric pressure compensation	Differential	Differential	Differential	Differential	
Construction				Phosphor bronze or stainless steel tubes and capsules	
Temperature range (°C)				0–100	
Accuracy					
Percentage of full scale deflection				±1	
Percentage of reading	± 0·003	± 0·1	± 0·1		± 2·5
Transducer method (Transducer types)					
Readout (Transducer types)	Analogue d.c. output	Meter ±1 V	Digital, 3 places plus 1 digit over-range		
Temperature range (Transducer types)					

MANUFACTURER:	Foster-Cambridge	Foxboro-Yoxall		Furness
Model no.	Clearway pressure gauge	11AL	11AM	MDC
Pressure range (torr)	0–760	0·1–0·4	0·4–0·1520	Up to 100
Pressure scales				Linear
Sensitivity				FSD 100 Torr; sensitivity 10^{-3} FSD 10 Torr, sensitivity 10^{-4}
Minimum readable pressure change (torr)				10^{-4}
Differential pressure of atmospheric pressure compensation	None			
Construction	Non-ferrous metal	Stainless steel	Stainless steel	Metal with rubber O-ring
Temperature range (°C)	−10–+50	−40–+120	−40–+120	10–30 (up to 50 available)
Accuracy				
Percentage of full scale deflection				
Percentage of reading		±0·25	±0·25–±1·0	1
Transducer types: Transducer method				Capacitance
Transducer types: Readout				Analogue
Transducer types: Temperature range				

MANUFACTURER:	General Engineering				Granville-Phillips	
Model no.	GVG11	GVG12	GVG13	GVG14	Series 212	Series 216
Pressure range (torr)	0–20	0–40	0–100	0–760	1000–5 x 10^{-3}	1000–10^{-11}
Pressure scales	Linear	Linear	Linear	Linear	1x, 0·3x, 0·1x 0·03x, 0·01x	<±1% of FSD of transducer input
Sensitivity			2 torr/division	20 torr/division		
Minimum readable pressure change (torr)			0·5	5	<5 x 10^{-3}	
Differential pressure of atmospheric pressure compensation	Compensated	Compensated	Compensated	Compensated		
Construction	Metal	Metal	Metal	Metal	Stainless steel head	
Temperature range (°C)					Bakeable to 450	Bakeable to 450
Accuracy						
Percentage of full scale deflection	2	2			± 2%	
Percentage of reading						
Transducer types: Transducer method						Pressure or gas flow
Transducer types: Readout						10 m V d.c. to 10 V d.c.
Transducer types: Temperature						

MANUFACTURER:	Hattersley Newman Hender	Leybold-Heraeus				
Model no.	Series B, C, A, G	160-40	890-50	160-35,	160-37,	Diavac range
Pressure range (torr)	20, 40, 100, 760	0–760	0–1500	0–760	0–100	1–760
Pressure scales		Linear	Linear	Linear	Linear	Non linear
Sensitivity						
Minimum readable pressure change (torr)						
Differential pressure of atmospheric pressure compensation		Differential Compensated ⟶				
Construction		Bourdon tube		Copper and beryllium diaphragm capsule	Copper and beryllium diaphragm capsule	Copper and beryllium diaphragm capsule or corrosion-proof design
Temperature range (°C)				Up to 70	up to 70	
Accuracy						
Percentage of full scale deflection			±1	±1·5	±1·5	±1
Percentage of reading						
Transducer method } Transducer types						
Readout } Transducer types						100 mV to 10 k ohm output
Temperature range } Transducer types						

MANUFACTURER:	MKS Instruments	Smith-Dennis	Wallace and Tiernan	
Model no.	77, 90, 100, 126, 144 145, 170, 210	A, B, C-CF, D-CF, G17	FA160/15	FA233
Pressure range (torr)	5000–10^{-5}	1–760	0·1–20 or 0–100 or 0–800	0–760
Pressure scales	Linear	Linear	Nearly linear	45 inch scale 1000 gradation
Sensitivity		0·5% FSD (industrial) 0·1% FSD (test)	0·2% of range	0·01%
Minimum readable pressure change (torr)	10^{-5}			
Differential pressure of atmospheric pressure compensation	Both	None		
Construction	Various, including stainless steel and Inconnel X		Aneroid capsule	Bourdon tube gauge
Temperature range (°C)	–125–+450			
Accuracy				
Percentage of full scale deflection				
Percentage of reading	Better than 0·05			
Transducer types: Transducer method	Thin taut diaphragm capacitance manometer			
Transducer types: Readout	0–10 V d.c. digital			
Transducer types: Temperature range	–125–+450			

MANUFACTURER:	**Wallace and Tiernan** *cont'd*		
Model no.	FA145	FA141700	FA141200
Pressure range (torr)		0–760	0–38
Pressure scales	45 inch scale, 1000 gradation	16 inch scale, 200 gradations	7 inch scale
Sensitivity	0·01%	0·2%	0·2%
Minimum readable pressure change (torr)			
Differential pressure of atmospheric pressure compensation			
Construction			
Temperature range (°C)			
Accuracy			
Percentage of full scale deflection			
Percentage of reading	0·1	0·33	0·33
Transducer types: Transducer method			
Transducer types: Readout			
Transducer types: Temperature range			

2.2.5 Penning gauges

Cold cathode ionization gauges in which ionization is produced by electrons emitted from an unheated cathode and in which a magnetic field is used to lengthen the electron path and so increase the number of ions produced. Magnetron and inverted magnetron gauges are Penning gauges with special cylindrical electrode arrangements.

MANUFACTURER:	AEI Scientific Apparatus		Aerovac	Alcatel	
Model no.	VC13	VC14	22GC200	5396	ACF10
Total pressure range (torr)	10^{-3}–10^{-6}	10^{-4}–10^{-13}	10^{-4}–10^{-14}	10^{-2}–10^{-6}	1×10^{-2}–1×10^{-6}
Number of ranges	1	Linear switched through to positive and log scales	10 decade ranges 10^{-4}–10^{-13}		
Number of pressure scales	1	2	2		1
Scale readings at FSD (torr)					
10%		3×10^{-8} (log)			
50%		4×10^{-6} (log)			
90%		3×10^{-4} (log)			
Number of measuring points	1	1	2		
Gauge head data					
Envelope material	Glass	Stainless steel	Stainless steel		
Anode/cathode; other electrodes	Nickel	Stainless steel	Tungsten filament		
Operating voltage (kV)	2	2			
Magnetic field (tesla)	0·1	0·1			
Maximum temperature (°C)		400	450		
Ancillary equipment					

MANUFACTURER:	**Alcatel** *cont'd*				
Model no.	ACF301	ACF104	ACF123	ACF12	UHV Penning gauge
Total pressure range (torr)	1×10^{-2}–1×10^{-6}	1×10^{-2}–1×10^{-7}			1×10^{-3}–1×10^{-9}
Number of ranges		4	3		
Number of pressure scales	1	4	3	1	
Scale readings at FSD (torr)					
10%					
50%					
90%					
Number of measuring points					
Gauge head data					
Envelope material					
Anode/cathode; other electrodes					
Operating voltage (kV)					
Magnetic field (tesla)					
Maximum temperature (°C)					
Ancillary equipment					

MANUFACTURER:	**Balzers**			**Cenco**	
Model no.	IKR010 (head)	HV2 (head)	HV5 (head)	94183-1	94183-8
Total pressure range (torr)	5×10^{-3}–5×10^{-8}	5×10^{-3}–10^{-6}	5×10^{-3}–5×10^{-8}	2×10^{-2}–10^{-6}	2×10^{-2}–10^{-6}
Number of ranges					
Number of pressure scales				1	1
Scale readings at FSD (torr)					
10%					
50%					
90%					
Number of measuring points					
Gauge head data					
Envelope material	Steel	Steel	Steel		
Anode/cathode; other electrodes					
Operating voltage (kV)	3	2	5		
Magnetic field (tesla)					
Maximum temperature (°C)	80	50	50		
Ancillary equipment					

MANUFACTURER:	Cooke	Edwards		General Engineering
Model no.	CC64	Penning 8		PEN1
Total pressure range (torr)	$2{\cdot}5 \times 10^{-2}$–1×10^{-7}	10^{-2}–10^{-7}	3×10^{-3}–10^{-6}	3×10^{-3}–10^{-6}
Number of ranges	3	3		1
Number of pressure scales	3	3		1
Scale readings at FSD (torr)				
10%				$1{\cdot}5 \times 10^{-3}$
50%				2×10^{-4}
90%				3×10^{-5}
Number of measuring points	1	Penning 6	Penning 5MFC	
Gauge head data				
Envelope material	Stainless steel	Stainless steel		
Anode/cathode; other electrodes	Molybdenum anode	Nickel cathode	Stainless steel anode Nickel cathode	
Operating voltage (kV)	1, 1·2 or 2·4			
Magnetic field (tesla)			70	
Maximum temperature (°C)	90	70		
Ancillary equipment	OA42 adjustable control			

MANUFACTURER:	Kinney			Leybold-Heraeus		
Model no.	KDG1		KDGN	PM30	PM4	Combitron CM30 Combines PM3 and TM202
Total pressure range (torr)	10^{-3}–2×10^{-7}		1–5×10^{-6}	10^{-2}–10^{-6}	10^{-2}–10^{-9}	
Number of ranges	3		1	1	1	
Number of pressure scales	2		2	1	1	
Scale readings at FSD (torr)	High	Low				
10%	6×10^{-4}	1×10^{-4}	3×10^{-5}	3×10^{-6}	5×10^{-9}	
50%	17×10^{-4}	5×10^{-4}	12×10^{-4}	1×10^{-4}	5×10^{-6}	
90%	5×10^{-3}	9×10^{-4}	9×10^{-3}	3×10^{-3}	5×10^{-4}	
Number of measuring points	1		1	1		
Gauge head data						
Envelope material	Stainless steel		Steel/brass	Nickel plated mild steel	Stainless steel	
Anode/cathode; other electrodes	Stainless steel		Stainless steel		Stainless steel	
Operating voltage (kV)						
Magnetic field (tesla)						
Maximum temperature (°C)	90		90			
Ancillary equipment						

MANUFACTURER:	NGN	Nanotech	Pennwalt-Stokes	Sadla-Vac	UHV	
Model no.	PGU3	NPE1 (control unit) NPE2 (head)		PEU-1	PNG-1	
Total pressure range (torr)	10^{-2}–10^{-7}	1×10^{-6}–4×10^{-3}	10^{-1}–10^{-7}	10^{-2}–10^{-6}	10^{-2}–10^{-6}	
Number of ranges	1	1		2	2	
Number of pressure scales	1	1		2	2	
Scale readings at FSD (torr)						
10%	4×10^{-6}	2×10^{-5}			$2{\cdot}5 \times 10^{-5}$	2×10^{-6}
50%	1×10^{-4}	2×10^{-4}			2×10^{-4}	10^{-5}
90%	4×10^{-3}	2×10^{-3}			1×10^{-3}	$1{\cdot}8 \times 10^{-5}$
Number of measuring points	1	1		1		
Gauge head data						
Envelope material	Stainless steel	Stainless steel				
Anode/cathode; other electrodes	Double anode cathode					
Operating voltage (kV)	2	2000				
Magnetic field (tesla)						
Maximum temperature (°C)	200	130				
Ancillary equipment		Built in variable setting electronic switch				

MANUFACTURER:	Varian		
Model no.	NRC821 Alphatron*	NRC 524-1	NRC 524-F
Total pressure range (torr)	760–10^{-3}	10^{-1}–1×10^{-9}	10^{-1}–1×10^{-9}
Number of ranges			
Number of pressure scales	6 (linear)		
Scale readings at FSD (torr)			
10%			
50%			
90%			
Number of measuring points	1		
Gauge head data			
Envelope material		Stainless steel	Stainless steel
Anode/cathode; other electrodes	Sealed 100 microcurie Radium 226	Aluminium cathode	Aluminium cathode
Operating voltage (kV)			
Magnetic field (tesla)			
Maximum temperature (°C)	60	Bakeable to 400	Bakeable to 400
Ancillary equipment		Control units NRC851, 852 and 853	Equipped with metal gasket flange for bakeout in uhv.

* Note: this gauge is a cold cathode ionization gauge in which ionization is

MANUFACTURER:	**Veeco**		
Model no.	DG 2 AR (rack) DG 2 10 (bench)	DG2 2T (DG2 + thermocouple)	DC71 (controller)
Total pressure range (torr)	10^{-2}–5×10^{-6}	1–5×10^{-6}	1×10^{-3}–5×10^{-7}
Number of ranges	2 (10^{-4} and 10^{-2} full scales)	3	1
Number of pressure scales			
Scale readings at FSD (torr)			
10%			
50%			
90%			
Number of measuring points			
Gauge head data			
Envelope material	Aluminium		
Anode/cathode; other electrodes	Ceramic shielded		
Operating voltage (kV)			
Magnetic field (tesla)			
Maximum temperature (°C)			
Ancillary equipment			

2.2.6 Hot cathode ionization gauges

A vacuum gauge in which pressure is measured in terms of the ion current produced by electrons emitted from a heated cathode under controlled conditions. Ionization gauges used in the high vacuum range are generally simple three-electrode (triode) structures. Special uhv ionization gauges use an electrode structure designed to minimize the residual current at the ion collector, particularly the X-ray induced component. Particular forms are the Bayard Alpert gauge in which a thin ion collector, is arranged axially in the cylindrical grid with the cathode mounted outside, and others using a suppressor, a modulator electrode, having an external or extractor ion collector arrangement, the hot cathode magnetron or the orbitron etc. The gauge heads are listed separately from the gauge control units.

Bayard Alpert gauge heads

MANUFACTURER:	Balzers			CVC		
Model no.	IMR120	IMR124	IMR125	GIC015	GIC016	GIC017
Lower pressure limit (torr) (X-ray limit)	5 x 10^{-11}	5 x 10^{-11}	5 x 10^{-11}	10^{-10}	10^{-10}	10^{-10}
Sensitivity ($torr^{-1}$)	5	5	5	10	10	10
at mA	0·2–2	0·2–2	0·2–2	10	10	10
Upper pressure limit of measuring range (torr)	10^{-2}	10^{-2}	10^{-2}	10^{-3}	10^{-3}	10^{-3}
Electrodes construction						
Electron collector	Tungsten	Tungsten	Tungsten			
Filament	Thoria coated iridium	Thoria coated iridium	Thoria coated iridium			
Ion collector	Tungsten	Tungsten	Tungsten	Tungsten	Tungsten	Tungsten
Envelope	Nude	Nude	Nude			
Operating data						
Electron collector (V)	+160	+160	+160			
Filament current (A)	2–3	2–3	2–3			
Cathode voltage (V)	+35/45	+35/45	+35/45			
Ion collector potential	0	0	0			
Other electrode						
Vacuum connection	Flange, viton seal	Flange, gold seal	Flange, copper seal	12·7 mm o.d.	19·05 mm o.d.	25·4 mm o.d.

MANUFACTURER:	**Cenco**					
Model no.	94176-1	94176-2	94176-3	94176-4	94176-5	94176-6
Lower pressure limit (torr) (X-ray limit)	10^{-10}	10^{-10}	10^{-10}	10^{-10}	10^{-10}	10^{-10}
Sensitivity ($torr^{-1}$) at mA						
Upper pressure limit of measuring range (torr)	10^{-3}	10^{-3}	10^{-3}	10^{-3}	10^{-3}	10^{-3}
Electrodes construction						
Electron collector						
Filament	Coated iridium	Coated iridium	Coated iridium	Tungsten	Tungsten	Tungsten
Ion collector						
Envelope	Glass	Glass	Glass	Glass	Glass	Glass
Operating data						
Electron collector (V)						
Filament current (A)						
Cathode voltage (V)						
Ion collector potential						
Other electrode						
Vacuum connection	19·05 mm o.d. nonex	19·05 mm o.d. pyrex	19·05 mm kovar	19·05 mm o.d. nonex	19·05 mm o.d. pyrex	19·05 mm o.d. kovar

MANUFACTURER:	Cenco *cont'd*			Cooke	Edwards	
Model no.	94195-1	94195-2	94195-3	BA	1G5G and 1G5GM	1G6G and 1G6GM
Lower pressure limit (torr) (X-ray limit)	10^{-10}	10^{-10}	10^{-10}	2×10^{-12}	10^{-10}	10^{-8}
Sensitivity ($torr^{-1}$)	1	1	1	10	12	6
at mA	5	5	5	0·1	0·1	0·1
Upper pressure limit of measuring range (torr)	10^{-3}	10^{-3}	10^{-3}	8×10^{-3}	10^{-3}	10^{-2}
Electrodes construction						
Electron collector				Molybdenum		
Filament				Tungsten or coated iridium	Tungsten	Tungsten
Ion collector				Tungsten		
Envelope	Glass	Glass	Glass	Nude	Glass (model G) or metal (model M) (nude)	Glass (model G) or stainless steel (model m) (nude)
Operating data						
Electron collector (V)				+150		
Filament current (A)					3·2	2·1
Cathode voltage (V)				3·5	4·2	4·3
Ion collector potential						
Other electrode						
Vacuum connection	19·05 mm o.d. nonex	19·05 mm o.d. pyrex	19·05 mm o.d. kover	Flange or 12·7, 19·05, 25·4 mm tubulation	Graded glass seal (G) Flange (M) with gold or aluminium wire seal or copper gasket	Graded glass seal (G) Flange (M)

MANUFACTURER:	Edwards *cont'd*	GEC Hirst Research	High Voltage Engineering
Model no.	IG4G and IG4M	Made to order	DVAC235
Lower pressure limit (torr) (X-ray limit)	10^{-11}		$<10^{-9}$
Sensitivity ($torr^{-1}$)	16		10
at mA	0·1		10
Upper pressure limit of measuring range (torr)	10^{-3}		
Electrodes construction			
Electron collector			
Filament	Tungsten		
Ion collector	Tungsten		
Envelope	Glass (model G) Metal (model m) (nude)		
Operating data			
Electron collector (V)			+150
Filament current (A)	3·6		
Cathode voltage (V)	3·9		4
Ion collector potential			
Other electrode			
Vacuum connection	Graded glass seal (G) Flange (M) with gold or aluminium wire seal or copper gasket		

MANUFACTURER:	Leybold-Heraeus			Mill Lane Engineering
Model no.	IR36GA, IE36 and IE34	IR411, IE421 and IE431	IR142, IE422 and IE432	ML30
Lower pressure limit (torr) (X-ray limit)	$<1 \times 10^{-8}$	$<1 \times 10^{-11}$	$<1 \times 10^{-11}$	10^{-10}
Sensitivity ($torr^{-1}$)	10	10	10	10
at mA	0·1–10	0·1–10	0·1–10	10
Upper pressure limit of measuring range (torr)	1×10^{-2}	1×10^{-3}	1×10^{-3}	10^{-4}
Electrodes construction				
Electron collector	Molybdenum	Tungsten	Tungsten	Nickel, tungsten and rhenium
Filament	Tungsten	Thoria coated iridium	Tungsten	Tungsten
Ion collector	Tungsten	Tungsten	Tungsten	Molybdenum
Envelope	Hard glass or nude	Hard glass or nude	Hard glass or nude	Glass
Operating data				
Electron collector (V)	+190	+160	+160	
Filament current (A)	1·5–2	2·9–3·8	1·5–2	
Cathode voltage (V)	+30	+50	+50	
Ion collector potential	0	0	0	
Other electrode		With modulator	With modulator	
Vacuum connection	Flange or hard glass tubulation	Flange or hard glass tubulation	Flange or hard glass tubulation	

MANUFACTURER:	RIBER		Torr Vacuum Products	UHV	Ultek
Model no.	JBA11V	JBA11I	1G4336	IGF1	804-7670
Lower pressure limit (torr) (X-ray limit)	2×10^{-10}	5×10^{-11}	10^{-10}	5×10^{-10}	2×10^{-11}
Sensitivity ($torr^{-1}$)				12	0·1
at mA				0·1	4
Upper pressure limit of measuring range (torr)	$1{\cdot}1 \times 10^{-3}$	$1{\cdot}1 \times 10^{-3}$	10^{-3}	10^{-4}	10^{-2}
Electrodes construction					
Electron collector				Tungsten	
Filament			Iridium coated or tungsten (x2)	Tungsten	
Ion collector				Tungsten	
Envelope			Glass	Glass, stainless steel	Nude
Operating data					
Electron collector (V)			+150	+120	+175
Filament current (A)			4–6	1·5	2·5–3·5
Cathode voltage (V)			3–5	3–4·5	3·5
Ion collector potential				0	0
Other electrode					
Vacuum connection			Pyrex, Nonex or Kovar tubulation		UHV copper gasketted flange

MANUFACTURER:	Vactronic		Vacuum Generators		Varian
Model no.	VAC-P-T VAC-N-T VAC-K-T	VAC-P-IR VAC-N-IR VAC-K-IR	VIG10	VIG20	NRC507
Lower pressure limit (torr) (X-ray limit)	10^{-10}	10^{-10}	2×10^{-11}	2×10^{-10}	
Sensitivity (torr^{-1})	10		25	25	16·7
at mA	1		0·1	0·1	0·1
Upper pressure limit of measuring range (torr)	10^{-4}	10^{-4}	10^{-3}	10^{-3}	10^{-3}
Electrodes construction					
Electron collector			Molybdenum	Molybdenum	Tungsten
Filament	Tungsten	Coated iridium	Tungsten or thoriated tungsten (x2)	Tungsten (x2) or thoriated iridium (x1)	Tungsten
Ion collector			Tungsten	Tungsten	
Envelope			Nude	Nude	Nonex
Operating data					
Electron collector (V)	+150	+150	+200	+200	+150
Filament current (A)	4·6	4·6			3·5–6·5
Cathode voltage (V)	3·5	3·5	+50	+50	3·5–6·5
Ion collector potential	0				
Other electrode					
Vacuum connection	Model P - Pyrex Model N - Nonex Model K - Kovar (tubulation)	Model P - Pyrex Model N - Nonex Model K - Kovar (tubulation)	FC flange	FC flange	18 mm tubulation

MANUFACTURER:	**Varian** *cont'd*						
Model no.	NRC518	NRC527	NRC538	NRC563	NRC571	UHV24	UHV12
Lower pressure limit (torr) (X-ray limit)				$1{\cdot}5 \times 10^{-10}$	2×10^{-10}	2×10^{-11}	2×10^{-11}
Sensitivity (torr^{-1})	16·7	16·7	16·7	11	11	25	25
at mA	0·1	0·1	0·1	0·1	0·1	0·1	0·1
Upper pressure limit of measuring range (torr)	10^{-3}	10^{-3}	10^{-3}	10^{-3}	10^{-3}	10^{-3}	10^{-3}
Electrodes construction							
Electron collector	Tungsten	Tungsten	Tungsten	Tungsten	Tungsten	Platinum/iridium alloy on molybdenum rods	Platinum/iridium alloy on molybdenum rods
Filament	Tungsten	Iridium	Iridium	Thoria coated iridium	Thoria coated iridium	Tungsten or thoria coated iridium	Tungsten
Ion collector				Tungsten	Tungsten		
Envelope	Pyrex	Nonex	Pyrex	Platinum coated glass		Nude	Glass
Operating data							
Electron collector (V)	+150	+150	+150	+150	+150	+180	+180
Filament current (A)	3·5–6·5	3·5–6·5	3·5–6·5			1–6	1–6
Cathode voltage (V)	3·5–6·5	3·5–6·5	3·5–6·5	2·4	2·4	0–11	0–11
Ion collector potential				0	0	0	0
Other electrode							
Vacuum connection	18 mm tubulation	18 mm tubulation	18 mm tubulation	Various tubulation including high conductance	Pyrex or Kovar tubulation	Flange	Pyrex or Kovar tubulation or Kovar tubulation on conflat flange

MANUFACTURER:	**Veeco**	
Model no.	RG75	TG75
Lower pressure limit (torr) (X-ray limit)	10^{-10}	10^{-10}
Sensitivity ($torr^{-1}$)	10	10
at mA	10	10
Upper pressure limit of measuring range (torr)	2×10^{-3}	10^{-4} (at 10 mA) 10^{-2} (at 0·1 mA)
Electrodes construction		
Electron collector	Molybdenum	Molybdenum
Filament	Thoria coated iridium	Tungsten (x2)
Ion collector		
Envelope	Hard glass	Hard glass
Operating data		
Electron collector (V)		+180
Filament current (A)		4–6
Cathode voltage (V)		3–5
Ion collector potential		0
Other electrode		
Vacuum connection	Pyrex, Nonex or Kovar tubulation	Pyrex, Nonex or Kovar tubulation

Triode gauge heads

MANUFACTURER:	Balzers						
Model no.	IMR110	IMR111	IMR112	IM15 52U	IMR103	IMR121	IMR122
Lower pressure limit (torr) (X-ray limit)	10^{-6}	10^{-6}	10^{-6}	5×10^{-12}	5×10^{-12}	3×10^{-10}	3×10^{-10}
Sensitivity ($torr^{-1}$)	2			25	25	5	5
at mA	0·05 ($<10^{-2}$) or 0·005 ($1-10^{-2}$)			0·004–10	0·004–10	0·2–2	0·2–2
Upper pressure limit of measuring range (torr)	1	1	1	10^{-3}	10^{-3}	10^{-3}	10^{-3}
Electrodes construction							
Electron collector	Molybdenum	Molybdenum	Molybdenum	Platinum/ iridium	Platinum iridium		
Filament	Thoria coated iridium	Thoria coated iridium	Thoria coated iridium	Tungsten or thorium/ tungsten or thorium/ iridium	Tungsten or thorium/ tungsten or thorium iridium	Tungsten	Tungsten
Ion collector	Tungsten	Tungsten	Tungsten	Platinum/ iridium	Platinum/ iridium		
Envelope	Nude	Nude	Nude	Nude	Nude	Nude	Nude
Operating data							
Electron collector (V)	+160	+160	+160	+150	+150	+150	+150
Filament current (A)	2–3	2–3	2–3	3	3	3	3
Cathode voltage (V)	+35/45	+35/45	+35/45	+50	+50		
Ion collector potential	0	0	0	0	0	0	0
Other electrode				Modulator	Modulator		
Vacuum connection	Flange, viton seal	Flange, gold seal	Flange, copper seal	Flange, gold wire seal	Flange, copper ring	Flange, gold wire seal	Flange, copper ring

MANUFACTURER:	Cenco	Cooke		Leybold-Heraeus	Mill Lane Engineering	Varian	Veeco
Model no.	VG1A/3	3P	VG1A	IR32G IR32GS	ML21	Millitorr	GS86
Lower pressure limit (torr) (X-ray limit)	10^{-8}	2×10^{-5}	10^{-8}	$<1 \times 10^{-6}$	10^{-8}	1×10^{-6}	10^{-5}
Sensitivity (torr^{-1})		0·4	10	5		0·5	
at mA		0·1	0·1	1		0·1	
Upper pressure limit of measuring range (torr)	10^{-3}	6×10^{-1}	8×10^{-3}	1×10^{-2}	10^{-4}		3
Electrodes construction							
Electron collector			Molybdenum	Nickel	Nickel, tungsten, thorium	Platinum/iridium wire on stainless steel	
Filament			Tungsten or iridium	Tungsten	Tungsten	Thoria coated iridium	Thoriated iridium
Ion collector	Platinum		Tungsten	Nickel	Molybdenum	Tungsten	
Envelope	Glass			Soft glass	Glass	Nude	Hard glass
Operating data							
Electron collector (V)			+150	+190		+180	
Filament current (A)				1–1·5		1–6	
Cathode voltage (V)			3·5	+30		0–11	
Ion collector potential				0		0	
Other electrode							
Vacuum connection			Nude + flange or 12·7, 19·05 25·4 mm tubulation				

Hot cathode ionization gauge control units

MANUFACTURER:	AEI		Balzers		Cenco
Model no.	VC9	VC20	IMG010	IMG U2	94197/8
Model number of associated gauge heads	Various	Various	IMR 110, 111, 112 120, 124, 125		
Pressure range (torr)	10^{-3}–10^{-7}	0·25–10^{-11}	1–10^{-9}	10^{-3}–10^{-12}	1 x 10^{-3}–2 x 10^{-10}
Number of scales	1	2	3 (linear and log)	2	
Scale factors between linear ranges	1 decade	1 decade			
Range of electron emission currents (mA)	0·5	0–10	0·005–0·05	0·01–10	
Stability of electron emission control	1%	1%	< 1% FSD	< 0·5%	
Ion current amplifier					
Current range	10^{-9}–5	10^{-10}–10^{-5}			
Response time (s)					
Zero drift					
Accuracy	1%	10%			
Readout	Meter or recorder	Meter or recorder			Meter or recorder
Outgassing method	None	Yes			
Mains input (V)	110–240	110–240	115, 220 or 240	115, 220 or 240	115
Dimension (m)	0·38 x 0·5 cabinet (or panel)	0·43 x 0·18 cabinet (or rack)	0·48 x 0·13 x 0·25	0·48 x 0·13 x 0·25	

MANUFACTURER:	Cooke			Denton
Model no.	IGC18	IGC20	IGC19	DVG5
Model number of associated gauge heads	SP80	BA60	SP80, BA60	BA
Pressure range (torr)	10^{-1}–10^{-5}	10^{-4}–10^{-10}	10^{-1}–10^{-10}	10^{-3}–2×10^{-10}
Number of scales	5	7	10	6
Scale factors between linear ranges	10	10	10	
Range of electron emission currents (mA)	0·1	0·01-10	0·01–10	
Stability of electron emission control	1%	1%	1%	
Ion current amplifier				
Current range				
Response time (s)	1	1	1	
Zero drift	½%	½%	½%	
Accuracy	1%	1%	1%	
Readout	Analogue (digital optional)	Analogue (digital optional)	Analogue (digital optional)	
Outgassing method	Resistance or electron bombardment	Resistance or electron bombardment	Resistance or electron bombardment	
Mains input (V)				
Dimensions (m)				

MANUFACTURER:	Edwards High Vacuum		Granville-Phillips	
Model no.	5M	Series 70, Ion 7	Series 260	Series 224
Model number of associated gauge heads	IG4G, IG4M	IG5, IG6	BA	WL7903
Pressure range (torr)	10^{-3}-10^{-11}	10^{-3}–10^{-10}	1×10^{-3}–2×10^{-10}	1–4 x 10^{-6}
Number of scales		3	6 (linear)	6 (linear)
Scale factors between linear ranges				
Range of electron emission currents (mA)	0·1, 1, 10	1	0·1-10	0·002–0·25
Stability of electron emission control	1% for up to ±10% supply variation	1% for ±10% supply variation		
Ion current amplifier				
Current range				
Response time (s)	0·25 (at best)	0·25 (at best)		
Zero drift	<3% per 24 hours	<3% per 24 hours		
Accuracy	10^{-4}–10^{-8} A within 4%, 10^{-9}–10^{-11} A within 5%	5% of scale reading		
Readout		0–10 mV		
Outgassing method	Electron bombardment	Electron bombardment		
Mains input (V)	100/125 or 200/250	100/125 or 200/250	105/125 or 210/250	105/125 or 210/250
Dimensions (m)	0·49 x 0·26 x 0·39	0·22 x 0·13 x 0·26	0·44 x 0·09 x 0·29	0·48 x 0·09 x 0·29

MANUFACTURER:	Leybold-Heraeus				Nuclide	RIBER
Model no.	Ionivac IM30	IM10	IM20	Ionivac IM50	IG-12	JBA10A
Model number of associated gauge heads	IR32G, IR32GS IR36GA, IE34 IE36, IR32GA	IR10, IE10	IR20, IE20	IR50, IE50	BA	JBA11V, JBA11I
Pressure range (torr)	10^{-2}–10^{-8}	1–5 x 10^{-7}	1 x 10^{-2}–2 x 10^{-10}	3 x 10^{-4}–2 x 10^{-12}		1·1 x 10^{-3}–2 x 10^{-10}
Number of scales	8 (7 linear 1 log)	7 (1 log)	7 linear	9 (8 linear 1 log)		
Scale factors between linear ranges				0·3		
Range of electron emission currents (mA)	0·7–1·3			1–5		0·25–1·1
Stability of electron emission control						
Ion current amplifier						
Current range						
Response time (s)						
Zero drift						
Accuracy						
Readout	0–10 V	0–10 V	0–10 V	0–10 V		Galvanometer
Outgassing method	Electron bombardment	Resistance heating ⟶		Electron bombardment		Electron bombardment
Mains input (V)	110 or 220/240	110, 240	110, 240	110 or 220		220 ± 10%
Dimensions (m)	0·26 x 0·13 x 0·27	0·26, 0·13, 0·27 ⟶		0·26 x 0·13 x 0·27		

MANUFACTURER:	RIBER *cont'd*	Torr Vacuum Products				
Model no.	JBA10N	PS100	PS200	PS400	PS500	PS1000
Model number of associated gauge heads	JBA11V, JBA11I	IG4336	IG4336	IG4336	IG4336	IG4336
Pressure range (torr)	$1{\cdot}1 \times 10^{-3}$–2×10^{-10}	10^{-3}–10^{-9}	10^{-3}–10^{-9}	10^{-2}–2×10^{-10}	10^{-3}–2×10^{-11}	760–10^{-11}
Number of scales		1 (log)	1 (log)	6 linear ranges	7 linear ranges	8 linear ranges or log scale
Scale factors between linear ranges						
Range of electron emission currents (mA)	0·25–1·1	10	10	1–10	1–10	1–10
Stability of electron emission control						
Ion current amplifier						
Current range						
Response time (s)						
Zero drift		$<2\%$	$<2\%$			
Accuracy						
Readout	Digital					
Outgassing method	Electron bombardment	Radiant heating	Radiant heating	Radiant heating	Radiant heating	Radiant heating or electron bombardment
Mains input (V)	220 ± 10%	115	115	115	115	115
Dimensions (m)						

MANUFACTURER:	Twentieth Century Electronics	UHV	Ultek	
Model no.	AIG50*	PCL1	DIG605-0010	60-760/761
Model number of associated gauge heads	AIG50 00 008 0002 00 008 0003 00 008 0004	IGF1		
Pressure range (torr)	10^{-3}-10^{-10}	10^{-3}–10^{-8}	10×10^{-3}–2×10^{-12}	10–10^{-11}
Number of scales	9 (linear)	1 (linear) + 6 range switches	Automatic range changing with digital display	7 linear + thermocouple
Scale factors between linear ranges	10	10	5 digits, 3 places and exponent	x1 and 3·5
Range of electron emission currents (mA)	0–1	1	0·01-10	0·01–10
Stability of electron emission control	1%	$\leqslant$ 2% for 24 hours	$<$1%	
Ion current amplifier				
Current range	10^{-4} A–10^{-12} A			10^{-5} A–$3{\cdot}5 \times 10^{-12}$ A
Response time (s)	0·2			
Zero drift		$\leqslant$2% for 24 hours		0·03% of FS/h or 5×10^{-13} A/h
Accuracy	1%			± 1%
Readout	Analogue 10 V positive and internal meter	Recorder	BCD output optional	Meter and 10 mV recorder output
Outgassing method	External bakeout heater	None	Electron bombardment	Electron bombardment
Mains input (V)	105/120 or 210/240	250	115 or 220	105/125 or 210/250
Dimensions (m)	0·3 x 0·2 x 0·2	Half rack	0·48 x 0·08 x 0·36	
Ancillary units	External scan unit		Oscillocorder process controller	

* This instrument is a combined total and partial pressure gauge. [See also under Residual gas analysers, quadrupole type instruments].

MANUFACTURER:	Vactorr	Vactronic	Vacuum Generators			
Model no.	10479	410	IGP3	TCS6	IGP4	TCS7
Model number of associated gauge heads		VG1A527	VIG10	VIG10	VIG20	VIG20
Pressure range (torr)	10^{-3}–10^{-5}	10^{-2}–10^{-11}	10^{-2}–10^{-11}	10^{-2}–10^{-11}	10^{-3}–10^{-9}	10^{-3}–10^{-9}
Number of scales	1	9 (8 linear 1 log)	Log	Log	Log	Log
Scale factors between linear ranges						
Range of electron emission currents (mA)			0·1, 1, 10	0·1, 1, 10	4	4
Stability of electron emission control			<1%			
Ion current amplifier						
Current range			10^{-14}–1 mA	10^{-14}–1 mA	10^{-10}–10^{-4} A	10^{-10}–10^{-4} A
Response time (s)						
Zero drift			Nil	Nil	Nil	Nil
Accuracy			Better than 10%	Better than 10%	Better than 10%	Better than 10%
Readout			Analogue 10 mA per decade →			
Outgassing method	120		Electron bombardment →			
Mains input (V)	120		110/120 or 220/240 →			
Dimensions (m)	0·28 x 0·09 x 0·24	0·21 x 0·13 x 0·28	Front panel 0·48 x 0·13 for 0·48 rack mount Control unit 0·44 x 0·13 x 0·27			
Ancillary units	Carries thermocouple gauge		'A' model controllers IGP3A, TCS6A, IGP4A, TCS7A GLD1 leak detector			

MANUFACTURER:	Varian		
Model no.	NRC830	NRC831 (ionization + thermocouple)	NRC836 (ionization + thermocouple)
Model number of associated gauge heads	NRC507, NRC563	NRC563, NRC531-thermocouple	
Pressure range (torr)	1×10^{-3}–1×10^{-8}	1×10^{-3}–2×10^{-9} (ionization) 1–1×10^{-3} (thermocouple)	1–2×10^{-9}
Number of scales	1 (log)	3 (linear or log)	3
Scale factors between linear ranges			
Range of electron emission currents (mA)			
Stability of electron emission control			
Ion current amplifier			
Current range			
Response time (s)			
Zero drift	± $\frac{1}{4}$ decade or ± 5% of FS per 10°C after 50 hours	Nil	
Accuracy	3%	±2% FS (meter) ±5% (resistors)	
Readout	0–10 mV	10 mV FS (thermocouple) 0–13 mV FS (ionization)	
Outgassing method		Resistance heating	
Mains input (V)	115 or 230	105/130 or 210/260	
Dimensions (m)	0·21 x 0·14 x 0·18	0·31 x 0·14 x 0·19	
Ancillary units	Holding and switching supply (allowing use of single control units for up to five gauges); NRC897 (three station selector)		

MANUFACTURER:	**Varian**			
Model number	NRC852	NRC853 (ionization + thermocouple)	NRC851	Dual range ionization gauge
Model number of associated gauge heads	NRC524-1	NRC524-1 NRC531-thermocouple	NRC524-1	Millitorr and UHV
Pressure range (torr)	1×10^{-9}–5×10^{-3}	2–1×10^{-9}	10^{-1}–5×10^{-5}	1–<2×10^{-11}
Number of scales	1	3		Mirrored linear scale
Scale factors between linear ranges				12 major division 50 μA meter movement
Range of electron emission currents (mA)				0·002–0·1 (millitorr) 0·01–10 (UHV)
Stability of electron emission control				± 10% of line voltage = ± 0·25% FS
Ion current amplifier				
Current range				
Response time (s)				
Zero drift				
Accuracy				
Readout	0–10 mV	0–10 mV		Meter reading – 1·2 V FS
Outgassing method				Electron bombardment
Mains input (V)	115 or 230	115 or 230		120 or 240
Dimensions (m)	0·21 x 0·14 x 0·17	0·21 x 0·14 x 0·17		0·48 x 0·09 x 0·38
Ancillary units	Holding and switching supply (allowing use of single control unit for up to five gauges); NRC897 (three station selector)			Has two process control channels

MANUFACTURER:	Veeco		
Model no.	RG81	RG83	RG84
Model number of associated gauge heads			
Pressure range (torr)	2×10^{-3}–1×10^{-10}	1×10^{-2}–2×10^{-11}	1×10^{-2}–2×10^{-11}
Number of scales	1 (log)	2 (log and linear)	2 (log and linear)
Scale factors between linear ranges	8 decades	8 decades each	8 decades each
Range of electron emission currents (mA)		0·2–4	0·2–4
Stability of electron emission control	± 2% FS per 24 hours ± 1% FS with line voltage ± 0·1% FS per °C	± 2% for FS per 24 hours ± 1% FS with line voltage ± 0·1% FS per °C	± 2% FS per 24 hours ± 1% FS with line voltage ± 0·1% FS per °C
Ion current amplifier			
Current range			
Response time (s)			
Zero drift			
Accuracy			
Readout			
Outgassing method	Resistance heating	Resistance heating	Electron bombardment
Mains input (V)	105/130 or 210/260	105/130 or 210/260	105/130 or 210/260
Dimensions (m)	0·21 x 0·13 x 0·28	0·21 x 0·13 x 0·28	0·21 x 0·13 x 0·28
Ancillary units	RG88 quarter rack controller centre, controls up to six external circuits		

MANUFACTURER:	**Veeco** *cont'd*	
Model no.	RG86	RGS7
Model number of associated gauge heads	GS86	RG75, TG75, BGT75
Pressure range	$3–1 \times 10^{-5}$	$1 \times 10^{-2}–2 \times 10^{-12}$
Number of scales	2 (log and linear)	1 linear
Scale factors between linear ranges	5 decade linear, 6 decade log	10 decades
Range of electron emission currents (mA)		$10^{-3}–10$
Stability of electron emission control	Linear: ± 2% FS per 24 hours; ± 1% FS with line voltage; ± 0·1% FS per °C Log: ± 10% FS per 24 hours; ± 10% FS with line voltage; ± 0·1% per decade per °C (output); ± 6% of reading per °C (calibration)	better than 1% over 50 hours
Ion current amplifier		
Current range		
Response time (s)		
Zero drift		
Accuracy		
Readout		100 mV full scale or less
Outgassing method	Electron bombardment	Resistance heating
Mains input (V)	105/130 or 210/260	100/130 V (convertible to 200/260)
Dimensions (m)	0·21 x 0·13 x 0·27	0·21 x 0·13 x 0·43
Ancillary units	RG88 quarter rack controller centre, controls up to six external circuits	

2.2.7 Residual gas analysers

Residual gas analysers are low resolution mass spectrometer instruments adapted to measure partial pressure of gases occurring in vacuum systems and vacuum processing plant. Instruments are generally of a number of types which may be classified according to the mass analysis principle used as: magnetic deflection instruments using 180°, 90° or 60° sector fields; cycloidal; time of flight; radio frequency; Omegatron; quadrupole; monopole. They are supplied in the form of an analyser head which may be mounted on the vacuum system or processing chamber with separate electronic control and measuring gear.

Mass spectrometer instruments with integral pumping systems as used for more general chemical analysis applications and particularly high resolution larger instruments are included here.

Magnetic deflection instruments

MANUFACTURER:	AEI		Aerovac		
Model no.	MS10	Minimass	AVA2	AVA200	AVA610
Sector angle	180°	180°	60°	60°	60°
Total mass range (amu)	0–200	0–200	2–70, 1–200 with 6 kg magnet	2–200	2–300
Number of mass ranges and coverage of each	5: 2, 3, 4, 12–45 36–200	4: 2, 4, 12–60 48–200	2–11, 12–70 (2)	2–11, 12–200	
Resolution (amu)	40–100 depending on slits		35, 50 with 6kg magnet	100, 170 with 6 kg magnet	150, 200 with 6 kg magnet
Sensitivity (A/torr)	5×10^{-5}	2×10^{-5}	10^{-10} for N_2	10^{-12} for N_2	10^{-13} for N_2
Minimum detectable partial pressure (torr)	5×10^{-13}	5×10^{-11}			

Partial pressure sensitivity					
Maximum pressure (torr)	10^{-3}				
Speeds of scanning	Variable	Variable	30 or 120 s/mass range	$2\frac{1}{2}$ or 10 min/mass range	15 or 25 min/mass range
Scan law					
Maximum temperature					
Operating (°C)	120	120			
Bakeout (°C)	400	400	400	400	400
Analyser head dimensions (mm)	20	10			
Analyser weight (kg)	6·8	0·91	38·5	52·2	52·2
Analyser dimensions			0·48 x 0·38 x 0·23 m	0·48 x 0·38 x 0·23 m	0·48 x 0·38 x 0·23 m
Ion path radius	50 mm radius	10 mm radius			
Magnet weight (kg)	15·88	9·98	8		
Magnet dimensions (m)					
Magnet strength (tesla)	0·148	0·4			
Ion source					
Nude or immersed operation	Immersed	Immersed			
Electron current range	10–300 μA	6–60 μA			
Cathode material	Rhenium	Rhenium	Tungsten	Tungsten	Tungsten
Detector					
Multiplier					
Gain	None	None			
Readout					
Built in	Yes	Yes			
Accessories	External meter and recorder	External meter and recorder			

MANUFACTURER:	Edwards	Nuclide	Thomson CSF	VG Micromass
Model no.	E180	3-60 Sectorr	THN 205 SE	Micromass 1
Sector angle	180°	60°	180°	180°
Total mass range (amu)	1–105	0–640	2–350	2–240
Number of mass ranges and coverage of each	5: 1, 2, 3, 4, 5–105	3: 0–64, 0-320, 0–640		5: 2, 3, 4, 12–60 48–240
Resolution (amu)	44		>60	22 for 120 μA
Sensitivity (A/torr)	Total pressure 6×10^{-4} Partial pressure 4×10^{-5}			3×10^{-4} for N_2
Minimum detectable partial pressure (torr)				3×10^{-12}
Partial pressure sensitivity	10^{-4}–10^{-10}			10 ppm
Maximum pressure (torr)				5×10^{-3}
Speeds of scanning	1 amu per 10 second	0–± 100 amu/min or 0–± 1000 amu/min	2·5 s to 25 min	1 amu/s or slower
Scan law				Linear
Maximum temperature				
Operating (°C)				200
Bakeout (°C)	350 with magnet 400 without magnet		400	400
Analyser head dimensions	0·22 x 0·18 x 0·16 m			10 mm radius
Ion path radius	15 mm			
Analyser weight (kg)	20·4 (head: 10)	9·98		0·9
Analyser dimensions	0·25 x 0·49 x 0·34 m	0·60 x 0·16 x 0·22 m		0·11 x 0·07 m
Magnet weight (kg)	6·1	22·2		10
Magnet dimensions		0·10 x 0·24 x 0·17 m		0·17 x 0·17 x 0·09 m
Magnet strength (tesla)	0·3	0·8		0·4
Ion source				
Nude or immersed operation	Immersed		Immersed	Immersed
Electron current range	8, 40, 120 μA	0–1000 μA		0·5–500 μA
Cathode material				W, Re, THO_2 or LaB_6
Detector				
Multiplier	No			No
Gain				
Readout				
Built in	10 mV recorder socket			Yes
Accessories	High sensitivity head Fast scan attachment Sampler system	Oscilloscope Potentiometric recorder		

MANUFACTURER:	VG Micromass *cont'd*		Veeco
Model no.	Micromass 2	Micromass 6	GA-4R
Sector angle	180°	90°	60°
Total mass range (amu)	2–240	2–500	1–760 (electromagnet) 2–30 (permanent magnet)
Number of mass ranges and coverage of each	5: 2, 3, 4, 12–60, 48–240	1	12–300 std permanent magnet 2–50 with magnet shunt 1–760 with electromagnet
Resolution (amu)	44 at 45 μA	50–400 adjustable at 200 μA	Unit resolution at mass 100
Sensitivity (A/torr)	3×10^{-5} (N_2)	10^{-4} (N_2)	
Minimum detectable partial pressure (torr)	3×10^{-11}	5×10^{-4} with multiplier 1 amu per 45	1×10^{-13} (multiplier) 1×10^{-10} (electrometer)
Partial pressure sensitivity	10 ppm	1 ppm	
Range of linearity	5×10^{-4}	3×10^{-5}	To 10^{-5} (multiplier) To 10^{-4} (electrometer)
Speeds of scanning	1 amu/s or slower	1 amu/10^{-3} s or slower (+multiplier)	100 ms–40 min
Scan law	Linear	Linear or exponential	
Maximum temperature			
Operating (°C)	200	200	
Bakeout (°C)	400	400	400 (tube), 300 (multiplier) 350 (electromagnet)
Ion path radius	10 mm radius	62 mm radius	
Analyser weight (kg)	0·9	1·4	
Analyser dimensions	0·11 x 0·07 m	0·20 x 0·30 x 0·12 m	0·61 x 1·84 x 0·58 m
Magnet weight (kg)	10	1·6–13	5·44 (electromagnet)
Magnet dimensions (m)	0·17 x 0·17 x 0·09		0·12 x 0·19 x 0·09
Magnet strength (tesla)	0·4	0·23–0·85	0·446, 0·183
Ion source			
Nude or immersed operation	No	Yes	Nude version available
Electron current range	0·5–45 μA	0·5–400 μA	
Cathode material	W, Re, THO_2 or LaB_6	W, Re, THO_2 or LaB_6	
Detector			
Multiplier	No	Optional	
Possibility and gain		Up to 10^6	
Readout			
Built in	Yes	Yes	3 decade linear scale
Accessories	Faraday plate available	Faraday plate available	

Quadrupole type instruments

MANUFACTURER:	Extranuclear Export	Ion Equipment	RIBER			
Model no.	Spectrel model 275-5	100B	QS50	QS100	QS200	QS300
Total mass range (amu)	1–1000	1–250	1–60	1–100	1–200	1–300
Number of mass ranges and coverage of each	1	1	1	1	1	1
Resolution	Unit resolution at mass 1000	500 at 250 amu	100	200	400	600
Sensitivity (A/torr)	1000 for N_2 M/AM = 56	10^{-3} for N_2; 300 with multiplier				
Minimum detectable partial pressure (torr)	1×10^{-16}	10^{-14}	10^{-11}	10^{-13}	10^{-13}	10^{-13}
Partial pressure sensitivity						
Maximum pressure (torr)						
Speeds of scanning	12 ranges	75 ms–120 s for complete mass range	1 s–15 min		50 ms–30 min	50 ms–30 min
Scan law	Linear					
Maximum temperature						
Operating (°C)	200					
Bakeout (°C)	350					
Analyser head dimensions	9·5 mm diameter 125 mm long	300 mm long 50 mm diameter				
Analyser weight (kg)		34				
Analyser dimensions						
Ion source						
Nude or immersed operation						
Electron current range (mA)	0–50		0·2–2	0·2–2	0·2–2	0·2–2
Cathode material	Tungsten or iridium					
Detector						
Multiplier	21 stage Cu/Be venetian blind					
Possibility and gain	10^6					
Readout						
Built in	Oscilloscope					
Accessories	Oscillograph, two channel strip chart, computer					

MANUFACTURER:	RIBER *cont'd*		Twentieth Century Electronics	
Model no.	QMM17	QMM51	Q806	AIG50*
Total mass range (amu)	1–300	1–750	2–200 (Optional 1–200)	2–50
Number of mass ranges and coverage of each	2	3	2: 2–100, 12–200	1: 2–50
Resolution	800	1200	100	M/ΔM > M
Sensitivity (A/torr)				
Minimum detectable partial pressure (torr)	10^{-15}	10^{-15}	$< 10^{-12}$ for N_2 at > 1 s/amu	10^{-9} for N_2
Partial pressure sensitivity			10 A/torr for N_2	10^{-4} A/torr for N_2
Maximum pressure (torr)			10^{-4}	10^{-3}
Speeds of scanning	10 ms–30 min	10 ms–30 min	1 ms per amu, variable to 10 min per spectrum scan	Manual or external
Scan law			Linear	Linear
Maximum temperature				
Operating (°C)			200 (RF unit water cooled)	60
Bakeout (°C)			400 (RF unit removed)	300
Analyser head dimensions			Quadrupole rods 6·35 mm diameter 127 mm long	43·7 mm diam 127 mm long
Analyser weight (kg)			5·66	0·15
Analyser dimensions			0·4 m (overall length) 110 mm o.d. flange	203 x 50·8 mm
Ion source				
Nude or immersed operation			Either	Either
Electron current range (mA)	0·3–3	0·3–3	0·1–1·5	0·1–1·5
Cathode material			Tungsten or Rhenium	Tungsten or Rhenium
Detector				
Multiplier			14 stage Be/Cu	Faraday cage
Possibility and gain			10^5	
Readout				Output meter
Built in				Yes
Accessories			Oscilloscope, UV recorder, X–Y recorder, potentiometric recorder	Scan drive unit, recorder

* Combined R.G.A. and Total Pressure Gauge.

MANUFACTURER:	Varian		
Model no.	QUAD250B	QUAD 1100A, QUAD 1110A	QUAD 1200, QUAD 1210
Total mass range (amu)	1–800	2–300	1–100
Number of mass ranges and coverage of each	4: 1–100, 5–250, 10–500, 500–800	1 scan (reduced coverage available)	1 scan (reduced coverage available)
Resolution	$M/\Delta M = \frac{1}{2}$ to 10 M adjustable	$\mu/\Delta M \geqslant 2M$ (ΔM at $\frac{1}{2}$ height)	$\mu/\Delta M \geqslant 2M$
Sensitivity (A/torr)	100 for N_2	100 for N_2 or, with multiplier, 2×10^{-4}	2×10^{-4} for N_2 at unit resolution
Minimum detectable partial pressure (torr)			
Partial pressure sensitivity	10^6 (1 ppm)		
Maximum pressure (torr)	10^{-4}	10^{-4}	
Speeds of scanning	30 ms–15 min 11 ranges	0·1, 0·2, 0·5, 1, 2, 5, 10, 20, 50, 100 200, 500, 1000 s	0·1–500 s 12 ranges
Scan law			
Maximum temperature			
Operating (°C)			
Bakeout (°C)	400	400	400
Analyser head dimensions			
Analyser weight (kg)			
Analyser dimensions			
Ion source			
Nude or immersed operation			
Electron current range (mA)			
Cathode material			
Detector			
Multiplier			
Possibility and gain			
Readout			
Built in			
Accessories			

Other types

MANUFACTURER:	Aerovac		
Model no.	Monitorr 702	Monitorr 710	Monitorr 712, 722
Type			
Total mass range	Up to 100 (300 with high mass tube)	Up to 100 (300 with high mass tube)	Up to 100 (300 with high mass tube)
Number of mass ranges and coverage of each		x1 and x10	
Resolution			
Sensitivity ($torr^{-1}$)			
Minimum detectable partial pressure (torr)	5×10^{-11} (N_2)		5×10^{-11} (N_2)
Partial pressure sensitivity	20 ppm		20 ppm
Maximum pressure (torr)			
Speeds of scanning		1–10 s/specimen	1 s/mass–10 s/mass (model 712)
Scan law	Automatic repetitive scan		Automatic repetitive scan (model 722)
Maximum temperature			
Operating (°C)			
Bakeout (°C)			
Analyser head dimensions (m)			
Analyser weight (kg)			
Analyser dimension (m)			
Magnet weight			
Magnet dimensions			
Magnet strength (tesla)			
Ion source			
Nude or immersed operation			
Electron current range (mA)			
Cathode material			
Detector			
Multiplier			
Possibility and gain			
Readout			
Built in	Analogue meter 0–100 mV, strip chart terminals (digital optional), BCD computer compatible terminal		
Accessories			

MANUFACTURER:	Leybold-Heraeus	Ultek	Veeco
Model no.	Topatron B	PPA	SPI-10
Type	Radio frequency	Cycloidal	Monopole
Total mass range	2-150	2–100	1–200
Number of mass ranges and coverage of each	1	1	3: 1–200, 1–50, 40–65
Resolution	30 at 10%	M/ΔM = 50	10% valley up to 65 amu
Sensitivity ($torr^{-1}$)			
Minimum detectable partial pressure (torr)	1×10^{-11}	5×10^{-12} (N_2)	1×10^{-10}
Partial pressure sensitivity	10^{-3}		
Maximum pressure (torr)	1×10^{-3}		5×10^{-10}–5×10^{-4}
Speeds of scanning	1 s to 20 min	300, 1, 3 ms/amu	3 ms–5 min
Scan law			linear over 1–50 amu
Maximum temperature			
Operating (°C)			
Bakeout (°C)	400	400	
Analyser head dimensions (m)	Generator: 0·13 x 0·09 x 0·04 Preamplifier: 0·05 diam., 0·14 long		
Analyser weight (kg)	20	11·34	1·8 (tube), 4·3 (+ RF head)
Analyser dimensions (m)	0·49 x 0·23 x 0·4	0·3 x 0·15 x 0·15	0·48 x 0·18 x 0·46 (control unit)
Magnet weight (kg)		9·07	
Magnet dimensions (m)		0·11 x 0·15 x 0·07	
Magnet strength (tesla)			
Ion source			
Nude or immersed operation	Yes		Nude
Electron current range			
Cathode material			W/Re filaments (2)
Detector			
Multiplier			
Possibility and gain			
Readout			
Built in			Linear scale, oscilloscope
Accessories			

2.2.8 Mass spectrometer leak detectors (with integral pumping systems)

The leak detectors included here are mass spectrometers adjusted to respond only to the search gas used and provided with an integral pumping system for direct vacuum testing of components and equipment, or for pressure testing using a sampling probe or sniffer to collect search gas from any leaks.

MANUFACTURER:	Aerovac		Alcatel	
Model no.	685 series	686 series	ASM4 series	ASM7 series
Search gas			Helium	Helium
Minimum detectable leak (+ response time) (torr litre/second)	10^{-13}	10^{-13}	4×10^{-12}	8×10^{-11}
Sensitivity (+ response time)				
Response and clean up time (s)			Response 1 Clean up 1	
Corresponding volume rate of flow at detector inlet (litre/second)				
Maximum volume rate of flow at detector inlet (litre/second)			4·5 for air	8
Auxiliary and roughing pump systems	2 inch and 1 inch diffusion 0·9 CFM mechanical	2 systems, each of 2 inch and 1 inch diffusion 2·1 CFM mechanical	Built in fore pump Roughing pump and secondary vacuum monoblock	
Minimum detectable concentration ratio	1 ppm	0·02 ppm		
Maximum operating pressure (torr)	5×10^{-10}	5×10^{-10}		10^{-4}
Sensitivity ranges				
Detector weight (total) (kg)	340·2	453·6	182	Vacuum unit 51 Electronics unit 9
Dimensions (m)			Base 0·6 x 0·55	Vacuum unit 0·44 x 0·24 x 0·45 Electronics unit 0·16 x 0·36 x 0·27
Auxiliary systems Calibrated leaks Pumping systems				

MANUFACTURER:	Edwards High Vacuum				Leybold-Heraeus	
Model no.	LT102	LT103	LT104	E180	Ultratest B	M
Search gas	Helium	Helium	Helium	Helium, hydrogen	Helium	Heluim
Minimum detectable leak (+ response time) (torr/litre/second)	10^{-11} or 10^{-12}	10^{-10} or 10^{-11}	10^{-11} or 10^{-12}	See 2.2 for details	4×10^{-12} < 1 s	2×10^{-12} < 1 s
Sensitivity (+ response time)	Response < 2 s for 63% FSD	Response < 2 s for 63% FSD	Response < 1 s for 63% FSD			
Response and clean up time (s)	Clean up < 3	Clean up < 2	Clean up < 2			
Corresponding volume rate of flow at detector inlet (litre/second)						
Maximum volume rate of flow at detector inlet (litre/second)	4	10	10		4	20
Auxiliary and roughing pump systems					Diffusion pump rotary backing pump	
Minimum detectable concentration ratio					0·05	0·5 ppm
Maximum operating pressure (torr)	5×10^{-4}	5×10^{-4}	5×10^{-4}		760	760
Sensitivity ranges					6	7
Detector weight (total) (kg)	55				235	Portable
Dimensions (m)	0·64 x 0·4 x 0·43	0·60 x 0·68 x 0·91	0·60 x 0·65 x 0·91		0·77 x 1·0 x 0·66	
Auxiliary systems						
Calibrated leaks		Additional high capacity roughing pump			Available	Available
P Pumping systems						

MANUFACTURER:	Twentieth Century Electronics		
Model no.	LD 401-001	LD 401-002	LD 401-004
Search gas	Helium	Helium 3 or 4	Helium/Argon
Minimum detectable leak (+ response time) (torr/litre/second)	10^{-11}	10^{-11}	10^{-11}
Sensitivity (+ response time)	50% FSD <1 s	50% FSD <1 s	50% FSD <1 s
Response (R) and clean up (C) time (s)	R < 1, C < 5	R < 1, C < 5	R < 1, C < 5
Corresponding volume rate of flow at detector inlet (litre/second)	<1	<1	<1
Maximum volume rate of flow at detector inlet (litre/second)	10	10	10
Auxiliary and roughing pump systems	Auxiliary of 100 l/min needed for vessels over 100 l	Auxiliary of 100 l/min needed for vessels over 100 l	Auxiliary pump of 100 l/min needed for vessels over 100 l
Minimum detectable concentration ratio	10^{-6}	10^{-6}	10^{-6}
Maximum operating pressure (torr)	10^{-4} (at head)	10^{-4} (at head)	10^{-4} at head
Sensitivity ranges	5	5	5
Detector weight (total) (kg)	163·29	163·29	163·29
Dimensions (m)	0·66 x 1·04 x 0·68	0·66 x 1·04 x 0·68	0·66 x 1·04 x 0·68
Auxiliary systems			
Calibrated leaks	Glass and metal leaks in range 10^{-9}–10^{-3} torr litre/second	Glass and metal leaks in range 10^{-9}–10^{-3} torr litre/second	Glass and metal leaks in range 10^{-9}-10^{-3} torr litre/second
Pumping systems			As required

MANUFACTURER:	Twentieth Century Electronics *cont'd*		Varian
Model no.	LD 800-01	LD800-02	925 Manumatic
Search gas	Helium	Helium 3 or 4	Helium 4
Minimum detectable leak (+ response time)	10^{-11}	10^{-11}	$1 \times 5 \times 10^{-11}$
Sensitivity (+ response time)	50% FSD <15	50% FSD <15	8×10^{-12} (2% FSD)
Response (R) and clean up (C) time (s)	$R < 1, C < 5$	$R < 1, C < 5$	R = 1, complete cycle: <20
Corresponding volume rate of flow at detector inlet (litre/second)	<1	<1	20 for helium at test port
Maximum volume rate of flow at detector inlet (litre/second)	15	15	10 for air at test port
Auxiliary and roughing pump systems	Built in roughing pump and sampling valve	Built in roughing pump and sampling valve	120 l/min mechanical pump
Minimum detectable concentration ratio			8×10^{-7}
Maximum operating pressure	10^{-4} (at head)	10^{-4} (at head)	2×10^{-4}
Sensitivity ranges	5	5	7 ($6 \times 10^{-5} \times 2 \times 10^{-11}$)
Detector weight (total) (kg)	139·71	139·71	272·16
Dimensions (m)	0·60 x 1·27 x 0·60	0·60 x 1·27 x 0·60	1·23 x 1·02 x 0·64
Auxiliary systems			
Calibrated leaks	Glass and metal leaks in range 10^{-9}–10^{-3} torr/litre/second	Glass and metal leaks in range 10^{-9}–10^{-3} torr/litre/second	
Pumping systems	As required	Only needed in special cases	

MANUFACTURER:	Varian *cont'd*	
Model no.	Portable NRC925-20	NRC925-50 (manual)
Search gas	Helium	
Minimum detectable leak (+ response time) (torr/litre/second)	8×10^{-12}	3×10^{-12}
Sensitivity (+ response time)		
Response (R) and clean up (C) time (s)	R < 2, <10 full recovery	
Corresponding volume rate of flow at detector inlet (litre/second)		
Maximum volume rate of flow at detector inlet (litre/second)		
Auxiliary and roughing pump systems	See below	HSA-150 150 litre/second diffusion Welch 1402-B 140 litre/second Optional roughing pump Welch 1400-B 21 litre/second Fore pump
Minimum detectable concentration ratio		
Maximum operating pressure (torr)		
Sensitivity ranges	3 direct reading scales, 1 arbitrary. Range selector – 7 positions	3 direct reading scales, 1 arbitrary. Range selector – 7 positions.
Detector weight (total) (kg)	Spectrometer 3·8, indicator 6·3, electronics unit 10·8	226
Dimensions (m)	Spectrometer 0·18 x 0·11 x 0·10 indicator 0·27 x 0·18 x 0·18 electronics unit 0·48 x 0·18 x 0·21	1·02 x 1·24 x 0·64
Auxiliary systems		
Calibrated leaks	Standard model 10^{-8} cc/s, others available	
Pumping systems	Mobile vacuum system 150 litre/second diffusion, 21 litre/second fore pump	

MANUFACTURER:	Veeco		
Model no.	MS90 series	MS90 UFT	MS12 series (split sector)
Search gas	Helium (argon-neon or helium-hydrogen available)	Helium	Helium
Minimum detectable leak (+ response time) (torr/litre/second)	8×10^{-11}	8×10^{-10} (5×10^{-10} at full production speed)	4×10^{-12}
Sensitivity (+ response time)			4×10^{-12} produces at least 2% FSD with time constant (to reach 63%) of < 2 s
Response and clean up time (s)		Complete cycle: 6	Complete cycle: 7 (MS12ABP) 12 (MS12ABR), 35 (MS12AB)
Corresponding volume rate of flow at detector inlet (litre/second)			
Maximum volume rate of flow at detector inlet (litre/second)			
Auxiliary and roughing pump systems	Diffusion pump 85 litre/second Roughing pump 5·6 or 10·6 CFM Fore pump 0·7 CFM	Diffusion pump 85 litre/second Roughing pump 10·6 CFM Fore pump 0·9 CFM	Diffusion pump 85 litre/second Roughing pump 1·2 CFM Fore pump 0·7 CFM (MS12ABP has additional 10 CFM roughing pump)
Minimum detectable concentration ratio	0·1 ppm	0·1 ppm	0·01 ppm
Maximum operating pressure (torr)			3×10^{-4}
Sensitivity ranges			
Detector weight (total) (kg)	MS90A - basic unit – 141·6 MS90AB - basic unit + pumping block – 191·5 MS90ABC – 90AB + back-filling block 275·4 MS90ABD – 90AB + automatic go/no go unit 236·9		6 decades direct readout
Dimensions (m)	90A – 0·66 x 0·93 x 0·70 90AB – 0·66 x 0·93 x 0·70 90ABC – 0·38 x 0·93 x 0·70 90ABD – 0·38 x 0·93 x 0·70	1·17 x 1·09 x 0·61	124·7 0·66 x 0·56 x 0·46
Auxiliary systems			
Calibrated leaks	Model SC4	Model SC4	Models SC4 or SC5
Pumping systems	Model MS90P roughing module (17·7 CFM)		MS12AB bench model MS12ABR roll-around model MS12ABP high speed production model

2.2.9 Leak detection instruments using search gas (other than mass spectrometer leak detectors)

Two cases are distinguished (a) instruments used for the location of leaks in pressurized equipment and (b) instruments which may be used for both proving of leak tightness and the location of leaks in evacuated equipment.

(a) For leak location in pressurized equipment

MANUFACTURER:	Analytical Instruments		
Model no.	Leakmeter	SF6B	Leakseeker TC
Method of operation	Hand held probe sniffing for electron capturing tracer gas	Hand held probe sniffing for electron capturing tracer gas	Thermal conductivity
Search gas	SF_6, Freons and others	SF_6, Freons and others	Hydrogen, helium and others
Minimum detectable concentration ratio	1 in 10^9 (SF_6)	1 in 10^{11} sampling, 1 in 10^9 continuous (SF_6)	1 in 10^4 (H_2)
Corresponding minimum detectable partial pressure of search gas	1 in 10^9 (SF_6) at atmospheric	1 in 10^{11} sampling, 1 in 10^9 continuous (SF_6)	0·076
Sensitivity to search gas (torr litre/second)	1 in 10^7 or $7{\cdot}6 \times 10^{-10}$	1 in 10^9 sampling $\equiv 7{\cdot}6 \times 10^{-9}$ 1 in 10^7 continuous $\equiv 7{\cdot}6 \times 10^{-7}$	8×10^{-6} (hydrogen)
Sensitivity to other gases			8×10^{-6} (helium). Other gases as ratio of thermal conductivity
Sensitivity ranges	Attenuation down to x50	Attenuation down to x50	Attenuation down to x500
Sampling probe			
Dimensions	Hand held gun	Hand held gun	Hand held gun
Weight	0·56		0·68
Overall			
Dimensions (m)	2·7 x 2·4 x 1·3 console	3·7 x 1·7 x 2·7 console	2·58 x 2·62 x 0·9 console
Weight (kg)	4·1	4·1	1·62
Electrical supply	Mains or battery (rechargeable)	Mains or battery (rechargeable)	Battery (rechargeable)

MANUFACTURER:	Cooke	Edwards High Vacuum	Furness Controls
Model no.	Models for use with either a sputter ion pump or ion gauge.	Handy-tector D141	FC020
Method of operation		Thermal conductivity	Differential pressure
Search gas	CO_2	Any gas with thermal conducting different from that of air	Any non-corrosive non-conducting gas
Minimum detectable concentration ratio			
Corresponding minimum detectable partial pressure of search gas		3×10^{-4} (helium)	
Sensitivity to search gas (torr litre/second)			
Sensitivity to other gases			
Sensitivity ranges		4 ranges (1:30) and audio signal	
Sampling probe Dimensions Weight			
Overall Dimensions (m) Weight (kg)			
Electrical supply		Battery and charging unit	

MANUFACTURER:	Leybold-Heraeus	Veeco
Model no.	Halogen Leak Detector Model 4	Accu Probe
Method of operation	Ion emission from heated anode	
Search gas	Halogen containing gases and vapours	
Minimum detectable concentration ratio	500 mg/year (Freon)	
Corresponding minimum detectable partial pressure of search gas		
Sensitivity to search gas (torr litre/second)		
Sensitivity to other gases		
Sensitivity ranges		
Sampling probe		
Dimensions		
Weight		
Overall		
Dimensions (m)		
Weight (kg)	15·5	
Electrical supply		

(b) For leak detection in vacuum equipment

MANUFACTURER:	Balzers		Hewlett Packard	Leybold-Heraeus
Model no.	HL2	HS1	4917A, 4905A, 4918A	Halogen leak detector 2
Method of operation	Halogen leak detector	Halogen leak detector	Ultrasonic leak detectors	
Search gas	Freon 12	Freon 12		Freon, Frigen or Kaltron
Minimum detectable partial pressure of search gas (torr litre/second)	10^{-6}	10^{-6}		5×10^{-7}
Pressure range (torr)	$760–10^{-6}$	$760–10^{-6}$		
Applicable to tightness proving	Yes	Yes		No
Sensitivity ranges (μA)	3: 1, 8, 50	3: 1, 8, 50		
Weight (kg)	9·3	17·8	Portable	

MANUFACTURER:	Leybold-Heraeus *cont'd*		Ultek
Model no.	Halogen leak detector 4	164-15	603-4000
Method of operation		Produces high frequency discharge in glass vacuum system; discharging appearance indicates pressure	Accessory to ionization gauge or ion pump to amplify pressure changes when probing leak with search gas
Search gas	Freon, Frigen or Kaltron		Oxygen, argon, helium, carbon dioxide NH_2
Minimum detectable partial pressure of search (gas torr litre/second)	5×10^{-7}		1% total gas load
Pressure range	$760–5 \times 10^{-7}$	$150–10^{-3}$	$10^{-4}–10^{-10}$
Applicable to tightness proving	Yes	No	Yes
Sensitivity ranges (μA)			
Weight (kg)	15·5		

2.2.10 Quartz crystal surface density monitors

Monitor crystal and holder

MANUFACTURER:	Airco Temescal	Balzers	Kronos	
Model no.	SHO-321	QSK 211, 181, 221, 113, 114	FTT4	FTT5
Type	AT cut	AT cut	AT cut	AT cut
Size (mm)				
Active area (mm)	12·7			
Electrode material (as supplied)		Aluminium, gold, silver	Gold	Gold
Resonant frequency (MHz)	6	5	5	5
Temperature range (°C)	−20–+105	5–40		
Maximum permissible temperature (°C)	105	70	125	Operating 125 Bakeout 300
Crystal holder dimensions and cooling method (mm)	7·94 thick x 34·9 diameter Water cooled	Water cooled	Water cooled	Water cooled

MANUFACTURER:	Kronos *cont'd*			Sloan	Ultek
Model no.	FTT6	XT7050	XT5544	Sloan QC sensor head	Monoprobe
Type	AT cut	AT cut	AT cut	AT cut	
Size (mm)		17·8 diameter	13·97 diameter	12·7 diameter	
Active area (mm)		15·2 diameter			
Electrode material (as supplied)	Gold	Gold over chrome	Gold over chrome	Gold or silver	
Resonant frequency (MHz)	5	5	4·4	5	5
Temperature range (°C)		0–350	0–350		
Maximum permissible temperature (°C)	125	350	350		
Crystal holder dimensions and cooling method (mm)	Water cooled	25·4 diameter 13·97 high Water-cooled	25·4 diameter 13·97 high Water cooled	25·4 diameter 38·1 long Water cooled	Water cooled

MANUFACTURER:	Varian		
Model no.	985–7101, 7102, 7103, 7104	988-0904	988-0907
Type			
Size (mm)	17·8 diameter		
Active area (mm)	12·7 diameter		
Electrode material (as supplied)	Gold		
Resonant frequency (MHz)	5		
Temperature range (°C)	0–125		
Maximum permissible temperature (°C)	300	250	250
Crystal holder dimensions and cooling method (mm)	23·4 x 33·0 Water cooled	29 x 18 Water cooled	29 x 18 Water cooled

2.2.11 Thickness monitors

MANUFACTURER:	Airco Temescal	Balzers	Edwards High Vacuum		Kronos
Model no.	DTM-321	QSG201	FTM2	FTM2D	QM300 series
Model number of crystal	SHO-321	QSK 211, 181, 221, 113, 114			FTT 4, 5, 6
Range					
Total frequency shift	1 MHz	0–30 kHz			
Number of ranges	2	6	10	4	3
Frequency shift to produce FSD for each			calibrated in Å, 100, 300, 1000, 3000, 10 000 for densities 1 to 10 g/cm^3		
Sensitivity: frequency shift per μg loading					
Stability: frequency shift per hour		5 Hz			
Accuracy	0·05%		± 2% FSD		± 0·1% of full scale plus resolution
Output indication	Digital		Meter	Digital	Analogue or digital
Output signal		10 mV to recorder	0–5 V	0–5 V	A: 0–5 V D: Buffered 5 V four digit parallel 8421 BCD code
Ancillary units			UHV crystal holder, additional oscillator, process terminator, source current stabilizer		

MANUFACTURER:	Sloan		Ultek	Varian	
Model no.	200	DTM-4	DDM	985-7001, to 985-7013	988-0900
Model number of crystal			Monoprobe	985-7150, 985-7151, 985-7152	988-0907
Range					
Total frequency shift		0–100 kHz		1 mHz	0–300 kHz
Number of ranges	3	5		3	6
Frequency shift to produce FSD for each				Calibrated in k Å 10 100, 1000	1, 3, 10, 30, 100, 300 kHz
Sensitivity: frequency shift per μg loading					20 Hz
Stability: frequency shift per hour				8 Hz	1 x 10^{-5}
Accuracy		± 2% FS		0·01%	± 2%
Output indication			Digital	Digital	Linear
Output signal	Recorder 0–5 V	Recorder 0–4·5 V		0–10 V	
Ancillary units	Remote crystal holder, perforated optic transmission shields, DDS digital dial set point			Rate controller programmer	

2.2.12 Rate meters

MANUFACTURER:	Airco Temescal	Balzers	Edwards High Vacuum	Electrotech
Model no.	DPC-333	QRG 201A	Deposition ratemaster	P1401
Associated thickness monitor model number	DTM321	QSG 201	FTM2, FTM2D	P1001
Ranges			3; 0–0·3, 0–1, 0–3%/s of selected thickness down to 0·005 Å/s	
Output signals	0–10 V	10 mV–100 mV recorder connections		0–10 V
Ancillary units	PCM341 Multi-material programmer			Programmable power supply P2101 for evaporation source

MANUFACTURER:	Kronos		Sloan	Ultek
Model no.	RI100	RO200	DRC	DDM
Associated thickness monitor model number	QM300 or ADS200	QM300 or ADS200	DTM4	
Ranges	3, 0–20, 0–200, 0–2000, Å/s	3, 0–20, 0–200, 0–2000 Å/s	10-300 Hz/s	
Output signals	Analogue 0-5 V	Analogue 0–5 V	0–9 V recorder	
Ancillary units				

MANUFACTURER:	Varian	
Model no.	985-7014	988-0901
Associated thickness monitor model number	985-7001 to 985-7009, 985-7011	988-0900
Ranges	0·0001, 0·1, 1, 10% FSD	18 ranges, 10,000 Hz/s
Output signals	0– –10 V	0––9 V, 0––100 mV
Ancillary units	Programmer	Deposition controller, signal amplifier

Note: 1 Å = 0·1 nm; 1 kÅ = 100 nm = 0·1 μm.

2.2.13 Vacuum microbalances

MANUFACTURER:	Cahn	Mettler		Stanton Redcroft
Model no.	Cahn RH model 2500	ME21	Thermoanalyzer 1	Thermobalance TG-750
Maximum capacity (g)	100	3·1	16-42	1
Maximum weight change (mg)		1, 10, 100	10, 100, 1000	
Smallest detectable weight change (μg)	200	0·1	10	
Sensitivity		0–10 V, 0–1 V, 0–100 mV	10 mV	Switchable from 1–250 mg for full scale
Zero suppression		± 100% of all electrical ranges	± 100% of all electrical ranges and weight set expanded ranges	
Range		± 1 mg ± 10 mg ± 100 mg	10 x 100 mg, 10 x 10 mg, 10 x 1 mg	
Accuracy		± 0·2 μg ± 1 μg ± 10 μg	± 0·5 mg ± 0·1 mg ± 0·03 mg expanded range	
Manual or automatic null balance		Electrically compensated balance	Electrically compensated balance	
Temperature range (°C)		25–45	– 150–+ 2400	Up to 1000
Microbalance (temperature °C)			Thermostatically controlled at 25	
Vacuum range		10^{-6}	8×10^{-6}	
Construction	Gold plate, glass	Glass, aluminium	Stainless steel, aluminium	
Suspension		Metal wires	Knife edges	
Transducer input		Inductive	Optical	

2.2.14 Optical film thickness monitors

MANUFACTURER:	Balzers	Edwards High Vacuum	Laser Optics
Model no.	GSM210	OFM2	401
Transmission or reflection measurement	Either	Either	Either
Maximum optical path length (m)			Up to 1·83
Scale expansion	1: 10 000	x1, x2, x5, x10 (ratio tolerance ± 2%)	0–10% and 90–100% plus variable scale expansion
Optical characteristics Light source type	Model WLL201		30W tungsten lamp
Wavelength characteristics	350 nm to 600 nm (blue) 450 nm to 1100 nm (red)	400–10 000 Å	4000-7000 Å
Detector type			Quartz window diode photocell, S4 response
If light source is modulated Method of modulation	Chopper motor with slotted disk		Chopper motor
Frequency of modulation		220 Hz (50 Hz supply) 265 Hz (60 Hz supply)	
Response time (or time constant) of detector (s)		Fast 0·5 Slow 2·5	

Note: 1 Å = 0·1 nm.

2.2.15 Ionization monitors

MANUFACTURER:	Sloan
Model no.	Iotron (monitor and controller)
Evaporation rates measured	1 to 1000 Å/s of aluminium
Method of residual gas background compensation	Dual ion compartments in head, background gas enters both, evaporant enters only one. Difference measured
Range of film thickness monitored	Up to 10^6 Å of aluminium
Associated deposition process controller	SCR-P 7·5, SCR-P 25, power controllers

Note: 1 Å = 0·1 nm.

2.2.16 Deposition process controllers

MANUFACTURER:	Edwards High Vacuum	Kronos	
Model no.	Automatic evaporation control system	ADS200	ADS100
Associated thickness and rate monitors	FTM2 Monitor and rate meter	FTT series Film thickness transducers built-in	QM310 thickness monitor and FTT4 film thickness transducers built-in

MANUFACTURER:	Sloan					
Model no.	Iotron monitor and controller	AB Programmer 116-002	SCR-P 7·5 power control module	SCR-P 25 power control module	OMNI IIA	Models 5000, 1000
Associated thickness and rate monitors			DRC		Incorporates QC sensing unit	Incorporates QC sensing unit

2.2.17 Vacuum switching or relay units

MANUFACTURER:	Balzers				Edwards High Vacuum	
Model no.	VW4	VWS101	VWG5	HWS11	VS6 Relay unit	VSK6 Vacuum switch
Model number of associated gauge			NV4, NVR4, TRP010	HV2		
Pressure range (torr)	10–400	Depends on gauge providing input signal	50–5 x 10^{-3}	5 x 10^{-3}–1 x 10^{-5}	0–20 and 0–100	0–20 and 0–100

MANUFACTURER:	Varian		
Model no.	NRC889 (dual set point)	NRC810 (single point)	NRC810-2 (dual point)
Model number of associated gauge	NRC800 series controllers	NRC531	NRC531
Pressure range (torr)		0–2	0–2

2.2.18 Vacuum process controllers

MANUFACTURER	Varian	Veeco	
Model no.	NRC703 Automatic valve control	TC-9	TCDG9
Pressure control functions	Controls main HV valve, roughing valve, foreline or air-release valve	Valve sequencing, system programming, pump control, alarm actuation	
Pressure range (torr)	$2{\cdot}5\text{–}10^{-3}$	$1\text{–}10^{-3}$	$1\text{–}5 \times 10^{-6}$

2.2.19 Surface analysis instruments

Electron diffraction including low energy (LEED), high energy reflection (RHEED) and scanning (SHEED) instruments.
Auger electron spectrometers (AES) for spectrometry of secondary electrons from electron bombardment.
Photoelectron spectrometers in which excitation is by UV or X-ray photons, including electron spectroscopy for chemical analysis (ESCA) and electron impact spectroscopy (EIS) in which energy losses are measured.
Scanning electron microscopy (SEM) and Field ion microscopy (FIM).
Secondary ion mass spectrometer (SIMS) in which secondary ions resulting from ion bombardment of surface are mass analysed.

Balzers

Electron diffraction unit
Eldigraph KD-G2 (for gases), Eldigraph KD-4.

Secondary ion probe
Model SIMS: for chemical analysis of monolayers at solid interfaces.
Cylindrical analyser for Auger electrons.
Model 10-234GC.

Physical Electronics Industries

Auger spectrometers
Model 5000: Components: cylindrical-Auger optics (10-150), Auger spectroscopy electronics (50-500). Electronics unit comprises: electron gun control (11-010), system control (11-500), oscilloscope (HP1206B), X-Y recorder (HP7035B), Lock-in amplifier (PAR122), electron multiplier supply (Keithley 246). Analyser energy resolution: 0·6% or better; beam voltage 0-5000 V; beam diameter 25 μm at specimen with 1 μamp beam current.
Model 3000: Components: cylindrical-Auger optics (10-130), Auger spectroscopy electronics (50-300). Electronics unit comprises: electron gun control (11-030), system control (11-500), oscilloscope (HP1206B), lock-in amplifier (PAR120), electron multiplier supply (Keithley 246). Analyser energy resolution: better than 0·8%; beam voltage 0–3000 V; beam diameter $<$0·1 mm at specimen.

LEED/Auger electron optics systems
Models 15-180, 15-120, 10-180, 10-120 (first figure gives the port internal diameter (cm), the second gives the angle subtended by the innermost grid). The units consist of an electron source and a 4 grid hemispherical retarding field electron energy analyser.

Additional equipment
Sputter ion gun (04-131), sputter ion gun control (20-005), electron bombardment specimen heater (04-121), heater control (20-025), specimen manipulator, alkali ion gun (04-141), gun control (20-015), specimen fracture attachment (10-520), photoemission energy analyser (10-185), photoemission control (20-500), grazing incidence electron gun (04-015).

Vacuum Generators

LEED Low Energy Electron Diffraction systems.
RHEED Reflection High Energy Electron Diffraction systems.
SHEED Scanning High Energy Electron Diffraction systems.
Auger electron spectrometers
(a) in conjunction with LEED using retarding grid electron energy analyser
(b) for fast scanning or low incident beam currents using hemi-cylindrical mirror analyser
(c) in conjunction with ESCA using 150° spherical sector analyser.

ESCA Photoelectron spectrometers
(a) using X-ray radiation
(b) using UV radiation
(c) using electron bombardment.
SIMS Secondary Ion Mass Spectrometry.
FIM Field Ion Microscopy including associated equipment such as time of flight mass spectrometer (Atom Probe) and proximity or magnetically focused image converters based on channel plate electron multipliers.
SEM Scanning Electron Microscopes. Simple microscopes for routine work at 500 Ångstrom resolution (Miniscan) to very high resolution (5 Ångstrom) ultra high vacuum microscopes based on very high brightness field emission tips (Crewe source).

Varian

Auger electron spectroscopy
Auger electron optics: model 981-2607 analyser, 981-0601 control unit. Accessories: electron gun, specimen manipulator.
Model 981-0603: complete system based on a cylindrical mirror analyser.

HEED systems
Model 120: 4 to 50 kV (100 kV to special order) compact table top model for reflection electron diffraction, uhv construction.
Model 240: as 120 but bench constructed system with cabinet mounted controls.
Model 360: as 120 with large diffraction chamber 0·46 m diameter, 0·61 m high bell jar mounted on hoist.

LEED/Auger electron optics
Model 981-0127: 4 grid assembly with fluorescent screen.
Model 981-0137: 4 grid assembly with externally driven Faraday cup.

LEED systems
Model 120: compact table top model.
Model 210: bench constructed model with cabinet mounted controls.
Model 360: Large diffraction chamber 0·46 m diameter, 0·61 m high bell jar mounted on hoist.

Veeco

Photoelectron spectrometer
Model ESCA-2: X-ray or uv sources, diffusion or ion pumped; resolution less than 1 eV, sensitivity greater than 18 000 counts/second for the gold 83 V line. Computer signal averaging system available.

Scanning electron diffractometer
Model ZF1: for SHEED and RHEED.

Combined LEED/Auger
Model AUG-1: available with 3 grid and 4 grid optics, hemi-cylindrical energy analyser available on special order. Ultimate pressure of 5 x 10^{-11} Torr.
Accessories: diffraction chamber (diameter from 0·15 to 0·46 m) UMD1 crystal manipulator, flange mounted electron optical and visual display unit, low energy electron gun LEG-2, gas handling system.

Field ion and field emission microscope
Model FIM-2.

Specimen manipulators
RD1, RD2 (rotary) LMD 25, LMD 50, LMD100 (linear), UMD1, UMD2 (Universal motion drive).

2.3 Vacuum process plant and vacuum systems

2.3.1 General pumping units

Environmental chambers
Standard: Airco Temescal, Andonian, Balzers, Barlow Whitney, Cooke, High Vacuum Equipment, International Vacuum, RIBER, Tenney, Torr Vacuum Products (controlled atmosphere welding chambers), Vacudyne, Karl Weiss.
Custom built: Airco Temescal, Andonian, Balzers, Barlow Whitney, Carrier Engineering, Cooke, Edwards, General Engineering, High Voltage Engineering, International Vacuum, Leybold, Mill Lane Engineering, Nanotech, RIBER, Tenney, UHV, Ultek, Vacudyne, Vactronic, Karl Weiss.

Exhaust and aluminizing carts for CRT tubes etc. (standard with custom built modifications)
Edwards, Eisler Engineering (also rotary pumping tables), Leybold, Shawfrank Engineering Corp.

High vacuum range (10^{-4} to 10^{-7})
Standard: AEI, Airco Temescal, Alcatel, Andonian, Balzers, Cenco, Centorr, Cooke, Denton, Edwards, General Engineering, International Vacuum, Kinney, Leybold, Microtol, Nanotech, Pope (all glass), Red Point, RIBER, Sadla-Vac, Sihi (10^{-3} only), Teltron, Ultek, Vactorr, Vactronic, Vacuum Barrier, Veeco.
Custom built: Airco Temescal, Andonian, Balzers, Cenco, Cooke, Denton, Edwards, General Engineering, International Vacuum, Kinney, Leybold, Microtol, Mill Lane Engineering, Nanotech, Pope (all glass), Red Point, RIBER, Ricor, UHV, Ultek, Vactorr, Vacuum Industries, Veeco.

Ultra high vacuum range (10^{-8} to 10^{-12} Torr)
Standard: AEI, Airco Temescal, Andonian, Balzers, Cenco, Denton, Edwards, Granville Phillips, International Vacuum, Leybold, RIBER, Ultek, Vactronic, Vacuum Generators, Veeco.
Custom built: Andonian, Balzers, Cenco, Centorr, Cooke, Denton, Edwards, Granville Phillips, International Vacuum, Leybold, Microtol, Mill Lane Engineering, RIBER, UHV, Ultek, Vactronic, Vacuum Generators, Veeco.

2.3.2 Chemical engineering applications of vacuum

Vacuum drying and freezing
Standard: Butner-Schilde-Haas, Denton, Edwards, Hotpack, Hull, International Vacuum, Leybold, Pennwalt-Stokes, Vactorr (Desiccator), Vacudyne, Varian, Virtis.
Custom built: APV Mitchell, H. Balfour (pumps bought-in), Butner-Schilde-Haas, Cooke, Denton, Edwards, General Engineering, Hull, International Vacuum, Leybold, Pennwalt-Stokes, UHV, Vacudyne, Virtis, Varian.

Vacuum impregnation (transformers and capacitors)
Standard: Barlow Whitney, Butner-Schilde-Haas, General Engineering, Hull, International Vacuum, Kinney, Leybold, Nanotech, Pennwalt-Stokes, Red Point, Vacudyne, Veeco.
Custom built: Balzers, Barlow Whitney, Butner-Schilde-Haas, Cooke, Edwards, General Engineering, Hull, International Vacuum, Kinney, Leybold, Nanotech, Red Point, UHV, Vactronic, Vacudyne, Veeco.

Molecular distillation
Standard: Leybold.
Custom built: Edwards, Leybold.

Vacuum autoclaves
Standard: Red Point.
Custom built: Red Point.

Vacuum degassing (oils)
Standard: APV Mitchell, Barlow Whitney, General Engineering, Kinney, Leybold, Nanotech, Pennwalt-Stokes, Vacudyne.
Custom built: APV Mitchell, Balzers, Barlow Whitney, Cooke, Edwards, General Engineering, Kinney, Leybold, Nanotech, UHV, Vacudyne.

Vacuum distillation
Standard: Butner-Schilde-Haas, Leybold, Vacudyne.
Custom built: H. Balfour (pumps bought-in), Butner-Schilde-Haas, Cooke, Edwards, Kinney, Leybold, J. Montz, Vacudyne.

Vacuum dry boxes
Standard: Blickman.
Custom built. Blickman.

2.3.3 Vacuum metallurgy

Brazing equipment
Standard: Balzers, Denton, Edwards, General Engineering, International Vacuum, Ipsen, Leybold, Metals Research, Thermo Electron, Torvac, Vacuum Industries, Varian, Wentgate, Wild Barfield.
Custom built: Balzers, Centorr, Denton, Edwards, General Engineering, International Vacuum, Ipsen, Leybold, Oxy-Metal, Torvac, Wentgate.

Electron beam welding equipment
Standard: Airco Temescal, Balzers, BTI, Industrial Electron Beams, International Vacuum, Leybold, Nuclide Corp., Torvac, Wentgate.

Custom built: Airco Temescal, Cooke, Industrial Electron Beams, International Vacuum, Leybold, Torvac, Wentgate.

Heat treatment equipment
Standard: Balzers, Centorr, Denton, Edwards, General Engineering, International Vacuum, Ipsen, Leybold, Metals Research, Thermo Electron, Torvac, Vacuum Engineering (Scotland), Vacuum Industries, Varian, Wentgate, Wild Barfield.
Custom built: Balzers, Centorr, Denton, Edwards, General Engineering, International Vacuum, Ipsen, Leybold, Oxy-Metal, Torvac, Ultek, Wentgate.

Melting and alloy preparation equipment
Standard: Airco Temescal, Balzers, Centorr, Edwards, Elphiac, General Engineering, International Vacuum, O. Jünker, Leybold, Metals Research, Nuclide, Vacuum Engineering (Scotland), Vacuum Industries, Varian, Wild Barfield.
Custom built: Airco Temescal, Balzers, Centorr, Edwards, General Engineering, International Vacuum, Leybold, Ultek.

Ultra high vacuum metallurgy equipment
Standard: Balzers, International Vacuum, Leybold, Varian.
Custom built: Balzers, Centorr, Cooke, International Vacuum, Leybold, Ultek, Vacuum Generators.

2.3.4 Vacuum furnaces

MANUFACTURER:	**Balzers**					
Model no.	VSG 02	VSG 02	VSG 10	MOV 025	MOV 3	MOV 10
Purpose of furnace	Melting and casting	Heat treatment	Melting and casting	Sintering, brazing, degassing and annealing →		
Heating method	Induction	Induction or resistance	Induction	Resistance →		
Furnace capacity	170 cm^3	70 cm^3	10 kg	0·24–0·34 litres	1·2–3·5 litres	6·65–10 litres
Furnace temperature (°C)	2200	3000	1600	1700–2400	1500–2400	1500–2200
Evacuation time (min)	1·5 to 1×10^{-3} 10 to 2×10^{-5}	1·5 to 1×10^{-3} 10 to 2×10^{-5}				
Cooling time (min)						
Working pressure (torr)	2×10^{-6}	2×10^{-6}	$<5 \times 10^{-6}$	5×10^{-6} →		

MANUFACTURER:	**Balzers** *cont'd*				
Model no.	MOV 20	MOV 50	MOV 80	MOV 150	MOV 250
Purpose of furnace	Sintering, brazing, degassing and annealing →				
Heating method	Resistance →				
Furnace capacity	21 litres	35·3–56·7	53–85 litres	155 litres	254 litres
Furnace temperture (°C)	1400–1700	1400–1700	1400–1700	1400	1400
Evacuation time (min)					
Cooling time (min)					
Working pressure (torr)	5×10^{-6} →				

MANUFACTURER:	**Balzers** *cont'd*					**Carbolite**
Model no.	IOV 16	IOV 50	IOV 100	VSG 30	VSG 50	
Purpose of furnace	Sintering heat treatment, brazing, out-gassing →			Melting and casting		Porcelain firing
Heating method	Induction →					Resistance
Furnace capacity	26 litres	70 litres	125 litres	4 litres	7 litres	
Furnace temperature (°C)	1600–2200	1600–220	1600–2200	1600	1600	500–1200
Evacuation time (min)						
Cooling time (min)						
Working pressure (torr)	2×10^{-3}–5×10^{-6}	2×10^{-3}-5×10^{-6}	2×10^{-3}–1×10^{-5}	5×10^{-6}	1×10^{-5}	20

MANUFACTURER:	Centorr					
Model no.	2B-20	Series 5 Single arc	Series 5 Reed tri-arc	Series 10 Graphite tube furnace	Series 14 Universal furnace	Series 18 Bell jar furnace
Purpose of furnace		High purity melting	Crystal growth high purity melting	Inert, reducing and oxidizing	High temperature, high vacuum, gases use	Converts B.J. unit to high temperature furnace
Heating method		Electric arc	Arc	Graphite tubular element	Refractory metals	Refractory metals
Furnace capacity		5 ml	5 ml	6 litres	1 litre	
Furnace temperature (°C)	800-900	3500	3500	3000	2500	3000
Evacuation time (min)					<15 to 10^{-4}	<10 to 10^{-4}
Cooling time (min)		0	0	60	<30	30
Working pressure (torr)					10^{-6}	10^{-7}

MANUFACTURER:	Centorr *cont'd*			Denton	Edwards-Abar
Model no.	Series 40 (vertical)	Series 50 (horizontal)	Series 60	Vertical and horizontal types available	Horizontal front loading HR20 HR26 HR34 HR34 x 48
Purpose of furnace	Heat treating brazing	Heat treating brazing	High temperature, physical testing	Sintering, outgassing, annealing, brazing	Heat treatment, brazing and degassing
Heating method	Refractory metal	Refractory metal rods	Refractory metal		Resistance using refractory metal bands
Furnace capacity	46 litres	130 litres	5 litres		68 litres 250 litres 340 litres 455 litres
Furnace temperature (°C)	3000	3000	3000	Up to 2500 (special furnaces over 3000)	1320 (service), 1370 (clean-up)
Evacuation time (min)	<10 to 10^{-4}	<20 to 10^{-4}	<10 to 10^{-4}		<25 to 1 x 10^{-4}
Cooling time (min)	<60	<15	60		<20 to 100°C (with load)
Working pressure (torr)	10^{-6}	10^{-6}	10^{-6}		10^{-5} range

MANUFACTURER:	**Edwards-Abar** *cont'd*							
Model no.	Horizontal front loading HR46	HR58	AL40	Vertical bottom loading VR36	VR36 x 72	VR48	VR60	VR96
Purpose of furnace	Heat treatment, brazing and degassing			Heat treatment, brazing and degassing				
Heating method	Resistance using refractory metal bands			Resistance				
Furnace capacity	1020 litres	1100 litres	250 litres	600 litres	1200 litres	1525 litres	2520 litres	11 025 litres
Furnace temperature (°C)	1320 (service), 1370 (clean-up)			1320 (service), 1370 (clean up)				
Evacuation time (min)	<25 to 1 x 10^{-4}			<25 to 1 x 10^{-4}				
Cooling time (min)	<20 to 100°C (with load)			<20 to 100°C (with load)				
Working pressure (torr)	10^{-5} range			10^{-5} range				

MANUFACTURER:	**Edwards-Abar** *cont'd*				**Elphiac**
Model no.	I-H series	RH series 1200	1600	T2200	Models VSG300, VSG600 VSG1200, VSG2500 VSG4000, VSG5000
Purpose of furnace	Alloy research and development	Melting, heat treatment	Melting, heat treatment	Sintering	Melting and casting
Heating method	Induction	Resistance	Resistance	Resistance	Induction
Furnace capacity	2, 7, 9, 5–15, 15–50, 100–750 kg				From 300 kg to 5 tons
Furnace temperature (°C)		1700	1200	1600	Up to 1700
Evacuation time (min)					
Cooling time (min)					
Working pressure (torr)		5 x 10^{-5} (ult.)			10^{-4}

MANUFACTURER:	General Engineering	International Vacuum	Ipsen			
Model no.	Over 60 models	Mark 14	Horizontal VFC series	Vertical VVFC series	CVFC-424	Bottom loading VFC(BL) series
Purpose of furnace	Heat treating and brazing, gas and oil quench	General purpose	Brazing annealing sintering	Brazing annealing sintering	Multi-chamber continuous heat-treatment	Brazing annealing sintering
Heating method		Resistance or induction	Resistance ⟶			
Furnace capacity		56·6 litres	10–700 kg	20–2000 kg		500–2000 kg
Furnace temperature (°C)	Up to 3000	Up to 2100	1100	1100	As specified	1300
Evacuation time (min)		15				
Cooling time (min)		Various. Liquid or gas quench	Fan assisted	Fan assisted		
Working pressure (torr)		10^{-7}				10^{-7}

MANUFACTURER:	Jünker				Metals Research	Nuclide
Model no.	IFT St. 600 vac	NFT St. 3000 vac	NFT St. 6000 vac	NFT St. 10 000 vac	VS2	H300
Purpose of furnace	Degassing	Degassing	Degassing	Degassing	Heat treatment, brazing, sintering, annealing	Melting
Heating method	Induction	Induction	Induction	Induction		Hollow cathode discharge
Furnace capacity	600 kg	3000 kg	6000 kg	10 000 kg	0·05 m diameter 0·153 m long	Chamber size 0·27 m diameter 0·91 m high Various size moulds
Furnace temperature (°C)	Maximum 1700	Maximum 1700	Maximum 1700	Maximum 1700	Maximum 1200	Up to 3000
Evacuation time (min)	7	8	10	10	10	
Cooling time (min)					240 (to 250°)	
Working pressure (torr)	0·2	0·2	0·2	0·2	10^{-6}	10^{-4} to 0·05

MANUFACTURER:	Oxy-metal	Ricor	Thermo Electron	Torvac	
Model no.	VCRF60-36	MF2A	Series 1000, series 1100 Series 1200, series 1300	S12	S16
Purpose of furnace	Brazing, heat treatment	Mössbauer furnace	Brazing, heat treatment annealing and outgassing	Brazing, hardening, heat treating sintering	
Heating method	Resistance (graphite)	11 Ω 2·5 A	Resistance	Molybdenum radiant elements	
Furnace capacity	1670 litres	16 mm diameter specimen	0·5 to more than 30 litres	5 kg	20 kg
Furnace temperature (°C)	1250	300–1000	Up to 2400	1400	1400
Evacuation time (min)	12/14 to 10^{-4}		4 to 1 x 10^{-4}	5	10
Cooling time (min)	35/40 (230 kg)		2000°C to 40°C in 45	1100°C to 100°C in 25	
Working pressure (torr)	$10^{-3}/10^{-4}$		10^{-5}–10^{-8}	10^{-5} band	10^{-5} band

MANUFACTURER:	Torvac *cont'd*		Vacuum Industries		Varian		
Model no.	S24	S36WW	Series 2100 System 7	Series 3600 Model 1-2300	2940	2941	2945
Purpose of furnace	Brazing, hardening, heat treating, sintering	Tungsten carbide sintering	General purpose	Hot press sintering	General	General	Sintering
Heating method	Molybdenum radiant elements	Induction	Various	H.F. 9·6 kHz, 15 kW	Resistance ⟶		
Furnace capacity	90 kg	50 kg			2·5 litres	2·5 litres	9 litres
Furnace temperature (°C)	1400	Maximum 1650	3000 (W) 2100 (graphite) Arc	2300	2300	2800	2200
Evacuation time (min)	20 (to 10^{-4})	15 (to 10^{-1})		30 (incl. heating)	10	20	20
Cooling time (min)	10 (1000–200°C)	120 (1450–200°C)	65°/min	180	45	150	180
Working pressure (torr)	10^{-5} band	10^{-1} band		10^{5}	10^{-5}	10^{-5}	10^{-5}

MANUFACTURER:	Varian *cont'd*		Veeco		Wentgate				
Model no.	2952	2966	C6	C11	0810	1218	1820	2424	3636
Purpose of furnace	Pre-sintering	General	Crystal growing	Crystal growing	Brazing and heat treatment →				
Heating method	Resistance	Resistance			Radiation–molybdenum or carbon →				
Furnace capacity	20 litres	35 litres			Model numbers denote diameter and depth in inches				
Furnace temperature (°C)	1000	2500	1100–1650	1100–1650	Up to 1400 on standard models → Up to 2100 on custom built models →				
Evacuation time (min)	10	30			5	7	7		
Cooling time (min)	120	60							
Working pressure (torr)	10^{-3}	10^{-5}	The performance of these basic units depends on the modules attached to them		5×10^{-5} →				

MANUFACTURER:	Wild and Barfield				
Model no.	Vertical VIP 0812 GQ	VIP 1218 GQ	VIP 1624 GQ	VIP 2030 GQ	VIP 2436 GQ
Purpose of furnace	Hardening high speed, tool and die steels →				
Heating method	Graphite resistor system →				
Furnace capacity	14 kg	45 kg	113 kg	227 kg	363 kg
Furnace temperature (°C)	Up to 1350 →				
Evacuation time (min)					
Cooling time (min)					
Working pressure (torr)	10^{-2}–1 with Roots pump, 10^{-2}–10^{-4} with vapour diffusion pump →				

MANUFACTURER:	Wild and Barfield *cont'd*				
Model no.	Horizontal VIH 0806/12 GQ	VIH 1209/18 GQ	VIH 1612/24 GQ	VIH 2015/30 GQ	VIH 2418/36 GQ
Purpose of furnace	Hardening high speed, tool and die steels →				
Heating method	Graphite resistor system →				
Furnace capacity	14 kg	113 kg	227 kg	227 kg	363 kg
Furnace temperature (°C)	Up to 1350 →				
Evacuation time (min)					
Cooling time (min)					
Working pressure (torr)	10^{-2}–1 with Roots pump, 10^{-2}–10^{-4} with vapour diffusion pump →				

MANUFACTURER:	Wild Barfield *cont'd*				
Model no.	Vertical VIP 0812C	Vertical VIP 1218C	Horizontal VIH 0806 12C	Horizontal VIH 1209 18C	VIP 0407
Purpose of furnace	Sintering tungsten carbide and titanium/tantalum grades →				Bright hardening of high speed tool and die steel, fluxfree brazing annealing of stainless steel, beryllium
Heating method	Graphite resistor system →				Tungsten sheet element
Furnace capacity	10–12 kg	30–40 kg	10–12 kg	30–40 kg	
Furnace temperature (°C)	Up to 1700 →				Up to 1350
Evacuation time (min)					
Cooling time (min)					
Working pressure (torr)	10^{-2}–1	10^{-2}–1	10^{-2}–1	10^{-2}–1	10^{-4}

2.3.5 Vacuum ovens

MANUFACTURER:	Barlow-Whitney	Red Point		Thelco		
Model no.	Series E250 Vaco H	VO 4236 H-S27	VO 2424H-C15	10	19	29
Chamber dimensions (m)	0·46–0·76 diameter 0·61–1·22 deep	1·06 diameter 0·91 deep	0·61 diameter 0·61 deep	0·2 diameter 0·2 deep	0·2 wide 0·2 High 0·3 deep	0·3 wide 0·3 high 0·45 deep
Oven temperature (°C)	250	750	750	150	200	200
Working pressure (torr)	0·5	10^{-1}	10^{-1}	Low pressure	Low pressure	Low pressure

2.3.6 Electron beam welding equipment

MANUFACTURER:	BTI	Balzers	Industrial Electron Beams (BOC)		International Vacuum	Leybold
Model no.		ESW5	201	301		ESW3000/6
Beam specification KV	25 and 50	30	50	150	60	150
MA	0–400		60	60	500	
Beam diameter (mm)	Under 1 and upwards				1·58	
Chamber size (mm)			200 300 60 cube	300 x 300 x 450		3 m^3
Vessel jigging	Single and multi-station rotary, linear tables		Rotary or linear		XYZ traverse	
Working pressure (torr)	$\leqslant 10^{-4}$ or $> 10^{-1}$		10^{-4}	5×10^{-2}	10^{-5}	

MANUFACTURER:	Nuclide	Torvac				Wentgate	
Model no.	Mark 6 EBW system	CVE62B	CVE66	BH range (fast cycle)	CVE31B	BW range	DW range
Beam specification KV	Variable 0–30	60	60	60/30	30	Up to 60	Up to 60
MA	Variable 0–300 (500 available)	33	100	33	33	10, 20, 40 or 100	10, 20, 40 or 100
Beam diameter (mm)	0·25 mm to more than 12·7 mm	0·15	0·15	0·15	0·15		
Chamber size (mm)		As required →				Up to 139·7	
Vessel jigging	Longitudinal, lateral and rotary work table	To suit application →				Rotary, linear or beam manipulation	
Working pressure (torr)	1×10^{-6} (ultimate)	5×10^{-4} or 5×10^{-2}	5×10^{-2} or 5×10^{-4}	5×10^{-2} 15 S working cycle	5×10^{-2} or 5×10^{-4}	5×10^{-4}	5×10^{-2} or 5×10^{-4}

2.3.7 Deposition plant: vacuum evaporation

Metal finishing of plastics
Standard: Airco Temescal, Balzers, Cenco, Cooke, Denton, Edwards, General Engineering, International Vacuum, Leybold, Veeco.
Custom built: Airco Temescal, Balzers, Cenco, Cooke, Edwards, General Engineering, International Vacuum, Leybold, Mill Lane Engineering, Nanotech, Vacuum Generators, Veeco.

Microelectronic components production
Standard: Airco Temescal, Balzers, Cenco, Cooke, Denton, Edwards, Electrotech (jigs and fittings only), Ion Equipment, International Vacuum, Leybold, Nanotech, Sloan, Ultek, Vacuum Industries, Varian, Veeco.
Custom built: Airco, Balzers, Cenco, Cooke, Denton, Edwards, Electrotech (jigs and fittings only), Hull, International Vacuum, Leybold, Mill Lane Engineering, Planer, Sloan, Ultek, Vactronic, Vacuum Generators, Vacuum Industries, Varian, Veeco.

Optical coatings (mirrors, lenses etc.)
Standard: Airco, Balzers, Cenco, Cooke, Denton, Edwards, Electrotech (jigs and fittings only), International Vacuum, Ion Equipment, Leybold, Sloan, Ultek, Varian, Veeco.
Custom built: Airco, Balzers, Cenco, Cooke, Edwards, Electrotech (jigs and fittings only), International Vacuum, Leybold, Mill Lane Engineering, Nanotech, Planer, Sloan, Ultek, Vactronic, Vacuum Generators, Varian, Veeco.

Specimen preparation for electron microscopy (shadowing, replication, etching)
Standard: Balzers, Cooke, Denton, Edwards, Electrotech (jigs and fittings only), General Engineering, International Vacuum, Ion Equipment, Kinney, Leybold, Nanotech, Sloan, Varian, Veeco.
Custom built: Cooke, Denton, Edwards, Electrotech (jigs and fittings only), General Engineering, International Vacuum, Kinney, Leybold, Planer, Sloan, Ultek, Vactronic, Vacuum Generators, Varian, Veeco.

MANUFACTURER:	Airco Temescal	Balzers				
Model no.	BJD1800	MICRO BA3	BAE 120	BAE121	BA360	BA510
*Plant type	VC, bench height top loading	VC	HC	VC	BJ	BJ
Vessel construction						
Diameter/width (m)	0·45	0·08	0·12	0·12	0·36	0·50
Height/length (m)	0·5	0·25	0·30	0·30	0·39	0·65
Material	SS	G	G	G	SS	SS
†Pumping system						
Pump type	D, C, TM	D	TM	D	D	D
Pump speed (litre/second)	2400 (D)	23	70	260	950	1800
Roughing pump	R	R	R	R	R	R
Exhaustion time						
Atmosphere to P torr (min)	10^{-6} (8) vessel loaded	10^{-5} (18)	10^{-5} (25)	10^{-5} (3)	10^{-5} (4)	10^{-5} (5)
P torr ultimate (mins or temp)	5×10^{-7} (15) vessel empty					5×10^{-8} (LN_2)
Vapour sources	SFIH-270 40 cc crucible STIM-270 4 x 15·4 cc crucibles					
Electron beam	ht 14 kW (CV-14) ht 8 kW (CV-8)				Yes	Yes
Ion plating facility						
Process monitors	hf crystal				Ionization	hf crystal Optical

Note: only special types of vapour sources (electron beam etc.) are listed; resistance types will normally be available.

* BJ Bell jar
BX Box
HC Horizontal cylinder
VC Vertical cylinder

† D Diffusion
C Cryopump
R Rotary
S Sorption
SI Sputter ion
TM Turbomolecular

MANUFACTURER:	Balzers *cont'd*					Cenco	Cooke
Model no.	BA511	BAK550	BA710	BAK750	BA1000K	Cenco vacuum coater	CV300
*Plant type	BJ	BX	VC	BX	BX	BJ	BJ
Vessel construction							
Diameter/width (m)	0·51	0·53 x 0·53	0·70	0·73 x 0·73	1 x 1	0·45	0·3
Height/length (m)	0·52	0·66	0·67	0·86	1·21	0·45	0·3
Material	SS	SS	SS	SS	SS	Aluminium	G
†Pumping system							
Pump type	TM	D	D	D	D	D	D
Pump speed (litre/second)	260	1800 or 5000	1700	11 500	13 000		285 Unbaffled 150 At vessel
Roughing pump	R	R	R	R	R	R	R
Exhaustion time							
Atmosphere to P torr (min)	10^{-5} (5)	10^{-5} (10)	10^{-5} (5–8)	10^{-5} (3)	10^{-5} (10)	10^{-3} (5)	10^{-5} (3)
P torr ultimate (mins or temp)		5×10^{-7} (30) (LN_2)	5×10^{-8} (LN_2)		1×10^{-7} (LN_2)	10^{-4} (8)	10^{-7} (15°C)
Vapour sources							
Electron beam			Yes				Yes
Ion plating facility							Yes
Process monitors			hf crystal Optical				Yes

* BJ Bell jar
BX Box
HC Horizontal cylinder
VC Vertical cylinder

† D Diffusion
C Cryopump
R Rotary
S Sorption
SI Sputter ion
TM Turbomolecular

MANUFACTURER:	Cooke *cont'd*		Davis and Wilder	Denton		
Model no.	CV600	CV1000	Model 630	DV502	DV503FP	DV504FP
*Plant type	BJ	BJ	BJ	BJ	BJ	BJ
Vessel construction						
Diameter/width (m)	0·6	1·2	0·45	0·3	0·46	0·61
Height/length (m)	0·45	1·2	0·75	0·3	0·76	0·76
Material	G	SS	G, SS	G	G	SS
† Pumping system						
Pump type	D	D	D, C	D	D	D
Pump speed (litre/second)	2200 Unbaffled 1400 At vessel	5200 Unbaffled 300 At vessel		250 litre/ second	750 litre/ second	1500 litre/second
Roughing pump	R	R	R	R	R	R
Exhaustion time						
Atmosphere to P torr (min)	10^{-5} (3)	10^{-5} (5)		10^{-4} (3) 5×10^{-5} (4)	10^{-6} (10)	10^{-6} (10)
P torr ultimate (mins or temp)	10^{-8} (15°C) 10^{-9} (LN_2)	10^{-7} (15°C) 10^{-8} (LN_2)		2×10^{-6} (water cool) 10^{-7} (LN_2)	10^{-7} (LN_2)	10^{-7} (LN_2)
Vapour sources			ht 8 kW, model 5714 multi crucible, model 5731 single crucible model 5700 two crucible	1 kVA filament	2 kVA filament 5, 10, 20, 40 V output	2 kVA filament 5, 10, 20, 40 V output
Electron beam	Yes	Yes	Yes		Yes	Yes
Ion plating facility	Yes	Yes				
Process monitors	Yes	Yes				

* BJ Bell jar
BX Box
HC Horizontal cylinder
VC Vertical cylinder

† D Diffusion
C Cryopump
R Rotary
S Sorption
SI Sputter ion
TM Turbomolecular

MANUFACTURER:	Denton *cont'd*				
Model no.		DV214	DV218	DV224	DV505FP
*Plant type		Twin BJ	Twin BJ	Twin BJ	BJ
Vessel construction					
Diameter/width (m)		0·35	0·45	0·61	0·91
Height/length (m)		0·61	0·76	0·76	0·01
Material		G	G	SS	SS
†Pumping system					
Pump type		D	D	D	D
Pump speed (litre/second)		750	1500	2400	4200
Roughing pump		R	R	R	R
Exhaustion time					
Atmosphere to P torr (min)					
P torr ultimate (mins or temp)		$<3 \times 10^{-7}$ (LN_2)	$<3 \times 10^{-7}$ (LN_2)	$<3 \times 10^{-7}$ (LN_2)	$<1 \times 10^{-7}$ (LN_2)
Vapour sources		2 x 2 kVA filaments 5, 10, 20, 40 V output	2 x 2 kVA filaments 5, 10, 20, 40 V output	2 x 2 kVA filaments 5, 10, 20, 40 V output	2 kVA filament 5, 10, 20, 40 V output
Electron beam		Yes	Yes	Yes	Yes
Ion plating facility					
Process monitors	DIP-1 (can be linked with DEG 810 EG gun)				

* BJ Bell jar
BX Box
HC Horizontal cylinder
VC Vertical cylinder

† D Diffusion
C Cryopump
R Rotary
S Sorption
SI Sputter ion
TM Turbomolecular

MANUFACTURER:	Denton cont'd				
Model no.	DV505UHV	DV503SC	DV504SC	DV505SC	DV515 automatic
*Plant type	BJ	BJ 'split chamber'	BJ 'split chamber'	BJ 'split chamber'	BJ
Vessel construction					
Diameter/width (m)	0·6	0·46	0·61	0·91	0·25 (0·30 optional)
Height/length (m)	0·76	0·76	0·61	0·66	0·30
Material	SS	G	SS	SS	C
†Pumping system					
Pump type	D	D	D	D	D
Pump speed (litre/second)	4200	1500	2400	4200	250
Roughing pump	R	R	R	R	R
Exhaustion time					
Atmosphere to P torr (min)					2×10^{-5} (5)
P torr ultimate (min or temp)	$<5 \times 10^{-9}$ (LN_2)	$<5 \times 10^{-7}$ (LN_2)	$<5 \times 10^{-7}$ (LN_2)	$<5 \times 10^{-7}$ (LN_2)	$<10^{-6}$ range
Vapour sources	2 kVA filament 5, 10, 20, 40 V output	2 kVA filament 5, 10, 20, 40 V	2 kVA filament 5, 10, 20, 40 V	$7\frac{1}{2}$ kVA filament 5, 10, 20, 40 V	
Electron beam	Yes	Yes	Yes	Yes	

MANUFACTURER:	Edwards High Vacuum						
Model no.	550P	6E4	24E5	36E4	E12E3	E1300	E19A3
*Plant type	HC	BJ	HC	HC	BJ	HC	VC
Vessel construction							
Diameter/width (m)	0·54		0·61	0·91	0·36	1·3	0·48
Height/length (m)	0·33		0·58	0·91	0·30	1·4	0·76
Material	SS	G	Mild steel	Mild steel	G	Mild steel	SS
†Pumping system							
Pump type	D	D	Vapour booster	D	D	D	D
Pump speed (litre/second)	1500		2800	6000	600	12 000	2500
Roughing pump	R	R	R	R	R	R	R
Exhaustion time							
Atmosphere to P torr (min)	10^{-6} (7)	1×10^{-4} (5)	1×10^{-4} (5)	5×10^{-4} (6)	$1 \times 1 \times 10^{-4}$ (4)	1×10^{-4} (4)	10^{-6} (7)
P torr ultimate (min or temp)					3×10^{-7} (LN_2)		
Vapour sources	4·5 kVA	200 VA	4 kVA	8 kVA	1 kVA	8 kVA	4 kVA
Electron beam					Yes		Yes
Ion Plating facility	As required						
Process monitors	hf crystal → ; Optical →						

* BJ Bell jar
BX Box
HC Horizontal cylinder
VC Vertical cylinder

† D Diffusion
C Cryopump
R Rotary
S Sorption
SI Sputter ion
TM Turbomolecular

MANUFACTURER:	Edwards High Vacuum *cont'd*			International Vacuum	
Model no.	E30E	E306	E500	IVI 5500, model 2108	IVI 4000 series
*Plant type	VC	VC or BJ	VC	BX	BX, BJ, HC
Vessel construction					
Diameter/width (m)	0·76	0·30	0·50	0·61	Up to 2·5
Height/length (m)	0·76	0·35	0·76	0·61	1·5 (BX)
Material	SS	G	SS	SS	G or SS
† Pumping system					
Pump type	D	D	D	D, C	D, C
Pump speed (litre/second)	4000	600	1500/2500	900 or 3000	Up to 50 000
Roughing pump	R	R	R	R	R
Exhaustion time					
Atmosphere to P torr (min)	2×10^{-5} (11)	1×10^{-4}(3)	2×10^{-5} (5)		
P torr ultimate (mins or temp)	5×10^{-7}	3×10^{-7}	1×10^{-7} (LN_2)	5×10^{-7} (15°C) 5×10^{-8} (LN_2)	5×10^{-7} (15°C) 5×10^{-8} (LN_2)
Vapour sources	4 kVA	1 kVA	4 kVA		
Electron beam	Yes		Yes	Yes 14 kW	Yes
Ion plating facility		As required		Yes	
Process monitors	hf crystal ⟶ Optical ⟶			hf crystal ⟶ Optical ⟶	

MANUFACTURER:	Ion Equipment				
Model no.	DPS-2	Series 1000	Series TTS50	Series TTS200	Series TTSB200
*Plant type	BJ	BJ	BJ	BJ	BJ
Vessel construction					
Diameter/width (m)		0·45	0·30	0·30	0·30
Height/length (m)		0·76	0·30	0·30	0·30
Material	G	G, SS	G	G	SS
†Pumping system					
Pump type	D	S1	S1	S1	S1
Pump speed (litre/second	80	450	50	200	200
Roughing pump	R	R + S	R or S	R or S	R or S
Exhaustion time					
Atmosphere to P torr (min)		5×10^{-7} (4) 5×10^{-8} (15)	5×10^{-7} (10) 5×10^{-9} (180)	5×10^{-7} (3) 5×10^{-9} (100)	5×10^{-10} (300)
P torr ultimate (mins or temp)			10^{-10}	10^{-10}	
Vapour sources					
Electron beam					
Ion plating facility					
Process monitors					

* BJ Bell jar
BX Box
HC Horizontal cylinder
VC Vertical cylinder

† D Diffusion
C Cryopump
R Rotary
S Sorption

MANUFACTURER:	**Kinney**	**Nanotech**		
Model no.	KSE-2A	Microprep 2505	Microprep 3005	Auto 300S
*Plant type	BJ	BJ	BJ	BJ
Vessel construction				
Diameter/width (m)		2·5	3·0	3·0
Height/length (m)		3·0	3·5	3·5
Material	G	G	G	G
†Pumping system				
Pump type	D	D	D	D
Pump speed (litre/second)	200	150 unbaffled	600	1000
		35 at vessel	150	200
Roughing pump	R	R	R	R
Exhaustion time				
Atmosphere to P torr (mins)	10^{-5} (10)	10^{-5} (10)	10^{-5} (4)	10^{-5} (3·5)
P torr ultimate (mins or temp)	5×10^{-7} (LN_2)	2×10^{-6} (15°C)	2×10^{-6} (15°C)	2×10^{-6} (15°C)
		5×10^{-7} (LN_2)	5×10^{-7} (LN_2)	5×10^{-7} (LN_2)
Vapour sources				
Electron beam				
Ion plating facility				
Process monitors		Optional	Optional	Optional

* BJ Bell jar
BX Box
HC Horizontal cylinder
VC Vertical cylinder

† D Diffusion
C Cryopump
R Rotary
S Sorption
SI Sputter ion
TM Turbomolecular

MANUFACTURER:	Pennwalt-Stokes			Planer	
Model no.	426-36	426-4	427-7		
*Plant type	HC	HC	HC		
Vessel construction					
Diameter/width (m)	0·92	1·22	1·83		
Height/length (m)	1·07	1·37	1·83		
Material	Mild steel	Mild steel	Mild steel		
†Pumping system					
Pump type	D	D	D		
Pump speed (litre/second)	4300 Unbaffled 3400 At vessel	10 000 8000	52 000 41 000		
Roughing pump	R	R	R		
Exhaustion time					
Atmosphere to P torr (min) P torr ultimate (mins or temp)	5×10^{-4} (3·5)	5×10^{-4} (4)	5×10^{-4} (5·5)		
Vapour sources					
Electron beam	12 kVA	20 kVA	30 kVA	Source EBS1/UH 'Unvala' type for uhv; power unit S7/1E 7 kV 200 mA	Source EBS1/V or H 'Unvala' type with baseplate (V) or source ring (H) mounting
Ion Plating facility					
Process monitors					

* BJ Bell jar
BX Box
HC Horizontal cylinder
VC Vertical cylinder

† D Diffusion
C Cryopump
R Rotary
S Sorption
SI Sputter ion
TM Turbomolecular

MANUFACTURER:	Planer *cont'd*		Sloan		
Model no.			SL180		SS2000
*Plant type			BJ		BX
Vessel construction					
Diameter/width (m)			0·45	0·45 or 0·60	0·78 x 0·78
Height/length (m)			0·76	0·76	0·83
Material			G	SS	SS
†Pumping system					
Pump type			D		D
Pump speed (litre/second)					
Roughing pump			R		R
Exhaustion time					
Atmosphere to P torr (min)					2×10^{-6} (10), 1×10^{-5} (4)
P torr ultimate (mins or temp)			1×10^{-8}		5×10^{-7}
Vapour sources	Source EB10 water-cooled Cu hearth, graphite crucible; power unit S10/50 Mk 1A, 10 kV 500 mA	Source EB4. Cermet crucible 5 kW. Source EBU/1 5 station crucible (1·4 kW)			
Electron beam			6 kW or 12 kW		Two EB guns or two resistance sources
Ion plating facility					
Process monitors			Ionization crystal		

* BJ Bell jar
BX Box
HC Horizontal cylinder
VC Vertical cylinder

† D Diffusion
C Cryopump
R Rotary
S Sorption
SI Sputter ion
TM Turbomolecular

MANUFACTURER:	Ultek					
Model no.	TNBX	MX14	TNB	HNB/C, VNB/C	RCS	TBK
*Plant type	BJ Table top	BJ Table top	BJ Table top	BJ Console	BJ Rapid cycling	BJ Bakeable 250°C uhv
Vessel construction						
Diameter/width (m)	0·3	0·3	0·3	0·45	0·45	0·3
Height/length (m)	0·3	0·3	0·3	0·75	0·75	0·3
Material	G or SS	G or SS	G or SS	G or SS	G or SS	G or SS
†Pumping system						
Pump type	SI or Ti/E‡	SI or Ti/E	SI or Ti/E	SI or Ti/E	SI or Ti/E	SI or Ti/E
Pump speed (litre/second)	75	75	50	400		50
Roughing pump	S§ or R	S or R	S or R	S or R	S or R	S or R
Exhaustion time						
Atmosphere to P torr (min)	10^{-7} (5-8)		10^{-7} (10)		10^{-7} (10)	
P torr ultimate (mins or temp)	10^{-11}		5×10^{-9} (180)	5×10^{-9} (180)		5×10^{-10} (360)
Vapour sources						
Electron beam	Yes	Yes	Yes	Yes	Yes	Yes
Ion plating facility						
Process monitors						

* BJ Bell jar
BX Box
HC Horizontal cylinder
VC Vertical cylinder

† D Diffusion
C Cryopump
R Rotary
S Sorption
SI Sputter ion
TM Turbomolecular

‡ The sublimation Ti-pump is termed 'Boostivac'.
§ A roughing station with 3 sorption pumps and a dry mechanical blower is available (model CFR).

MANUFACTURER:	Vacuum Industries		Varian		Veeco
Model no.	Series 2100, Model 1804	Series 2100 Model 1806	VE 10	NRC3117	VE400 (manual) VE401 (auto)
*Plant type	BJ		BJ	BJ	
Vessel construction					
Diameter/width (m)	0·45	0·45	0·25-0·35	0·45	0·35 or 0·50 baseplate diameter
Height/length (m)	0·45	0·75	0·30	0·76	
Material	G	G	G	G or SS	
†Pumping system					
Pump type	D	D	D	D	EP4W
Pump speed (litre/second)	(130 mm diameter)	(175 mm diameter)	80 unbaffled 30 at vessel	1500 or 2400 unbaffled 500 or 800 at vessel	425
Roughing pump	R	R	R	R	
Exhaustion time					
Atmosphere to P torr (min)	10^{-6} (7)	10^{-6} (4)	10^{-5} (10)	10^{-5} (18 or 14)	
P torr ultimate (mins or temp)	10^{-7} (LN_2)	10^{-7} (LN_2)	10^{-6} (LN_2)	10^{-7} or 10^{-8} (LN_2)	
Vapour sources					
Electron beam	5 kV, 0·4 A	5 kV, 0·4 A	LT 1·5 kVA	2 kVA	Type VeB-6: 6 kW electron gun, 3 mm diameter beam
Ion plating facility					
Process monitors				crystal	

* BJ Bell jar
BX Box
HC Horizontal cylinder
VC Vertical cylinder

† D Diffusion
C Cryopump
R Rotary
S Sorption
SI Sputter ion
TM Turbomolecular

MANUFACTURER:	**Veeco** *cont'd*				
Model no.	VS400 (manual) VS401 (auto)	VE776M (manual) VE776A (auto)	VA776M (manual) VA776A (auto) High vacuum pumping station	BE1200 Box coater	820 series Fast cycle
*Plant type					
Vessel construction					
Diameter/width (m)	0·45	0·45		0·80 long 0·75 wide	0·45
Height/length (m)	0·75	0·75		0·87	0·75
Material	Pyrex	Pyrex		Stainless steel	Pyrex
†Pumping system					
Pump type	EP4W	EP776 + water-cooled chevron baffle and LN_2 trap	EP776 + water-cooled chevron baffle and LN_2 trap	0·3 m diameter diffusion	Sputter ion + electron beam Ti-evaporation getter
Pump speed (litre/second)	425	2000 (550 at vessel)	2000 (550 at vessel)		
Roughing pump		17·7 CFM	10·6 or 7·7 CFM		Carbon vane + LN_2-sorption
Exhaustion time					
Atmosphere to P torr (min)		3×10^{-7} (15 min)		$\sim 10^{-6}$ (4 min)	$<5 \times 10^{-7}$ (4 min)
P torr ultimate (mins or temp)		$1{\cdot}5 \times 10^{-7}$ (1 h)			$<5 \times 10^{-9}$
Vapour sources		2 kW			
Electron beam	Type VeB-6: 6 kW electron gun, 3 mm diameter beam →				
Ion plating facility					
Process monitors					

* BJ Bell jar
BX Box
HC Horizontal cylinder
VC Vertical cylinder

† D Diffusion
C Cryopump
R Rotary
S Sorption
SI Sputter ion
TM Turbomolecular

2.3.8 Vacuum evaporation special systems, vessel accessories and materials

Coating materials (evaporation): oxides, sulphides, fluorides, tellurides, selenides etc.
Alcatel, Balzers, BDH, GTE Sylvania, Leybold, Mill Lane Engineering, Printa, Ventron.

Continuous coating systems: for deposition onto strip or roll stock using metal, polymer or paper base materials.
Airco, Cooke, Edwards, General Engineering, International Vacuum, Leybold, Mill Lane Engineering, Torvac.

General purpose evaporation plants.
AEI, Airco, Balzers, Cooke, Edwards, Hull, International Vacuum, Leybold, Mill Lane Engineering, Nanotech, UHV, Ultek, Vactronic, Vacuum Industries.

Reactive evaporation: gas pressure or deposition rate control systems.
Airco, Balzers, Cooke, Edwards, Electrotech, International Vacuum, Leybold, Mill Lane Engineering, Nanotech, Ultek.

Substrate cleaning systems (e.g. glow discharge).
Airco, Balzers, Cooke, Edwards, Electrotech, International Vacuum, Leybold, Mill Lane Engineering, Nanotech, Ultek, Veeco.

Ultra high vacuum systems: using sorption or expulsion pumps for pure metal film production.
AEI, Airco Temescal, Balzers, Cooke, Edwards, International Vacuum, Leybold, Mill Lane Engineering, RIBER, UHV, Ultek, Vacuum Generators.

Vapour sources: refractory metal filaments and boats, ceramic crucibles, electron beam types (electrostatic, focused gun and magnetic bent beam).
Airco Temescal, Alcatel, Balzers, Borax Consolidated, Denton, Edwards, Electrotech, Evaporation Apparatus, GTE Sylvania, Leybold, Mill Lane Engineering, Nanotech, Union Carbide, Vactronic, Vacuum Generators.

2.3.9 Sputtering plant: applications

Element and compound films: for research purposes.
Standard: Balzers, Cooke, Edwards, International Vacuum, Ion Beam Systems, Materials Research, Nordiko, Ultek.
Custom built: Balzers, Cooke, ERA, Edwards, International Vacuum, Ion Beam Systems, Materials Research, Mill Lane Engineering, Nordiko, Ultek, Vacuum Generators.

Ion beam thinning and milling.
Alba Engineers, Balzers, Commonwealth Scientific, Edwards, International Vacuum, Materials Research, Mill Lane Engineering, Nordiko, Planer, Technics, Ultek, Vacuum Generators.

Microelectronics: thin film passive devices, capacitors, resistors, connecting leads.
Standard: Balzers, Cooke, Edwards, International Vacuum, Ion Equipment, Leybold, Materials Research, Ultek, Vacuum Industries.
Custom built: Airco Temescal, Balzers, Cooke, ERA, Edwards, International Vacuum, Ion Equipment, Leybold, Materials Research, Mill Lane Engineering, Nordiko, Ultek, Vactronic, Vacuum Generators, Vacuum Industries.

Optical coatings: beam splitters of metals or dielectrics.
Standard: Airco Temescal, Balzers, Cooke, Edwards, International Vacuum, Leybold, Materials Research, Ultek.
Custom built: Airco Temescal, Balzers, Cooke, ERA, Edwards, International Vacuum, Leybold, Materials Research, Mill Lane Engineering, Nordiko, Ultek, Vactronic, Vacuum Generators.

Semiconductor coatings: insulating films and conducting electrodes.
Standard: Airco Temescal, Balzers, Cooke, Edwards, International Vacuum, Ion Equipment, Leybold, Materials Research, Ultek, Vacuum Industries.
Custom built: Airco Temescal, Balzers, Cooke, ERA, Edwards, International Vacuum, Ion Equipment, Leybold, Materials Research, Mill Lane Engineering, Nordiko, Ultek, Vactronic, Vacuum Generators, Vacuum Industries.

Sputter-etching: profile and structure.
Alba Engineers, Balzers, Commonwealth Scientific, ERA, Edwards, International Vacuum, Ion Beam Systems, Materials Research, Mill Lane Engineering, Nordiko, Planer, Technics, Ultek, Vacuum Generators, Vacuum Industries, Veeco.

MANUFACTURER:	Alcatel	Cooke	Denton	Edwards High Vacuum
Model no.	SCM450, SCM600	C70-6	DUSS universal sputtering system and Denton low pressure sputtering apparatus	E306/E500 STD Sputtering access
Plant type*		VC		VC
Type of sputtering	RF or DC diode, triode	RF, DC diode or triode	RF/DC triode	DC, AC, RF
Vessel dimensions				
Diameter (m)		0·6		0·306 0·450
Height (m)		0·46		0·190 0·225
Pumping system†		D or TM		D
Target characteristics				
Diameter (m)		0·025 to 0·6	0·34 diameter on basic system	0·075–0·204 (RF) Up to 0·300 (DC)
Number		1–4		Up to 4
Substrate table diameter		0·025 to 0·6		Up to 0·0250 (water cooled)
Power supply				
h.t. (kV)			5 kV, 400 mA	3 kV 500 mA
d.c. (kW)		Yes		1·5 kW
r.f. (kW)		Yes		1 to 5 kW
Target material				
Deposition rate				
Gas used				
Å/s		Dielectrics 2–20 Metals 6–40		Dielectrics up to 30 Metals up to 60

* BJ Bell jar
BX Box
HC Horizontal cylinder
VC Vertical cylinder

† D Diffusion
C Cryopump
R Rotary
S Sorption
SI Sputter ion
TM Turbomolecular

MANUFACTURER:	Electrical Research Association	International Vacuum	Ion Beam Systems		
Model no.	E RA Mark III	IVI-5000	E30	IBS30	A50
Plant type*	Horizontal box	BJ, HC, VC, BX	Ion beam/target vessel systems →		
Type of sputtering	DC or RF diode, double electrode, RF/DC, RF with DC or RF substrate bias, DC with RF or DC substrate bias. Evaporation with RF or DC substrate bias	RF or DC diode	Extracted ion beam/high/ultra high vacuum vessel		
Vessel dimensions					
Diameter (m)	1·12 x 0·5	Up to 1·2	0·30	0·15	0·30
Height (m)	0·07	Up to 0·6	0·40	Cube	0·40
Pumping system†	D or TM	R, D or TM	D, SI, TM	D, SI, TM	D, SI, TM
Target characteristics					
Diameter (m)	Various E.G. 0·34, 0·13	Up to 0·02			
Number	2, 6	Up to 4	Multi (4)	1	1
Substrate table diameter (m)	0·34 diameter on basic system	Up to 0·9	To customer requirements	To customer requirements	To customer requirements
Power supply					
h.t. (kV)	Yes				
d.c. (kW)	Yes	Up to 7·5	0–30	0–30	0–50
r.f. (kW)	Yes	Up to 20	0·5	0·05	Optional
Target material	Metal, semi conductor, dielectric. Power or solid	Metals and dielectrics	Optional →		
Deposition rate					
Gas used	Argon	Argon, O_2, N_2			
Å/s	Up to 17 for silica	250–400 S, O_2 1500-2500 Au			

* BJ Bell jar
BX Box
HC Horizontal cylinder
VC Vertical cylinder

† D Diffusion
C Cryopump
R Rotary
S Sorption
SI Sputter ion
TM Turbomolecular

MANUFACTURER:	Ion Beam Systems *cont'd*		Ion Equipment Corporation	
Model no.	A180	SES	SM218 sputter down SM118 sputter up	SM224 sputter down SM124 sputter up
Plant type*	Ion beam/target vessel system	VC	Sputtering modules	Sputtering modules
Type of sputtering	Extracted ion beam/high/ultra high vacuum vessel	RF plasma etching	RF/DC	RF/DC
Vessel dimensions				
Diameter (m)	0·30	0·15	0·45	0·6
Height (m)	0·40	0·20		
Pumping system†	D, SI, TM	D		
Target characteristics				
Diameter (m)		0·12	0·12 or 0·2	0·12 or 0·2
Number	1	1	1–3	1–3
Substrate table diameter (m)	Jigging to customer requirements	0·12	0·12 or 0·2. Heaters, rotary and lifting pedestal available	0·12 or 0·2. Heaters, rotary and lifting pedestal available
Power supply				
h.t. (kV)	180 kV			
d.c. (kW)			5 kV, 300 mA, 5kV 600 mA	5 kV, 300 mA, 5 kV 600 mA
r.f. (kW)	Optional	0·75	0·5–5	0·5–5
Target material	Optional	Optional	A range of refractory compounds hot pressed without binders available	
Deposition rate				
Gas used	Any. Argon or oxygen pref.	Any. Argon or oxygen pref.		
Å/s		5		

* BJ Bell jar
BX Box
HC Horizontal cylinder
VC Vertical cylinder

† D Diffusion
C Cryopump
R Rotary
S Sorption
SI Sputter ion
TM Turbomolecular

MANUFACTURER:	Ion Equipment Corporation *cont'd*	Materials Research	Nordiko	
Model no.	CL500B	System 90, System 89; 8500, 8550, 8600, 8610, 8620, 8630, 870	System 1	System 2
Plant type*	Sputtering target	Glass or SS cylindrical	Multi target with rotary substrate table and evaporator	RF two targets of dissimilar materials
Type of sputtering	RF/DC	RF/DC, RF bias, etching and heating		
Vessel dimensions				
Diameter (m)				
Height (m)				
Pumping system †		D or TM		
Target characteristics				
Diameter (m)	Cylindrical 0·25 high, 0·12 diameter	0·05 to 0·3		
Number		1–4		
Substrate table diameter (m)				
Power supply				
h.t. (kV)				
d.c. (kW)	5 kV, 300 mA, 5 kV, 600 mA			
r.f. (kW)	0·5–5	1·25, 2, 3, 5		
Target material	A range of refractory compounds hot pressed without binders available			
Deposition rate				
Gas used				
Å/s		1·6–6·6 dielectrics 3·3–33·3 conductors		

* BJ Bell jar
BX Box
HC Horizontal cylinder
VC Vertical cylinder

† D Diffusion
C Cryopump
R Rotary
S Sorption
SI Sputter ion
TM Turbomolecular

MANUFACTURER:	**Nordiko** *cont'd*		**Ultek**
Model no.	System 3	System 4	Randex, model 2400 multi-target
Plant type*	Cylindrical target, concentric substrate holder, d.c. supply	Rectangular parallel twin cathodes for coating both sides of substrate	VC
Type of sputtering			RF with RF and DC bias and sputter-etching
Vessel dimensions			
Diameter (m)			0·6
Height (m)			0·2
Pumping system†			D, TM, SI
Target characteristics			
Diameter (m)			0·15
Number			3 (number up)
Substrate table diameter (m)			0·15 to 0·53
Power supply			
h.t. (kV)			SCR controlled l.t. supply 2 kVA model 603-0200
d.c. (kW)			
r.f. (kW)			1 or 2 (13·56 MHz)
Target material			
Deposition rate			
Gas used			
Å/s			

* BJ Bell jar
BX Box
HC Horizontal cylinder
VC Vertical cylinder

† D Diffusion
C Cryopump
R Rotary
S Sorption
SI Sputter ion
TM Turbomolecular

MANUFACTURER:	Vacuum Industries			Varian	
Model no.	Series 2200 Model 2305/6	Series 2250 Model 3408	Series 2200 Model 2021	Unit A	Unit B
Plant type*	VC	VC	VC	VC	VC
Type of sputtering	RF diode vessel module	RF diode	RF diode	RF diode	RF diode
Vessel dimensions					
Diameter (m)	0·46	0·75	0·61	0·30	0·45
Height (m)	0·36	0·37	0·25	0·25	0·45
Pumping system†		D and R, TM	D and R	As specified	As specified
Target characteristics					
Diameter (m)	0·13	0·20	0·53	0·20	0·15
Number	1	2 or 4	1	1	3
Substrate table diameter (m)	0·08	0·20		0·18	0·37
Power supply					
h.t. (kV)					
d.c. (kW)					
r.f. (kW)	1 (13·56 MHz)	2 (13·56 MHz)	10 (13·56 MHz)	1 or 2 (13·56 MHz)	1 or 2 (13·56 MHz)
Target material					
Deposition rate					
Gas used					
Å/s					

* BJ Bell jar
BX Box
HC Horizontal cylinder
VC Vertical cylinder

† D Diffusion
C Cryopump
R Rotary
S Sorption
SI Sputter ion
TM Turbomolecular

MANUFACTURER:	Varian *cont'd*		Veeco		
Model no.	Unit C	Coaxial 1-CSS-30	500 series	501	511
Plant type*	VC	VC	Targets, work holders sputtering accessories	VC	BJ
Type of sputtering	RF diode	RF diode		RF sputter-etching and evaporation	RF cylindrical diode axial target
Vessel dimensions					
Diameter (m)	0·60	0·45		0·46	0·46
Height (m)	0·45	0·76		0·76	0·76
Pumping system†	As specified	TM			
Target characteristics					
Diameter (m)	0·20	0·76 long		Note; the target is also the evaporation receiver	Inner cylinder
Number	4	1			
Substrate table diameter (m)	0·50	7·870 (cm^2)			Cylindrical jig with indexed rotary arms
Power supply					
h.t. (kV)					
d.c. (kW)					
r.f. (kW)	1 or 2 (13·56 MHz)	1 or 2 (13·56 MHz)			
Target material			Range available of: borides, carbides, fluorides, nitrides, silicides, sulphides, selenides, metals and their oxides		
Deposition rate					
Gas used					
Å/s					

* BJ Bell jar
BX Box
HC Horizontal cylinder
VC Vertical cylinder

† D Diffusion
C Cryopump
R Rotary
S Sorption
SI Sputter ion
TM Turbomolecular

MANUFACTURER:	Veeco				
Model no.	515	525	528	535	538
Plant type*	VC (pyrex)	VC (pyrex)	VC (stainless steel)	VC	BJ
Type of sputtering	RF planar-diode	RF planar-diode	RF	RF/DC planar-diode	RF/DC planar-diode
Vessel dimensions					
Diameter (m)	0·46	0·46	0·61	0·46	0·61
Height (m)	0·25			0·25	0·25
Pumping system†	10 CFM mechanical pump	Model VE 766A	Model VE 766A	Model VE 766A	Model VE 766A
Target characteristics					
Diameter (m)	0·125	0·125	0·2	0·125	0·2
Number	1 (offset)	2 (offset)	2 (offset)	3 (offset)	3 (offset)
Substrate table diameter (m)		Elevating, rotary, indexing	Elevating, rotary indexing	Elevating, rotary, indexing	Elevating, rotary, indexing
Power supply					
h.t. (kV)					
d.c. (kW)					
r.f. (kW)	1 (Crystal controlled)	1 (Crystal controlled)	2 (Crystal controlled)	0·5 (Crystal controlled)	3 (Crystal controlled)
Target material	Range available of: borides, carbides, fluorides, nitrides, silicides, sulphides, selenides, metals and their oxides				
Deposition rate					
Gas used					
Å/s					

* BJ Bell jar
BX Box
HC Horizontal cylinder
VC Vertical cylinder

† D Diffusion
C Cryopump
R Rotary
S Sorption
SI Sputter ion
TM Turbomolecular

2.4 Manufacturers' names and addresses

AEI SCIENTIFIC APPARATUS LTD., P.O. Box 1, Edinburgh Way, Harlow, Essex, England.

APV MITCHELL (DRYERS) LTD., Denton Holme, Carlisle, England.

AEROQUIP CORPORATION, Aerospace-Marman Division, 11214 Exposition Boulevard, Los Angeles, California 90064, U.S.A.

AEROVAC CORPORATION, P.O. Box 448, Troy, New York 12181, U.S.A.

AIRCO TEMESCAL (A division of Air Reduction Co. Inc.), 2850 7th Street, Berkeley, California 94710, U.S.A.

ALBA ENGINEERS, Asinères, France.

ALCATEL VACUUM TECHNIQUES, 41 Rue Perier, 92 Montrouge, France.

ALLEY COMPRESSORS LTD., 149 Newlands Road, Cathcart, Glasgow S.4, Scotland.

ANACONDA AMERICAN BRASS CO., Anaconda Metal Hose Division, Waterbury, Connecticut 06720, U.S.A.

ANALYTICAL INSTRUMENTS LTD., Fowlmere, Royston, Hertfordshire SG8 7QS, England.

ANDONIAN CRYOGENICS INC., 26 Farwell Street, Newtonville, Massachusetts 02160, U.S.A.

APIEZON PRODUCTS LTD., 8 York Road, London, SE1, England.

ATLAS COPCO (Great Britain) Ltd., Maylands Avenue, Hemel Hempstead, Hertfordshire. England.

CHARLES AUSTEN PUMPS LTD., 100 Royston Road, Byfleet, Surrey, England.

BDH CHEMICALS LTD., Broom Road, Poole, Dorset, England.

BTI: BRAD THOMSON INDUSTRIES INC., 83–810 Tamarisk Street, Indio, California 92201, U.S.A.

HENRY BALFOUR & CO. LTD., Durie Foundry, Leven, Fife, Scotland.

BALZERS A.G., FL-9496 Balzers, Liechtenstein.

BARLOW WHITNEY LTD., Watling Street, Bletchley, Buckinghamshire. England.

BINGHAM WILLAMETTE CO., 2800 North West Front Avenue, Portland, Oregon 97210, U.S.A.

S. BLICKMAN INC., 536 Gregory Avenue, Weehawken, New Jersey 07087, U.S.A.

BORAX CONSOLIDATED LTD., Borax House, Carlisle Place, London, S.W.1, England.

BOSCH & NOLTES APPARATEN, Anjelier Straat 224–230, Amsterdam C., Holland.

SAS ING. BRIZIO BASI & CO., 20128 Milano, Viale Monza 198–200, Italy.

BURCKHARDT ENGINEERING WORKS, Basle, Switzerland.

BÜTTNER-SCHILDE-HASS AKTIENGESELLSCHAFT, D-5630 Remscheid-11, Postfach 110125.

CVC: CONSOLIDATED VACUUM CORPORATION, 1775 Mount Read Boulevard, Rochester, New York 14603, U.S.A.

CAHN DIVISION, VENTRON CORPORATION, 7500 Jefferson Street, Paramount, California 90723, U.S.A.

CAJON CO., 32550 Old South Miles Road, Cleveland, Ohio 44139, U.S.A.

CARBOLITE COMPANY LTD., Bamford Mill, Bamford, Sheffield, England.

CARRIER ENGINEERING CO. LTD., Carrier House, Warwick Row, London S.W.1, England.

CENCO, CENTRAL SCIENTIFIC CO., 2600 South Kostner Avenue, Chicago, Illinois 60623, U.S.A.

CENTORR ASSOCIATES INC., Canal Street, Suncook, New Jersey, U.S.A.

CERAMASEAL INC., New Lebamon Centre, New York 12126, U.S.A.

COMMONWEALTH SCIENTIFIC CORPORATION, 500 Pendleton Street, Alexandria, Virginia 22314, U.S.A.

COOKE VACUUM PRODUCTS INC., 13 Merritt Street, South Norwalk, Connecticut 06854, U.S.A.

THE CRAWFORD FITTING CO., 29500 Solon Road, Solon, Ohio 44139, U.S.A.

DATAMETRICS, 127 Coolidge Hill Road, Watertown, Massachusetts 02172, U.S.A.

DAVIS AND WILDER INC., 1115 East Arques Avenue, Sunnyvale, California 94086, U.S.A.

DAWSON MCDONALD AND DAWSON LTD., Compton Works, Ashbourne, Derbyshire, England.

DENTON VACUUM INC., Cherry Hill Industrial Centre, Cherry Hill, New Jersey 08034, U.S.A.

DITRIC CORPORATION, 170 Maple Street, Marlboro, Massachusetts 01752, U.S.A.

DIVERSEY CORPORATION, 100 West Monroe Street, Chicago, Illinois 60603, U.S.A.

DRESSER INDUSTRIES INC., Industrial Products Division, 900 West Mount Street, Connersville, Indiana 47331, U.S.A.

DUNELLEN LTD., Number 8 Factory, Dunmurry Trading Estate, Dunmurry, Co. Antrim, N. Ireland.

DURCO, THE DURIRON CO. INC., 3490 South Dixie Avenue, Kettering, Ohio 45439, U.S.A.

EDWARDS HIGH VACUUM, Manor Royal, Crawley, Sussex, England.

EISLER ENGINEERING CO., 750 South 13th Street, Newark, New Jersey 07103, U.S.A.

ELECTRICAL RESEARCH ASSOCIATION, Cleeve Road, Leatherhead, Surrey, England.

ELECTROTECH ASSOCIATES, Prince of Wales Industrial Estate, Abercarn, Monmouthshire, England.

ELPHIAC S.A., Rue PJ Antoine 79, B 400 Herstal, Belgium.

EVAPORATION APPARATUS, 1440 Crystal Street, Los Angeles, California 90031, U.S.A.

EXTRANUCLEAR EXPORT CORPORATION, P.O. Box 11412, Pittsburgh, Pennsylvania 15238, U.S.A.

FERROFLUIDICS CORP., 144 Middlesex Turnpike, Burlington, Massachusets 01803, U.S.A.

FISONS SCIENTIFIC APPARATUS, Bishop Meadow Road, Loughborough, Leicestershire, LE11 ORG, England.

FLEXTUBE LTD., Mark Road, Hemel Hempstead, Hertfordshire, England.

FOSTER CAMBRIDGE LTD., Sydney Road, Muswell Hill, London, N10 2NA, England.

FOXBORO-YOXALL LTD., Redhill, Surrey, England.
FURNESS CONTROLS LTD., 21 Endell Road, Bexhill-on-Sea, Surrey, England.
GEC ELLIOTT CONTROL VALVES LTD., Airport Works, Rochester, Kent, England.
GEC LTD., Hirst Research Centre, East Lane, Wembley, Middlesex, England.
GTE SYLVANIA INC., Lighting Products Group, Special Products Division, Emissive Products, Portsmouth Avenue, Exeter, New Hampshire 03833, U.S.A.
OFFICINE GALILEO, Via C. Bini 44, Firenze, Italy.
GAST MANUFACTURING CORP., P.O. Box 117, Benton Harbour, Michigan, U.S.A.
APPARATEBAU GAUTING GMBH., Gauting, Ammerseestrasse 31, Germany.
GELMAN INSTRUMENT CO., 600 South Wagner Road, Ann Arbor, Michigan 48106, U.S.A.
GENERAL ELECTRIC COMPANY, 570 Lexington Avenue, New York, New York 10022, U.S.A
THE GENERAL ENGINEERING CO.(RADCLIFFE) LTD., Station Works, Bury Road, Radcliffe, Manchester M26 9UR, England.
S.A.E.S., Getters SpA Via Gallarate 215, 20151 Milano, Italy.
GRANVILLE-PHILLIPS CO., 5675 East Arapahoe Avenue, Colorado 80303, U.S.A.
GRESHAM & CRAVEN LTD., Chippenham, Wiltshire SN15 1JD, England.
HASTINGS RAYDIST, P.O. Box 1275, Hampton, Virginia 23361, U.S.A.
HATTERSLEY NEWMAN HENDER LTD., Burscough Road, Ormskirk, Lancashire, England.
HEWLETT PACKARD LTD., The Graftons, Altrincham, Cheshire, England.
HICK HARGREAVES AND CO. LTD., Soho Works, Bolton BL3 6DB, Lancashire, England.
HIGH VACUUM EQUIPMENT CORP., 2 Churchill Road, Hingham, Massachusetts 02063, U.S.A.
HIGH VOLTAGE ENGINEERING LTD., 8 Kildare Close, Eastcote, Ruislip, England.
HOKE INTERNATIONAL LTD., 1 Tenahill Park, Cresshill, New Jersey 07626, U.S.A.
HOTPACK CORP., 5022 Cottman Avenue, Philadelphia, Pennsylvania 19135, U.S.A.
HULL CORP., Hatboro, Pennsylvania 19040, U.S.A.
HUNTINGTON MECHANICAL LABORATORIES INC., 1988 Leghorn Street, Mount View, California 94040, U.S.A.
INDUSTRIAL ELECTRON BEAMS (BOC), Manor Royal, Crawley, Sussex, England.
INGERSOLL RAND CO. LTD., Bowater House, Knightsbridge, London SW1X 7LU, England.
INTERNATIONAL VACUUM INC., (617) 826–3195 Oak Street, Pembroke, Massachusetts, 02359, U.S.A.
ION BEAM SYSTEMS, CRL Edwards High Vacuum, Manor Royal, Crawley, Sussex, England.
ION EQUIPMENT CORP., 1805 Walsh Avenue, Santa Clara, California 95050, U.S.A.
IPSEN INDUSTRIES, P.O. Box 500, Rockford, Illinois 61105, U.S.A.
THE JET-VAC CORP., 73 Pond Street, Waltham, Massachusetts 02154, U.S.A.

JIGTOOL PRODUCTS PTY. LTD., Melbourne, Australia.
O. JÜNKER GMBH, 5101 Lammersdorf, Durenerstrasse, Germany.
KINNEY VACUUM CO., 3529 Washington Street, Boston, Massachusetts 02130, U.S.A.
THE KRAISSL CO. INC., 299 Williams Avenue, Hackingsack, New Jersey, U.S.A.
KRONOS INC., 1647–7 West Sepulveda Boulevard, Torrance, California 90501, U.S.A.
LKB-PRODUKTER AB, Fredsforssitgen 22–24 Mariehäll, Germany.
LARSON ELECTRONIC GLASS, Box 371, Redwood City, California, 94063, U.S.A.
LASER OPTICS LTD., Mill Plain Road, Danbury, Connecticut 06810, U.S.A.
LEYBOLD-HERAEUS GMBH & CO. KG, 5000 Köln 51, Bonner Strasse 504, Germany.
LUCIFER AG, 8610 Uster-Zurich, Ackerstrause 42, Switzerland.
CH. LUMPP & CIE, 12 Rue Jouffroy-d'Abbans, Lyon 9, France.
MKS INSTRUMENTS INC., 25 Adams Street, Burlington, Massachusetts 01803, U.S.A.
MATERIALS RESEARCH CORPORATION, Orangeburg, New York 10962, U.S.A.
METALS RESEARCH LTD., Melbourn, Royston SG8 6EJ, Hertfordshire, England.
METTLER INSTRUMENTS AG, CH 8606 Greifensee – Zurich, Switzerland.
MICROTOL ENGINEERING CORPORATION, Box 154, State College, Pennsylvania 16801, U.S.A.
MILL LANE ENGINEERING CO. INC., 10 Gasfield Circle, Burlington, Massachusetts, U.S.A.
MONSANTO CO., 800 North Lindbergh Boulevard, St. Louis, Missouri 63166, U.S.A.
J. MONTZ GMBH, 401 Hilden, Hofstrasse 82, Germany.
NGN LTD., Church Bank Works, Church, Nr. Accrington, England.
NANOTECH (THIN FILMS) LTD., Sedgely Trading Estate, Bury New Road, Prestwich, Manchester, England.
THE NASH ENGINEERING CO., Norwalk, Connecticut 06856, U.S.A.
NESLAB INSTRUMENTS INC., 871 Islington Street, Portsmouth, New Hampshire 03801, U.S.A.
NEWMAN & ESSER, 51 Aachen, Classenstrasse 11, Schliessfach 1049.
NEW JERSEY MACHINE CORP., 16th Street and Willow Avenue, Hoboken, New Jersey 07030, U.S.A.
NEYRPIC, Usine de Beauvert, 75 Rue General Mangin, 38 Grenoble, France.
NORDIKO LTD., Brockhampton Lane, Havant, Hants PO9 1JB, England.
NORTHEY ROTARY COMPRESSORS LTD., Alder Road, Parkstone, Poole, Dorset BH12 4BB, England.
NUCLIDE CORP., 642 East College Avenue, State College, Pennsylvania, U.S.A.
NUPRO CO., 15635 Saranac Road, Cleveland, Ohio 44110, U.S.A.
ORMANDY & STOLLERY LTD., Station Road, Brightlingsea, Colchester, Essex, England.
OXY METAL FINISHING (GREAT BRITAIN) LTD., Sheerwater, Woking, Surrey, England.
PENBERTHY DIVISION, HOUDAILLE INDUSTRIES INC., Prophetstown, Illinois 61277, U.S.A.
PENNWALT CORP., Stokes Vacuum Components Dept., 5500 Tabor Road, Philadelphia, Pennsylvania 19210, U.S.A.

A. PFEIFFER VAKUUMTECHNIK GMBH, 6330 Wetzlar, Postfach 147.
PHÖNIX ARMATUREN-WERKE, 6 Frankfurt (Main) 94, Postfach 9400230.
PHYSICAL ELECTRONICS INDUSTRIES INC., 7317 South Washington Avenue, Edina, Minnesota 55435, U.S.A.
GV PLANER LTD., Windmill Road, Sunbury-on-Thames, Middlesex TW16 7HD, England.
POPE SCIENTIFIC INC., N90 W 14337 Commerce Drive, Menomenee Falls, Wisconsin 53051, U.S.A.
PRECISION SCIENTIFIC CORP., 3737 West Cortland Street, Chicago, Illinois 60647, U.S.A.
PRINTA INKS & PAINTS LTD., Neogene Works, Alfred Road, London W.2, England.
RALET-DEFAY, 63–66 Boulevard Poincaré, 1070 Brussels, Belgium.
RED POINT CORP., 105 West Spazier Avenue, Burbank, California 91502, U.S.A.
RIBER, BP 65 95 Boulevard de l'Hôpital Stell, Rueil-Malmaison, Paris, France.
RICOR LTD., En-Harod (Ihud), Israel.
SADLA VAC LTD., Halliwell Mill, Weymouth Street, Bolton, Lancashire, England.
SARGENT-WELCH SCIENTIFIC CO., 7300 North Linder Avenue, Skokie, Illinois 60076, U.S.A.
SCHUF ARMATUREN GMBH, D 623 Frankfurt/Main, 80 Farkenstrasse 96, Germany.
SCHUTTE AND KOERTING CO., Cornwells Heights, Bucks County, Pennsylvania 19020, U.S.A.
SHAWFRANK ENGINEERING CORP., 6 North River Road, Des Plaines, Illinois 60016, U.S.A.
SIGMA LABORATORIES INC., 171 Sherwood Place, Englewood, New Jersey 07631, U.S.A.
SIHI GMBH AND CO., KG, 221 Itzehoe/Holstein, Germany.
SLOAN TECHNOLOGY CORP., 535 East Montecito Street, Santa Barbara, California 93103, U.S.A.
SMITH-DENNIS LTD., Bridge Road West, Nottingham NG7 5HZ, England.
SPIRAX SARCO LTD., Charlton House, Cheltenham GL53 8ER, England.
G. SPRINGHAM AND CO. LTD., 35 South Road, Temple Fields, Harlow, Essex, England.
STANTON REDCROFT (Division of Loertling Ltd.), Copper Mill Lane, London SW17 0BN England.
TECHNICS INC., 80 N. Gordon Street, Alexandria, Virginia 22304, U.S.A.
TELTRON LTD., 32–36 Telford Way, London W.6., England.
TENNEY ENGINEERING INC., 1090 Springfield Road, Union, New Jersey 07083, U.S.A.
THELCO, Precision Scientific Corporation, 3737 West Cortland Street, Chicago, Illinois 60647, U.S.A.
THERMO ELECTRON ENGINEERING CORP., 85 First Avenue, Waltham, Massachusetts 02154, U.S.A.
THOMSON CSF, 51 Boulevard de la Republique, BP17, 78 Chatou, France.
TORR VACUUM PRODUCTS, 7121 Hayvenhurst Avenue, Van Nuys, California 91406, U.S.A.

TORVAC LTD., Histon, Cambridge, England.

TWENTIETH CENTURY ELECTRONICS LTD., King Henry's Drive, New Addington, Croydon CR9 0BG, England.

U.H.V. LTD., Thor Works, Berinsfield, Oxford, England.

ULTEK DIVISION, Perkin Elmer Corp., Palo Alto, California, U.S.A.

UNION CARBIDE CORP., Carbon Products Division, 270 Park Avenue, New York 10017, U.S.A.

VG MICROMASS LTD., Nat Lane, Winsford, Cheshire CW7 3BX, England.

VACTORR, 3737 West Cortland Street, Chicago, Illinois 60647, U.S.A.

VACTRONIC LABORATORY EQUIPMENT INC., East Northport, New York 11731, U.S.A.

VACUDYNE CORP., 375 E Joe Orr Road, Chicago Heights, Illinois 60411, U.S.A.

VACUUM ACCESSORIES CORP. OF AMERICA, P.O. Box 322, 193 West Hills Road, Huntington Station, New York 11746, U.S.A.

VACUUM BARRIER CORP., 4 Barten Lane, Woburn, Massachusetts 01801, U.S.A.

VACUUM ENGINEERING (SCOTLAND) LTD., Motherwell House, Motherwell, Lanarkshire, Scotland.

VACUUM GENERATORS LTD., Charlewoods Road, East Grinstead, Sussex, England.

VACUUM INDUSTRIES INC., 34 Linden Street, Massachusetts 02143, U.S.A.

VACUUM RESEARCH MANUFACTURING CO., 3100 Crow Canyon Road, San Ramon, California 94583, U.S.A.

VARIAN, 611 Hansen Way, Palo Alto, California 94303, U.S.A.

VEECO INSTRUMENTS INC., Terminal Drive, Plain View, New York 11803, U.S.A.

VENTRON CORP., Alfa Products, 2098 Pike Street, San Leandro, California 94577, U.S.A.

J. M. VOITH GMBH, Heidenheirn (Brenz), W. Germany.

VIRTIS CO., Route 209, Gardiner, New York 12525, U.S.A.

WALLACE & TIERNAN LTD., Priory Works, Tonbridge, Kent, England.

K. WEISS-GIESSEN, D. 6301 Lindenstruth, W. Germany.

WENTGATE ENGINEERS LTD., Industrial Estate, St. Ives, Huntingdonshire, England.

WESTINGHOUSE, 93-Freinville-Sevran, BP2, France.

WHITEY CO., 5679 Landregan Street, Oakland, California 94608, U.S.A.

WIEGAND APPARATEBAU GMBH, 75 Karlsruhe-West, Postfach 4469, Andreas-Hofer Strasse 3, Germany.

WILD BARFIELD LTD., Otterspool Way, Watford By Pass, Watford WD2 8HX, England.

3 — Recent Developments in Vacuum Science and Technology

3.1 Vacuum pumps: recent developments

3.1.1 Introduction

A comprehensive account of vacuum pumps of all types is given by Power (1966) in his book 'High Vacuum Pumping Equipment'. Apart from some account needed for clarification, the present article therefore deals mainly with developments since 1965/6. These developments are almost all in the form of improvements on existing types of pump. Only one radically new method of pumping has been introduced: the accommodation pump (Hobson, 1970; Hobson and Pye, 1972) and this will not be considered further here because it does not promise to become a commercially viable pump for transporting significant quantities of gas.

Little need be said about vapour pumps and sorption pumps. The former have not changed significantly in design though much attention has been paid to the development of pump fluids in the form of oils which give rise to minimum backstreaming and decomposition products. Sorption pumps based on molecular sieve materials (particularly the zeolites 5A and 13X) have not changed in design either, though a fair amount of work has been devoted to the study of their performance, particularly from the point of view of the effect of a pre-sorbed gas on the sorption of a second gas.

3.1.2 Positive displacement pumps

Rotary pumps of this class have been developed with direct motor drive to operate at rotational speeds of as much as 1500 rev/min so that the pump volume rate of flow (speed) is twice or more that obtained with a pulley or gear driven pump of similar physical dimensions but operating at 400 to 700 rev/min.

Oil-sealed rotary pumps give rise to backstreaming oil and decomposition products of the oil at the inlet, with consequent organic contamination of the vacuum system. This is deleterious to several processes undertaken *in vacuo* and, in particular, when the oil-sealed rotary pump is used to create a fore-pressure, results in the malfunctioning of a sputter-ion pump and spoilation of the performance of a vapour-diffusion pump, especially when the latter is used to create an ultra-high vacuum.

The two main approaches to solving this problem which provide an oil-free vacuum by a rotary pump have been (a) to construct a 'dry' or oil-free pump and (b) to introduce a suitable trap for organic contaminants in the line to the oil-filled rotary pump inlet.

(a) the construction of an oil-free rotary pump has been attempted by making the vanes and bearings of a sliding vane rotary (Gaede type) pump of self-lubricating materials. The most suitable material is polytetrafluoroethylene (PTFE) with additives intended to strengthen the PTFE, reduce seep and minimize wear. Such additives to PTFE may be carbon, lead oxide, glass fibre, asbestos, carbon fibre or molybdenum disulphide. Holland (1970) and Baker, Holland and Stanton (1972) allege that the two-vane Gaede type rotary pump with PTFE containing carbon and graphite (Graflon) is the most effective design. Such a pump with a mild steel stator electroplated with nickel, Graflon vanes and drive shaft vacuum sealing by means of two lip seals with the interspace filled with a lubricant paste compounded with a low vapour pressure lubricant and fine PTFE powder, operated for 500 hours at a rotational speed of 1450 rev/min at a steady temperature of 20°C above ambient. Such a single-stage oil-free rotary pump is used frequently with a backing stage of the conventional oil-sealed type (see also Brand, 1970).

The alternative approach (b) of using a trap in the inlet line to an oil-filled pump has been chiefly by means of a trap containing molecular sieve (Zeolite 13X). Explored by Hennings and Schütze (1966) and Holland, Fulker and Laurenson (1967), a disadvantage is that the Zeolite 13X deteriorates somewhat after water vapour is sorbed. Hence activated alumina is preferred. Fulker (1968) has shown that a 50 mm column of activated alumina balls traps 99·7% of the backstreaming vapours from a rotary pump filled with Edwards 16 oil (a hydrogen oil without special additives). Baker and Staniforth (1968) allege that activated alumina balls each of 5 to 10 mm diameter are the most effective. Baker, Holland and Stanton (1972) made an effective trap in the form of an in-line cylinder 50 mm long by 50 mm diameter containing 6 x 4 mm pellets of alumina on which active nickel was deposited chemically, and containing an electrical heater for temperature control. The large surface area of active nickel so presented offers the advantage in trapping that hydrocarbons are chemisorbed without preferential physisorption of water vapour.

Sadler (1969) has described an oil-filled 'planetary piston pump' which has the advantages of a rotary piston pump (as compared with the sliding vane rotary pump, chiefly avoidance of wear as the vanes scrape the inside wall of the stator) plus double the free air displacement (the pump volume rate of flow at atmospheric pressure) of conventional rotary piston pumps of the same overall physical dimensions.

The most outstanding recent development, however, is probably the positive displacement rotary pump based on the principle of the Wankel engine. Following the work of Wutz (1970), Baechler and Knobloch (1972) report on the performance of a trochoid rotary pump capable of attaining compression ratios of 10^5. This makes use of a circulating piston (rotor) of elliptical cross-section within a stator of which a section of the inner wall is a cardioid (Fig. 3.1). There is consequently one sealing line between the points where the rotating elliptical piston touches the inner wall of the stator. The pump inlet and the discharge outlet (to the atmosphere) are on either side of this line. At these 'sealing points' it is essential for the inevitable gaps to be exceedingly small if a low ultimate pressure is to be attained. As the problems of making the difficult shapes involved forbid sufficiently small gaps, a so-called 'moving seal ledge' is fitted into a slot in the inner wall of the stator.

Fig. 3.1 A trochoidal rotary pump based on the Wankel principle. (Courtesy of Leybold–Heraeus GmbH and Co. Ltd.)

The face of this ledge is shaped to bear snugly and continuously against the elliptical surface of the rotating piston. The volume rate of flow (speed) against pressure characteristic of this type of pump is compared with that of a conventional rotary piston pump in Fig. 3.2. Advantages afforded compared with the conventional pump are a shorter inlet channel, improved static and dynamic balance enabling rotational speeds without undue vibration and with quiet running, and lower drive power demands.

The Roots pump, defined as 'a positive displacement pump in which a pair of interlocking and synchronized lobed rotors rotate in opposite directions, moving past each other and the housing wall with a small clearance and without touching' (BS 2951, Part 1, 1959) is frequently used as a 'mechanical booster' to obtain very high volume rates of flow in the pressure region from 1 to 10^3 Pa. It discharges to a backing pump (e.g. an oil-sealed rotary pump or a liquid ring pump) where the capacity ratio of the Roots pump to the backing pump is normally between 5 and 15. Usually, the backing pump reduces the

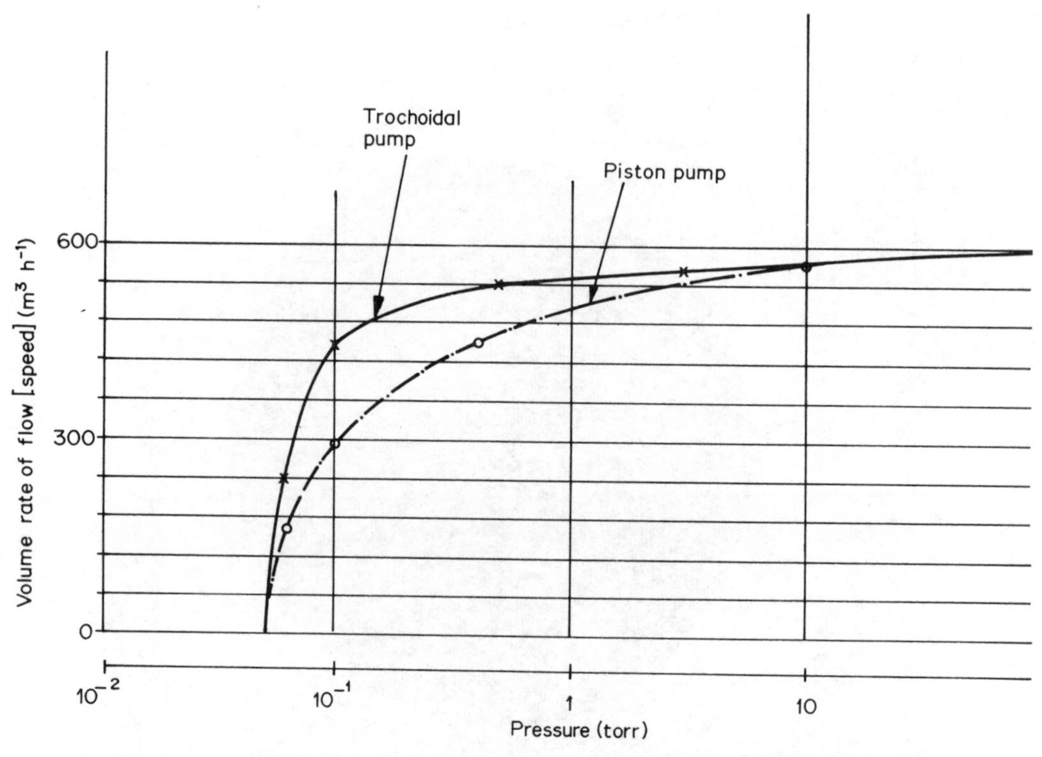

Fig. 3.2 Volume rate of flow against pressure characteristic of the trochoidal rotary pump.

pressure to about 10^3 Pa before the Roots pump is brought into operation. To save pump-down time on large vacuum systems, Wycliffe and Salmon (1971) describe the advantages of a hydrokinetic drive (fluid coupling drive) which, by providing a controllable torque, enables adjustable slip to occur so that the Roots pump rotates slowly when the gas load is high (at and near atmospheric pressure) and accelerates to full rotational speed as the pressure falls progressively.

3.1.3 Molecular drag pumps; the turbomolecular pump

Originally introduced in 1912 by Gaede, the molecular drag pump underwent checkered development until 1958 when Becker (1958) introduced his turbomolecular pump, which has now become a valuable device, especially for producing a 'clean' vacuum at pressures below 10^{-4} Pa (Becker, 1966; Henning, 1971).

Osterstrom and Shapiro (1972) describe a new form of turbomolecular pump aimed to provide an ultimate total pressure of 10^{-6} Pa with a compression ratio for hydrogen of 50 and utilizing only 8 rotor discs and 7 stator discs (as against 19 and 20 respectively in the Becker design) at each end of the double-ended machine (Fig. 3.3). The first pumping element (nearest the pump inlet) is a rotor disc instead of a stator disc as in the Becker

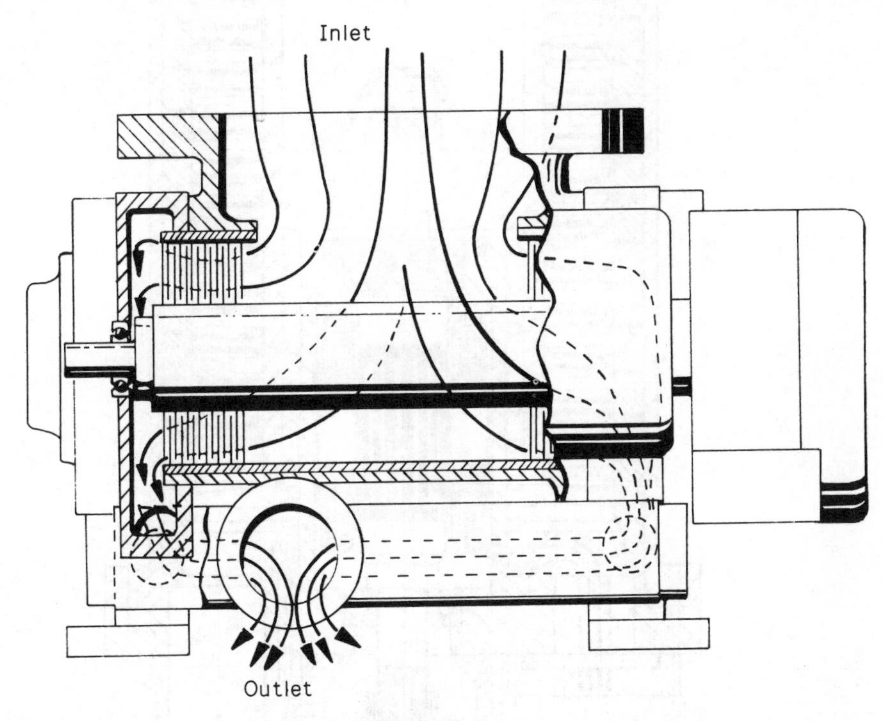

Fig. 3.3 The turbomolecular pump of Osterstrom and Shapiro.

pump. The reduced number of rotor and stator discs forbids as high a compression ratio but enables a much larger centre space in the pump and so a much larger pump inlet port (optionally vertical at the side of the housing or horizontal at the top). Combined with double-cut rotor and stator discs at the inlet side, the pump volume rate of flow for nitrogen at a rotor speed of 42 000 rev/min (as compared with 16 000 for the Becker pump) is alleged to be 1600 litre/second as compared with 140 litre/second for a Becker pump of approximately the same physical dimensions. This new pump also incorporates an integral liquid-nitrogen cooled panel having a pump volume rate of flow for water vapour of 16 000 litre/second.

Mirgel (1972) describes a turbomolecular pump of vertical design (Fig. 3.4) with the rotor drive by a squirrel cage motor within the pump (i.e. in the vacuum) which provides, at a rotational speed of 24 000 rev/min, a pumping speed of 370 litre/second for air with a compression ratio of 10^6 for air and 10^2 for hydrogen, and a pressure operating range of 1 to 10^{-6} Pa. The firm SNECMA in France make* a vertical turbomolecular pump of similar design which gives a pump volume rate of flow for air of 650 litre/second and an ultimate pressure (on bakeout) of 6×10^{-9} Pa in conjunction with a mechanical rotary pump having a volume rate of flow of 500 litre/min with a diffusion pump of 100 litre/second.

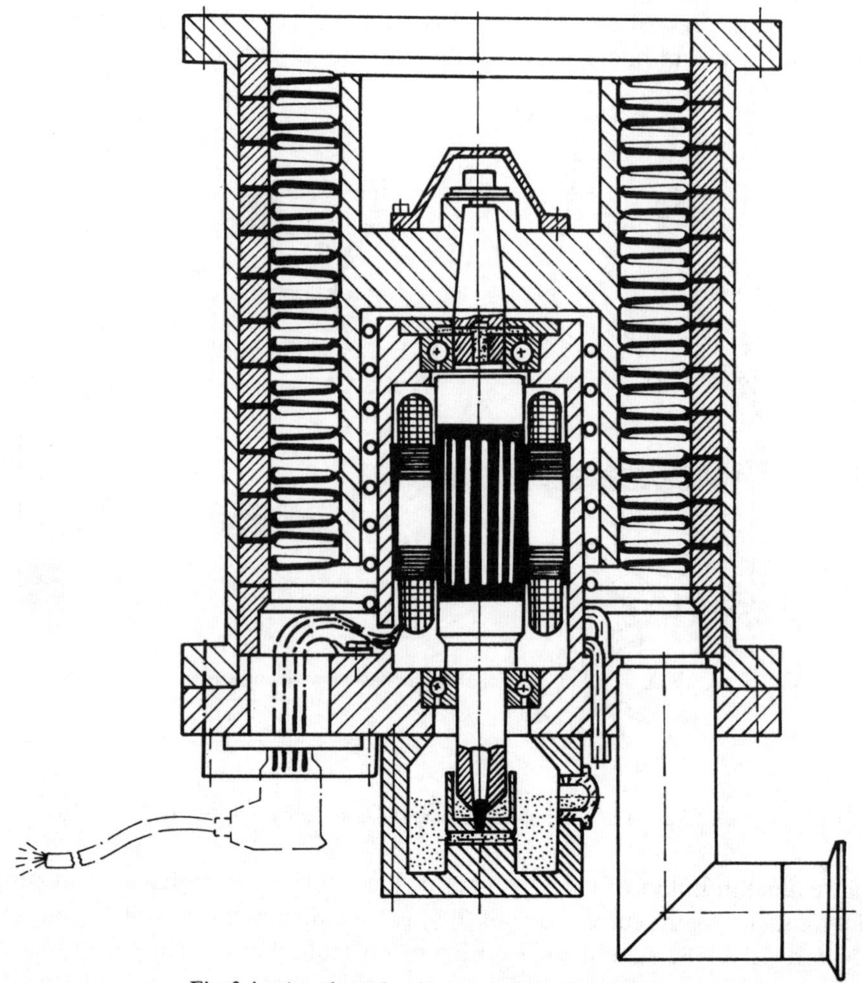

Fig. 3.4 A turbomolecular pump of vertical design.

Following the introduction by Beams (1961), turbomolecular pumps are now often supplied with an integral electrical heater which enables degassing of the pump at temperatures of up to 400°C under vacuum.

3.1.4 Getter pumps and getter-ion pumps

To improve the pump volume rate of flow for the inert gases, Tom and James (1966) introduced a 'differential' sputter-ion pump having two cathodes of different metals, one

* The suppliers are Leybold-Heraeus Ltd.

titanium and the other tantalum, apparently leading to enhanced inert gas ion burial on the cathode of lower sputtering rate. A fuller consideration of the operation of this pump is given by Jepsen (1968) using his so-called 'energetic-neutral' hypothesis. The 'differential' sputter-ion pump using combinations of the cathode materials aluminium, titanium and

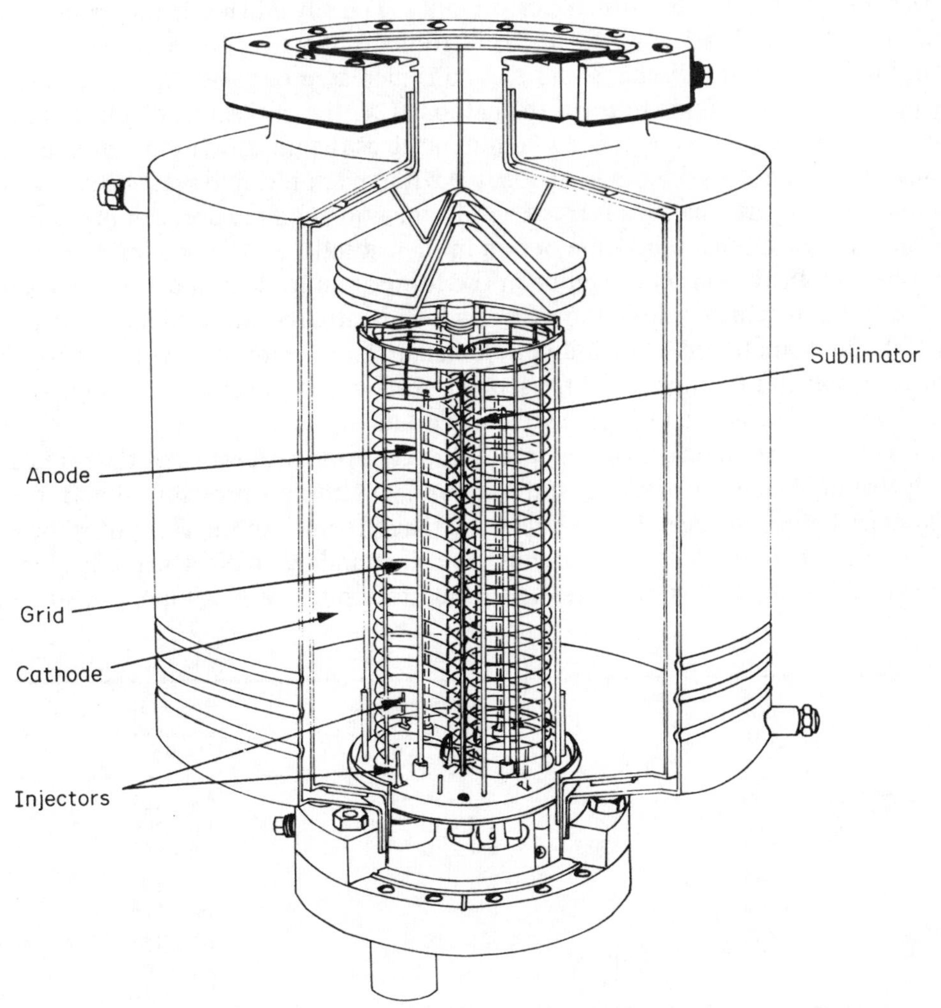

Fig. 3.5 The triode electrostatic getter-ion pump.

tantalum has also been explored experimentally by Baker and Laurenson (1972). Tom (1972) describes a sputter-ion pump in which one of the cathodes is of magnesium lined with spaced, parallel, tungsten rods. Magnesium provides the required getter material but also, having a high vapour pressure compared with titanium, evaporates much more readily under localized heating in the discharge. The relatively abundant magnesium atoms made available ensure a higher ionization current (and hence higher pump volume rate of flow) for a given pressure of gas in the pump. The volume rate of flow is alleged to be increased by 50% compared with the use of titanium for both cathodes and, moreover, the magnesium vapour (which condenses readily on cold surfaces in the pump) does not contribute to the ultimate total gas pressure obtainable.

Electrostatic getter-ion pumps of triode geometry have been introduced by Bills (1972) whilst Denison (1972) describes their performance. The advantages of the triode design (Fig. 3.5) over the diode pattern (e.g. the orbitron pump) are (a) the number of energetic electrons injected into the discharge per second is increased because the capacitance per unit length of the anode is greater, (b) the cathode (i.e. the outermost electrode) diameter may be increased (so increasing the gas conductance as the inlet) without influencing the ion formation within the grid-anode region, (c) with an appreciable bias voltage between the cathode and the grid, all ions formed within the grid-anode region can be accelerated to sufficiently high speeds to become buried in the cathode, and (d) the grid-anode cell need not be concentric and may even be in the form of multiple grid-anode arrangements within the cathode cylinder along the axis of which is situated the sublimator of titanium. With a 150 mm diameter inlet, a triode pump had a volume rate of flow for nitrogen of 1700 litre/second and for argon 25 litre/second. A new resistance heated sublimator of titanium (Warren, Denison and Bills, 1967) was used.

Denison (1968) has studied experimentally the gettering properties of the rare earth element yttrium in an electrostatic getter-ion pump. Yttrium is probably superior to titanium in gettering hydrogen at low pressures. The yttrium (in the form of ribbon 0·96 x 2·6 mm) on a sheath of molybdenum foil surrounding an electrically heated beryllia core, when sublimated at low pressures was found to produce an ultimate pressure in which

TABLE 3.1

Filament type	Current (A)	Temp (K)	Average evaporation rate over first 15 min (g cm^{-1} s^{-1})	Energy for evaporation (J g^{-1})	Available titanium (mg cm^{-1})
85% Ti, 15% Mo*	50	1780	$2{\cdot}3 \times 10^{-4}$	$4{\cdot}8 \times 10^{6}$	55
	55	1860	$1{\cdot}0 \times 10^{-5}$	$1{\cdot}6 \times 10^{6}$	
	57	1900	$2{\cdot}1 \times 10^{-5}$	$7{\cdot}5 \times 10^{6}$	
Variant†	45		5×10^{-6}	3×10^{6}	60
	48		$1{\cdot}3 \times 10^{-5}$	$1{\cdot}3 \times 10^{6}$	
Pure titanium‡	500		$4{\cdot}6 \times 10^{-5}$	$1{\cdot}8 \times 10^{6}$	700
		1750		$8{\cdot}3 \times 10^{6}$	
		2450		$1{\cdot}75 \times 10^{4}$	

* Hairpin: length 240 mm; diameter 2·10 mm.
† Tungsten wire 0·76 mm diameter overwound with two 0·76 mm diameter titanium wires and a 0·38 mm diameter molybdenum wire.
‡ Tantalum wire core overwound with one layer of 0·76 mm diameter niobium and two layers of 0.76 mm diameter titanium (Clausing, 1966).

the partial pressures of hydrogen and methane were about 0·1 and 0·33 of those obtained with titanium and with excellent retention of argon. Unfortunately, the yttrium surface is readily poisoned by most active gases. It is therefore recommended to use titanium sublimation to reach the 10^{-9} Pa range and then employ yttrium sublimation subsequently.

Titanium sublimation pumps (titanium getter pumps) have become more widely used in recent years. McCracken and Pashley (1966) stress the advantages from the points of view of good mechanical strength (even after evaporation), reliability, high resistivity and self-regulation (so avoiding burn-out) of 85% Ti, 15% Mo alloy as a filament material for titanium sublimation. Their tests on this alloy in comparison with some other titanium sublimators lead to the results summarized in Table 3.1.

Lawson and Woodward (1967) also report on the 85% Ti, 15% Mo alloy wire, showing that about 40% of the titanium in the alloy can be evaporated whilst at constant voltage, 24% can be evaporated at a constant, predictable rate and that hydrogen is the chief gas present in this alloy.

Harra and Snouse (1972) describe a titanium sublimator in the form of an approximately spherical shell of titanium (32 mm o.d. of wall thickness 3·8 mm) supported on four molybdenum rods and heated by radiation from an internal tungsten heater. It can dispense about 37 g of titanium at a rate of up to 0·5 g h^{-1}.

The pump volume rate of flow is considerably increased if the titanium is evaporated on to a surface at liquid nitrogen temperature (–196°C). The sticking probabilities for nitrogen, oxygen, carbon monoxide and carbon dioxide are then between 0·8 and 1·0, and for hydrogen between 0·2 and 0·5. Consequently a pump volume rate of flow of about 10 litre per second per cm^2 of titanium surface is achievable, though the inert gases are not pumped. With such chilled titanium films, saturation is at about 10^{16} molecules/cm^2 and as much as 10^{19} molecules/cm^2 for hydrogen, which diffuses rapidly into the titanium. Prevot and Sledziewski (1972) report that titanium monoxide (TiO) at –186°C gives better results than titanium, with a capacity for air four times greater than that of titanium, and also that the chilled TiO pumps argon effectively, with a sticking probability of 0·17 (pumping volume rate of flow of 1·7 litre/second per cm^2).

In a survey of developments in the pumping of large space-simulation chambers, Barnes and Pinson (1968) recommend titanium sublimation in which titanium rods of as much as 20 mm diameter are used to sublime 1 g per hour on to the chilled surfaces of the thermal shroud of a liquid helium cryopump. High speed for hydrogen (one of the major outgassing components of steel) is achieved. Robertson (1968) describes a titanium sublimator in which electron bombardment heating is used (regulated power from 0·6 to 3 kW) and capable of dispensing 8 g per hour. The Russian workers Butussow, Masslennikow, Miljagin, Tschuchin and Schewtscenko (1968) describe a titanium sublimator in drop form heated by electron bombardment (the electrons are confined to fairly definite paths by a magnetic field) capable of dispensing 3 to 6 g per hour.

Non-evaporable getter pumps based on the gettering properties of a zirconium–aluminium alloy (ST 101 from SAES – Getters Ltd.) have been developed in Italy. They are used chiefly as appendage pumps (Pisani and della Porta, 1967) particularly in sealed-off vacuum tubes. A high gettering rate for all the active gases is obtained (della Porta and Ferrari, 1968). The Zr–Al alloy is deposited in a thin layer on both sides of a metal strip (without organic binders) suitably pleated so that large active areas are provided in a small volume. So that the getter material may develop its pumping properties, activation by heating *in*

Fig. 3.6 Non-evaporable getter pumps.
(Courtesy of SAES-Getters, Milan)

vacuo to 700–800°C for some minutes is practised. Subsequently, the getter surface is at room temperature or up to 400°C, where heaters within the pump serve both to activate the getter and maintain it at any desired operating temperature (Giorgi and Ricca, 1963). These convenient non-evaporable getter pumps are now available in a variety of sizes and forms (Fig. 3.6).

3.1.5 Cryopumps

The chief recent advances have been in the development of cryopanels cooled with either liquid hydrogen (b.p. 20·35 K) or liquid helium (4·25 K). Liquid hydrogen, in addition to being cheaper, has the advantage over liquid helium of a latent heat of vaporization of 7640 calorie per litre as compared with only 650 calorie per litre but the disadvantages of fire risk (avoidable by satisfactory design of plant as, for example, shown by Gomer, 1972) and that hydrogen itself has a high vapour pressure at 20 K, make liquid helium the better choice.

Stern, Hemstreet and Ruttenbur (1966) have developed 'cryosorption' panels for the pumping of hydrogen gas at 20 K which consist of as much as 40 g of molecular sieve 5 A (preferably mixed with 5% of fine aluminium flakes to enhance thermal conductivity and diminish thermal emissivity) bonded to a cross-milled aluminium plate. After suitable heat treatment for activation *in vacuo*, at 20 K (liquid hydrogen cooling) a pumping volume rate of flow of 6600 litre/second per ft^2 of panel is obtained, corresponding to a sticking coefficient of hydrogen on the panel of 0·16. However, cryosorption methods, initially promising, have been abandoned except for relatively small chamber pumping because their performance deteriorates considerably because of 'poisoning' of the sorbents by high freezing point gases (Barnes and Pinson, 1968). Cryotrapping (e.g. the trapping of hydrogen and/or helium by means of CO_2 or argon pre-condensed on to cryopanels at 20 K) has also become deprecated because of inadequate efficiency. For large space-simulation chambers, in particular, the present practice is to make use of titanium sublimation, cryopumping with liquid helium and sputter-ion pumping of inert gases.

Turner and Hogan (1966) have developed a small cryopump which is based on a cryopanel of surface area 1000 cm^2 chilled to 30 K or below by a modified Taconis-cycle refrigerator (the working fluid is helium gas) developed by Challis and Hogan (1966). In conjunction with a 50 litre/second sputter-ion pump, a pumping volume rate of flow for air of 3200 litre/second is achieved at pressures below 6×10^{-2} Pa and ultimate pressures of 6×10^{-7} Pa are obtained in an unbaked stainless steel chamber of volume 150 litre. This ultimate pressure can be reduced to about 3×10^{-8} Pa by subliming titanium on to a liquid-nitrogen chilled chevron baffle. Eder, Hengevoss and Woessner (1968) describe a cryopump with an integral helium refrigerator based on the simple Claude cycle without making use of a Joule-Thomson expansion stage. Eder and Wossner (1971) have developed further the integration of cryopanels with a refrigeration unit. Pumping volume rates of flow in production systems are obtainable with small 1 watt refrigerator type cryopumps (the amount of heat which has to be removed from the condensation surface and baffle is about 1 W at 20 K and 15 W at 80 K for a pumping speed of 10 000 litre/second). Possible refrigerators are those based on (a) the Brayton helium cycle including two expansion engines at 20 K and 80 K with heat exchangers, (b) the Claude hydrogen cycle with only one expansion engine at medium temperature and partial liquefaction of the working gas and (c) Gifford – McMahon cycles applying thermal regenerators and displacers without external work.

Interest has developed in cryopumps in which chilled helium gas is circulated (Bailey and Chuan, 1959). Akiyama, Nakayama and Saito (1971) discuss a space simulation chamber consisting of a 2 m diameter cylinder with doors, pumped initially by a chevron-baffle shielded diffusion pump. The cryopanel consists of a set of aluminium pipes through which is circulated helium gas at 10 K and just inside a liquid-nitrogen chilled cylindrical shroud against the inside wall of the chamber. Attached to this shroud are fins acting as radiation shields for each pair of helium gas chilled pipes. A Monte Carlo analysis of the molecular flow and sorption within this system enabled the pumping volume rate of flow to be estimated to be 48 000 litre/second for nitrogen as compared with an experimentally determined value of 47 500 litre/second. Furthermore, this method of calculation enabled the transmission probability of the gas molecules to be related to the geometry and configuration of the shroud fins so that an optimum arrangement could be predicted.

Holkeboer (1968) describes a 'cryosublimation' trap. A welded stainless-steel casing encloses a copper baffle cooled by liquid nitrogen on which titanium is deposited. This baffle is designed to ensure that any molecule passing through it makes at least two collisions with the active surface. To prevent this baffle structure from being by-passed by gas molecules, it is connected to the casing by a thin stainless steel diaphragm. Placed between the vacuum chamber and the oil vapour trap to an oil-diffusion pump, the use of this 'cryosublimation' trap is alleged to enable ultimate pressures of 4×10^{-12} Pa to be obtained and as low as 10^{-12} Pa if the cooling is by liquid helium instead of liquid nitrogen.

Cryopumping with liquid helium boiling under reduced pressure to achieve temperatures between 4·2 and 1·0 K has been investigated chiefly by Forth (1968).

REFERENCES

AKIYAMA, Y., NAKAYAMA, K., and SAITO, M. (1971) *Vacuum,* **21**, 167.
BAECHLER, W. G., and KNOBLOCH, D. (1972) *J. Vac. Sci. Technol.,* **9**, 402.
BAILEY, B. M., and CHUAN, R. L. (1959) *1958 5th Natl. Symp. Vac. Tech. Trans.,* 262.
BAKER, M. A., HOLLAND, L., and STANTON, D. A. G. (1972) *J. Vac. Sci. Technol.,* **9**, 412.
BAKER, M. A., and LAURENSON, L. (1966) *Vacuum,* **16**, 633.
BAKER, M. A., and LAURENSON, L. (1972) *J. Vac. Sci. Technol.,* **9**, 375.
BAKER, M. A., and STANIFORTH, G. H. (1968) *Vacuum,* **18**, 17.
BARNES, C. B., and PINSON, J. (1968) *Proc. 4th Intl. Vac. Congress,* 219, (Institute of Physics, London).
BEAMS, J. W. (1961) *Trans. 7th AVS Natl. Symp. 1960,* 1.
BECKER, W. (1958) *Vakuum-Technik,* **7**, 149.
BECKER, W. (1966) *Vacuum,* **16**, 625.
BILLS, D. G. (1967) *J. Vac. Sci. Technol.,* **4**, 149.
BRAND (1970) *Brit. Pat.* 1, 178, 265.
BUTUSSOW, W. I., MASSLENNIKOW, J. A., MILJAGIN, M. J., TSCHUCHIN, I. A., and SCHEWTSCENKO, T. A. (1968) *Proc. 4th Intl. Vac. Congress,* 377 (Institute of Physics, London).

CHELLIS, F. F., and HOGAN, W. H. (1964) *Advan. Cryog. Eng.*, **9**, 545.
CLAUSING, R. E. (1960) *Thermonuclear Semiannual Report* ORNL 2926.
DENISON, D. R. (1967) *J. Vac. Sci. Technol.*, **4**, 156.
DENISON, D. R. (1968) *Proc. 4th Intl. Vac. Congress,* 377 (Institute of Physics, London).
EDER, F. X., HENGEVOSS, J., and WOESSNER, H. (1968) *Proc. 4th Intl. Vac. Congress,* 409 (Institute of Physics, London).
EDER, F. X., and WÖSSNER, H. (1971) 'Ergebnisse der Hochvakuumtechnik und der Physik dünner Schichten' (M. Aüwarter, Ed.), 348 (Wissenschaftliche Verlag, Stuttgart).
FORTH, H. J. (1968) *Proc. 4th Intl. Vac. Congress,* 425 (Institute of Physics, London).
FULKER, M. J. (1968) *Vacuum,* **18**, 445.
GIORGI, T. A., and RICCA, F. (1963) *Suppl. Nuovo Cim.*, **5**, 261.
GOMER, R. (1972) *Vacuum,* **22**, 521.
HARRA, D. J., and SNOUSE, T. W. (1972) *J. Vac. Sci. Technol.*, **9**, 552.
HENNING, J. (1971) *Vacuum,* **21**, 523.
HOBSON, J. P. (1970) *J. Vac. Sci. Technol.*, **7**, 231.
HOBSON, J. P., and PYE, A. W. (1972) *J. Vac. Sci. Technol.*, **9**, 252.
HOLKEBOER, D. M. (1968) *Proc. 4th Intl. Vac. Congress,* 405 (Institute of Physics, London).
HOLLAND, L., FULKER, M. J., and LAURENSON, L. (1967) *Suppl. al Nuovo Cimento,* **5**, (1), 242.
HOLLAND, L. (1970) *Vacuum,* **20**, 175.
JEPSEN, R. L. (1968) *Proc. 4th Intl. Vac. Congress,* 317 (Institute of Physics, London).
LAWSON, R. W., and WOODWARD, J. W. (1967) *Vacuum,* **17**, 205.
MCCRACKEN, G. M., and PASHLEY, N. A. (1966) *J. Vac. Sci. Technol.*, **3**, 96.
MIRGEL, K. H. (1972) *J. Vac. Sci. Technol.*, **9**, 408.
OSTERSTROM, G. E., and SHAPIRO, A. H. (1972) *J. Vac. Sci. Technol.*, **9**, 405.
PISANI, C., and DELLA PORTA, P. (1967) *Suppl. Nuovo Cim.*, **5**, 261.
DELLA PORTA, P., and FERRARIO, B. (1968) *Proc. 4th Intl. Vac. Cong.*, 369 (Institute of Physics, London).
POWER, B. D. (1966) 'High vacuum pumping equipment' (Chapman and Hall, London).
PREVOT, F., and SLEDZIEWSKI, Z. (1972) *J. Vac. Sci. Technol.*, **9**, 49.
ROBERTSON, D. D. (1968) *Proc. 4th Intl. Vac. Congress,* 373 (Institute of Physics, London).
SADLER, P. (1969) *Vacuum,* **19**, 17.
STERN, S. A., HEMSTREET, R. A., and RUTTENBUR, D. M. (1966) *J. Vac. Sci. Technol.*, **3**, 99.
TOM, T., and JAMES, B. D. *AVS 13th Natl. Vac. Symp. Abstracts,* 21.
TOM, T. (1972) *J. Vac. Sci. Technol.* **9**, 383.
TURNER, F. A., and HOGAN, W. H. (1966) *J. Vac. Sci. Technol.*, **3**, 252.
WARREN, K. A., DENISON, D. R., and BILLS, D. G. (1967) *Rev. Sci. Instrum.*, **38**, 1019.
WUTZ, M. (1968) *Proc. 4th Intl. Vac. Congress,* 283 (Institute of Physics, London).
WYCLIFFE, H., and SALMON, A. (1971) *Vacuum,* **21**, 223.

3.2 Vacuum instruments for the analysis of surfaces

3.2.1 Introduction

Solid surfaces may be subjected to irradiation with photons, electrons, ions or neutral atoms or molecules. Alternatively the surface may be stimulated thermally or by the application of electric fields. The resulting physical effects observed may be in the form of the emission of photons, electrons, ions or neutral particles. The various instrumental developments arising from these techniques are summarized in Table 3.2.

It is seen that a number of instruments are now available suitable for specific investigations and measurements of solid surfaces. Small parts of a specimen surface may be

TABLE 3.2 Many instrumental techniques have been developed depending on excitation by photons, electrons and ions. The methods discussed in more detail in the text are shown here in capitals together with their respective paragraph numbers (eg. (6) corresponds to 3.2.6)

Detection	Excitation		
	Photons	Electrons	Ions
Photons	(2) ELLIPSOMETRY X-ray fluorescence X-ray diffraction X-ray microscopy Light microscopy Interferometry Spectrophotometry	(6) ELECTRON MICROPROBE (often using scanning electron microscope, with X-ray analyser attachment) Cathodoluminescence (usually with scanning electron microscopy)	Protons or ion impact X-ray analysis or u.v. analysis
Electrons	(7) ESCA (8) PHOTOELECTRON SPECTROSCOPY Photoelectron emission microscopy U.V.-electron emission microscopy	(3) ELECTRON DIFFRACTION; LEED, HEED SCANNING E.D. (4) AUGER ELECTRON SPECTROSCOPY Electron microscopy (5) SEM Electron mirror, Electron reflection microscopy, Appearance potential spectroscopy,* Ionization loss spectroscopy, Electron impact spectroscopy	Ion neutralization spectroscopy Ion impact with electron spectrometer
Ions	Laser beam mass spectrometry Photo-ionization mass spectrometry	Electron impact ionization mass spectrometry	(9) ION SCATTERING SPECTROMETRY (10) ION SCATTERING MICROSCOPE (12) STATIC SIMS (13) ION IMAGING SIMS (14) ION MICROPROBE

* Included here, since X-ray emission is usually detected by resulting photoelectrons (see also Tracy, J. C. (1972) J. App Phys. *10,* 4164–71).

studied by probing over a selected area or the microscopic distribution may be detected by scanning with a fine beam. Generally one is concerned with the properties of the solid-vacuum interface, and conventional vacuum and usually ultra high vacuum techniques are used to maintain surfaces sufficiently free from adsorption of impurities from the background gas during the measurement. However, all measurements tend to disturb the system being investigated. The degree of disturbance depends on the type of probe, its intensity and duration. Thus photons and electrons can give rise to desorption and decomposition effects, the associated heating can lead to sorption and diffusion phenomena while ions cause sputtering, ion damage and possibly surface roughening. The intrinsic sensitivity of a particular technique is often proportional to its destructiveness so that the relatively destructive ion beam may only disturb a very small percentage of the surface to obtain useful information.

To obtain solid surfaces for analysis, which have not been affected by exposure to the atmosphere or possible contamination by the vacuum system, facilities are often provided to generate clean surfaces (or thin films) within the ultra high vacuum system of the instrument itself. Measurements can then be made on an 'atomically' clean surface before appreciable interaction of the residual gas with the surface could have occurred. A number of different procedures have been used for the generation of the clean surfaces:

(a) cleavage *in vacuo* usually of single crystals
(b) deposition of films by evaporation or sputtering
(c) glow discharge cleaning
(d) ion bombardment cleaning
(e) sputter etching in a gas discharge or
(f) removal of surface layers by sputtering with ion beams directed at the surface.

The principal characteristics of the instruments to be discussed are summarized in Table 3.3 at the end of this section.

3.2.2 Ellipsometry

In this technique a beam of linearly polarized light is directed obliquely at the surface being measured and the polarization of the reflected beam is analysed. The underlying principle of such an instrument is shown in Fig. 3.7. E_p and E_s represent the amplitudes of the reflected electric vector parallel and perpendicular to the plane of incidence respectively and Δ is the phase difference between these two reflected components. In practice the commonly used ellipsometric parameters are ϕ and Δ obtained from a measurement of

$$E_p/E_s = \tan \phi,$$

and the phase difference

$$\Delta = \Delta_p - \Delta_s.$$

In the case of an absorbing sample the reflected beam is elliptically polarized and its analysis allows the determination of the optical constants n, k of the material to be determined. Similarly when a thin film covers a surface the analysis of the reflected beam gives data on film thickness and refractive index. The technique can be used for surfaces *in vacuo* when changes in mean film thickness corresponding to fractions of a mono layer can be detected. To derive film properties from the measured results, computers are often applied to obtain solutions to the fundamental equations.

For high sensitivity and to simplify the operation automatic ellipsometric instruments have been designed. Modulation techniques are used in which the polarization of the incident light is varied and the polarizer and analyser are automatically adjusted to obtain the minimum intensity position. In this way Δ and ϕ are simply obtained from a direct readout and possibly as a digital display.

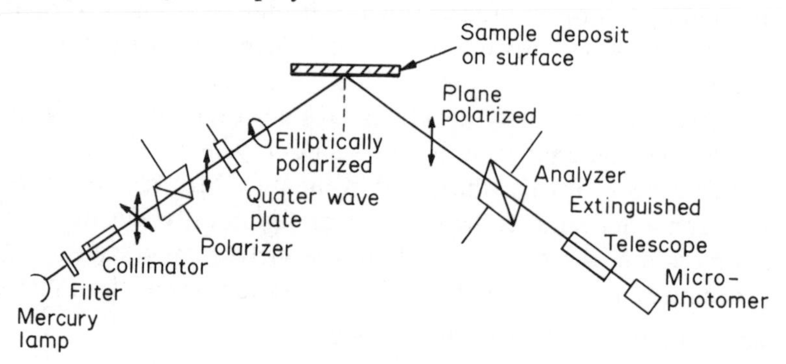

Fig. 3.7 Principle of ellipsometry instrument.

Ellipsometry has been most extensively applied to surfaces covered with thin films and used in the study of the growth of oxide layers on metals and semiconductors as well as sorption effects, contamination and coatings produced by evaporation and sputtering. In the case of very thin films interesting comparisons with other techniques such as Auger analysis and LEED have been made, but ellipsometry does not give any information on chemical composition but merely gives data on average optical thickness over the area 1 to 4 mm diameter covered by the beam from a bright discharge tube or laser source.

For further information on recent work with ellipsometer instruments and their applications, the papers in the conference proceedings edited by Passaglia et al. (1963), and that of Bashara et al. (1969) are of interest. Bootsma and Meyer (1969) discuss surface analysis of the submonolayer region. Vrakking and Meyer (1971) compared ellipsometry data with Auger analysis.

3.2.3 Electron diffraction (LEED and HEED)

In an electron diffraction instrument a point source of electrons is focused by an electron optical system on to a fluorescent screen. The electron beam is arranged to intercept the specimen to produce a diffraction pattern which may be observed on the fluorescent screen for the direct examination of surfaces. The reflected electron diffraction pattern from the surface is observed either at normal incidence or by inclining the surface at a small angle relative to the beam depending on the primary electron energy used.

A distinction is also usually made between low energy electron diffraction (LEED) employing electrons of the high energy range 5 to 10 000 eV and high energy electron diffraction (HEED) using electrons ranging from 10 to 100 keV. In the latter case one sometimes also refers to RHEED for reflection and THEED for the case of transmission electron diffraction.

In the interaction of a beam of monoenergetic electrons of energy E with a solid surface the elastically reflected electrons which have not suffered appreciably energy losses give rise to the diffraction pattern. The wavelength of the electrons is given by the relation due to De Broglie

$$\lambda = (150/E)^{1/2} \quad \text{(in Å)},$$

where E is their kinetic energy in eV.

At typical LEED energies, λ is of the order of 1 Å so that diffraction of the electrons by the crystal lattice of the surface layers occurs. The penetration of primary electrons is limited mainly by inelastic scattering and in the case of LEED is estimated at some 3 to 10 Å,

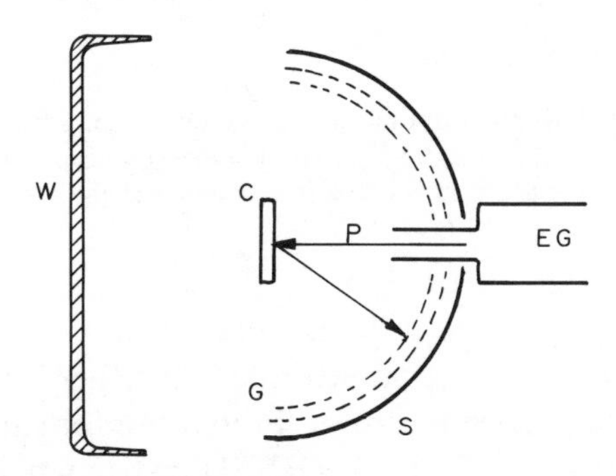

Fig. 3.8 Schematic of LEED instrument.
W = viewing window P = primary beam EG = electron gun and lenses
S = fluorescent screen C = crystal specimen G = grids

showing that only a few atomic layers parallel to the surface give rise to the elastic components of the emitted electrons. At higher energies, e.g. from 10 to 100 keV, the corresponding wavelengths ranges from 0·12 to 0·037 Å, and there is considerable penetration and forward scattering of the electrons. Therefore in RHEED experiments the specimen surface is arranged at a glancing incidence relative to the primary electron beam.

A typical LEED apparatus is shown schematically in Fig. 3.8. The specimen in the form of a single crystal is mounted on a crystal holder which must allow movement in three dimensions and usually has provision for heating and sometimes cooling of the crystal. The primary electrons are produced by an electron gun which floods the crystal surface by a nearly monoenergetic well-collimated electron beam. In the display type system the deflected beam reflected from the crystal surface gives rise to a spot on the fluorescent screen and the entire spot pattern can be viewed and photographed through a window in the vacuum chamber. High transparency grids are provided in front of the spherical screen. A typical arrangement is that the crystal and first grid is at ground potential, the second grid is used to reject inelastic electrons which have lost energy. A positive potential of several kV is applied to the screen to accelerate the electrons and improve the brightness of the display. To ensure well defined surface conditions it is necessary to operate the whole equipment in the ultra high vacuum range (10^{-9} Torr or better).

The geometrical construction of the diffraction conditions (Bragg's law) from a row of atoms is illustrated in Fig. 3.9.

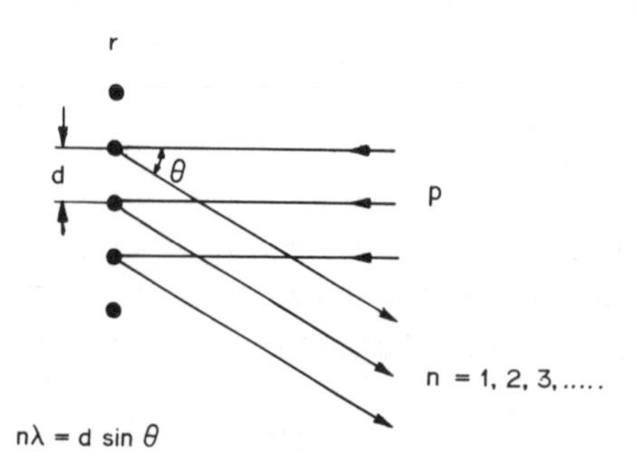

Fig. 3.9 Geometrical construction of diffraction conditions from row of atoms in crystal.
r = row of surface atoms
λ = wavelength of primary beam $\simeq (150/V)^{1/2}$ Å, where V = primary beam voltage
d = spacing between atoms
θ = angles at which diffraction occurs given by Bragg relation $n\lambda = d \sin \phi$, where, $n = 0, 1, 2, 3 \ldots$

Conventional electron microscopes or high voltage electron diffraction instruments can be used to obtain electron diffraction information either by transmission or reflection, but their conventional vacuum systems are of limited use for surface studies. However special ultra high vacuum version HEED or RHEED instruments have been designed.

Although considerable work has been done with electron diffraction both theoretically and experimentally, it appears that the complete interpretation of diffraction patterns is highly complex. It is a simple matter to determine the dimensions and spacing of the unit cell of a crystal lattice, but this is not so for any adsorbed atoms or surface compounds. The technique is therefore now usually combined with other experimental techniques, for example Auger electron spectroscopy.

A recent review of surface studies by electron diffraction was given by Estrup and McCrae (1971). A high voltage scanning electron diffraction system and its possible applications to surface analysis was described by Grigson (1968). In this case fast electromagnetic scanning of the pattern over a fixed energy filter and sensitive collector system produces a synchronous display of the intensity profiles.

3.2.4 Auger electron spectroscopy

Consider the bombardment of the surface with a beam of electrons (the primary electrons) of energy E then an analysis of the energy distribution of the secondary electrons and reflected electrons is of the form shown in Fig. 3.10(a) from which one can distinguish two distinct regions. The low energy region up to about 50 eV has a broad peak and is the main component of true secondary electron emission. Further electrons appearing at higher energies are the reflected electrons, showing a peak at the energy of the primary beam, and corresponding to the inelastic and elastic reflection effects at the surface. The peak at the upper end, corresponding to elastic reflection, is the one contributing to the

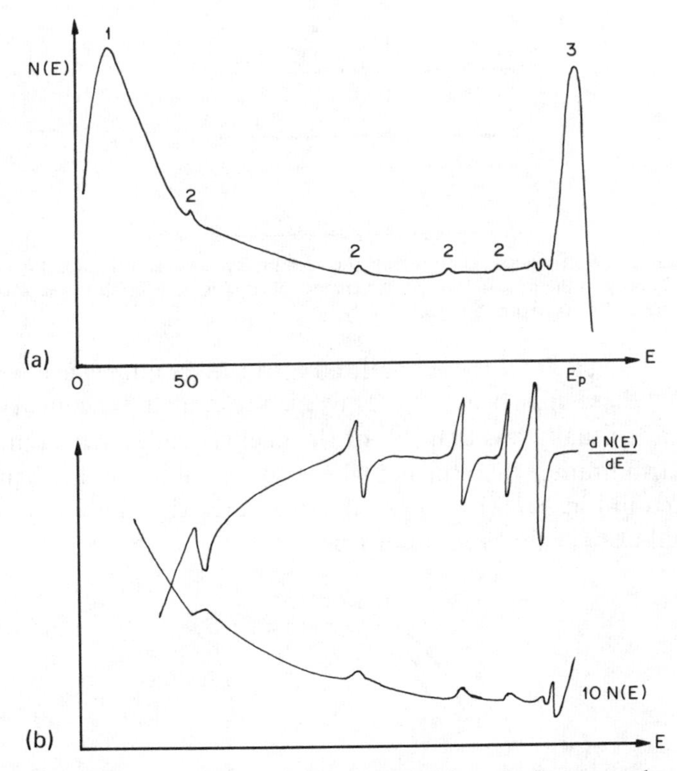

Fig. 3.10 Secondary electron emission energy distribution. (a) (1) Main peak of true secondary electron emission; (2) Small subsidiary peaks due to Auger electron emission process; (3) Peak at primary beam energy corresponding to elastic reflection of electrons (this is the peak contributing to the LEED pattern). (b) Magnified (x10) and differentiated energy distribution to show up the Auger peaks.

LEED pattern. Close examination of the intermediate region reveals the occurrence of very small peaks indicating the existence of preferred energies. Two types of these can be distinguished; the energy loss peaks due to primary electrons that have lost discrete amounts of energy, and a second group corresponding to the emission of Auger electrons produced by an ionization process of deep lying energy levels in the atom. Auger spectroscopy is concerned with revealing these small peaks in the energy spectrum.

When a beam of electrons with energies of 1 to 2 keV is used to bombard the surface the process is shown diagramatically in Fig. 3.11. Level 3 lying a few hundred volts below the vacuum level 0 is ionized by the primary electron. An electron from level 2 can occupy the hole in level 3, and the energy E_{2-3} released in this way can be transferred to an electron in the adjacent level 1. This electron will then leave the atom with an energy $E_{2-3} - E_{0-1}$. The emission of this electron is the Auger effect, and it is noted that the energy of the Auger electron depends on the energy level diagram of the atom involved, but is independent of the energy of the incoming primary electron beam.

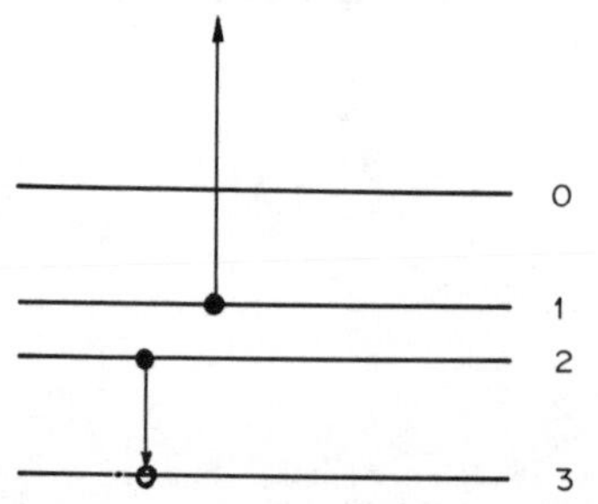

Fig. 3.11 Auger electron emission process. O represents the vacuum level. Level 3 is ionized by the primary electron. An electron from level 2 occupies the hole in level 3. The energy released is transferred to the adjacent level 1 resulting in the emission of the Auger electron from this level.

A fraction of the Auger electrons are emitted into the vacuum space without suffering any energy loss, and these appear in the energy spectrum of the secondary electrons as small characteristic maxima usually less than 1% of the background. By measuring the derivative of the energy spectrum rather than the actual energy distribution, the background is greatly reduced as indicated in Fig. 3.10(b). In practice the derivative is obtained electronically by a.c. modulation and fast synchronous detection.

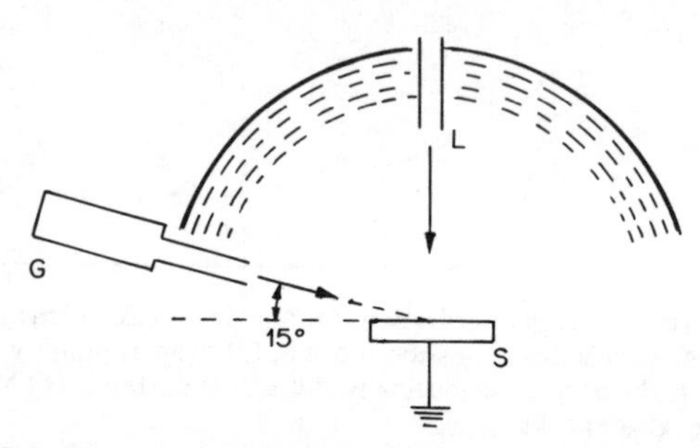

Fig. 3.12 LEED instrument adapted for Auger electron spectrometry.
G = electron gun, typically 3 kV, 100 μA S = sample L = LEED electron beam.

At present energy analysis is mostly carried out using retarding field analysers based on multi-grid LEED systems as indicated in Fig. 3.12. In this case the signal corresponds to the integrated secondary electron emission and to obtain the actual energy spectrum a first derivative needs to be obtained. Again in practice this is done by a.c. modulation of the retarding voltage applied to one or two of the grids. To obtain the Auger spectrum, it is necessary to have the derivative of the energy distribution, which in this case is the second derivative of the retarding potential curve. In practice this is obtained by measuring the second harmonic component of the collected current.

Greater sensitivity can be obtained by using deflection type energy analysers and Fig. 3.13 shows a scheme used for measuring an Auger spectrum with a coaxial cylinder (or electrostatic mirror type analyser). Since deflection type energy analysers give the energy distribution directly by scanning the voltage of a suitable electrode (the outer cylinder in this case) the Auger spectrum is obtained from the first derivative of the energy distribution, i.e. in practice again by a.c. modulation and synchronous detection of the fundamental frequency.

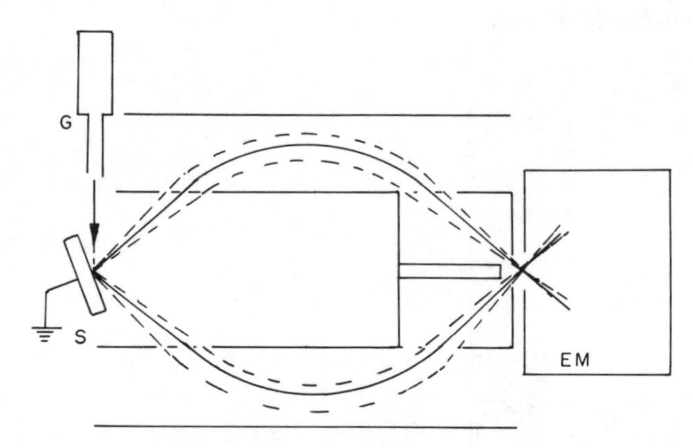

Fig. 3.13 Deflection type analyser for Auger electron spectrometry.
G = electron gun, typically 3 kV, 100 μA S = sample EM = electron multiplier detector

The method turns out to be very sensitive, and specific components at only 1% equivalent mono-layer coverage have been detected. In some instances however the spectra may be difficult to analyse. A considerable amount of accumulated Auger energy spectral data for different materials have been catalogued. The method is particularly sensitive for the detection of light elements, i.e. typically strong peaks occur at the characteristic Auger levels of oxygen, carbon and sulphur. This is in contrast to the insensitivity of X-ray fluorescent techniques with which it is complimentary.

Recent reviews were given by Chang (1971) and Gallon and Matthew (1972).

3.2.5 Scanning electron microscopy (SEM)

A finely focused beam of high energy electrons (greater than 5 keV) is scanned over the surface under investigation and the variation in intensity of the reflected and secondary electrons emitted is observed. The technique is often combined with other detectors to obtain a better characterization of the surface, as shown schematically in Fig. 3.14. The instrument can be used as an electron microprobe, in which the characteristic X-rays at the specimen surface are analysed by a suitable X-ray analyser combined with the instrument (as discussed more fully under this heading). The SEM can also be combined with a photo detector for cathodoluminescence analysis. Thin specimens may be investigated by observation of transmitted or forward scattered secondary electrons and an instrument used in this way is sometimes referred to as a scanning transmission electron microscope (STEM). Special STEM instruments based on a very fine focus electron gun using field electron emission (Crewe et al. 1968) have recently become available. In all scanning type instruments, the phenomena observed resulting from the scanning of the primary beam (i.e. electrons, X-rays or luminescence) are detected to give an electrical signal which can be displayed as an intensity variation on the screen of a synchronously scanned cathode ray tube. Such displays serve to give a very useful image of the topography and constitution of the surface under investigation.

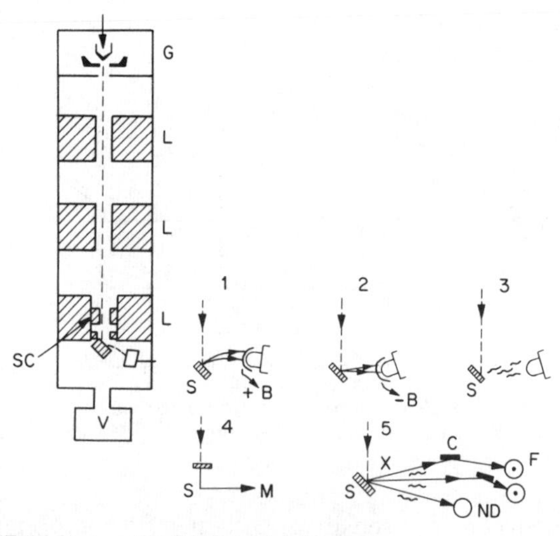

Fig. 3.14 Diagram showing SEM instrument and different detectors.
G = electron gun L = lenses SC = scanning coil S = specimen
1 = secondary + backscattered electrons detected by scintillator and multiplier, with bias electrode B positive
2 = backscattered electrons only detected with bias voltage B negative
3 = cathode-luminescence detected by light guide and photomultiplier only
4 = absorptive current detected (or conductive mode, for semiconductors) with meter M or display
5 = X-ray detection with analyser crystals C and flow proportional counters F, or by non-dispersive system ND using Li drifted silicon or Ge detectors with multichannel analysers.

Stereo scanning techniques can be used by the superimposition of two rasters taken with the electron beam tilted at two different angles in relation to the same region of the specimen. The present status and recent developments in the technique of SEM have been the subject of several conferences. For earlier work see Oatley, Nixon and Pease (1965) and the book by Thornton (1968); a brief review was given by Kammlott (1971), applications were reviewed in the book by Hearle, Sparrow and Cross (1972).

3.2.6 Electron microprobe

In this technique, electrons from a finely focused beam of sufficiently high energy are made to bombard the target surface. Inelastic scattering of the electrons in the target material leads to ionization and to the emission of X-rays. The chemical species present in the vicinity of the surface are identified by the characteristic X-rays emitted. A typical instrumental arrangement is shown in Fig. 3.15 with facilities for fine focusing a high intensity electron beam and X-ray analyser using in this case a full focusing crystal spectrometer of Johannsen geometry. Other dispersive detectors have also been used, e.g. lithium drifted silicon detectors combined with a multi-channel analyser.

For a full qualitative analysis a wide range of X-ray wavelengths needs to be separated and detected requiring the use of a range of detector crystals. Spatial resolution of better than 0·3 μm and quantitative measurements to within 5% or better are readily achieved.

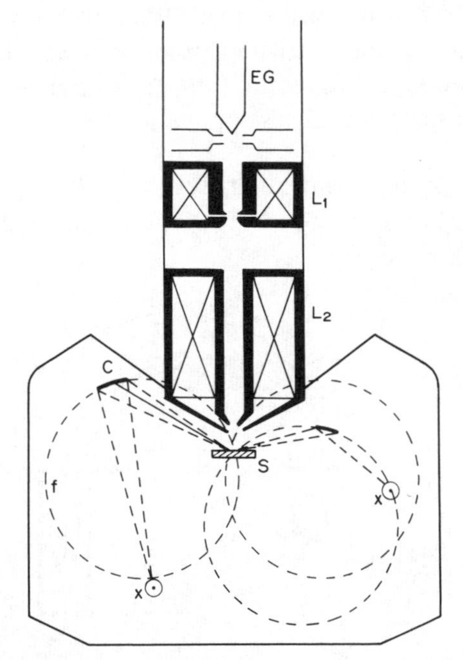

Fig. 3.15 Schematic of electron probe microanalyser, with crystal detectors.
EG = electron gun L_1 = condenser lens S = specimen L_2 = objective lens
In left hand X-ray spectrometer the crystal C is set at a high Bragg angle.
f = focusing circle x = X-ray detector.

However, the entire depth of penetration of the electron beam is sampled, which at typically 30 kV bombarding energy is usually greater than 1 μm. The electron microprobe is not really a true surface analysis tool, although it is one of the most widely used instruments for general analysis of solid samples. The probability of X-ray production P_x shows a strong dependence on the atomic number Z and may be expressed by

$$P_x = (1 + a/Z^4)^{-1}$$

where a is a constant depending on the energy level of the ionized electron. This is in contrast to Auger electron emission (discussed under the section dealing with Auger electron spectroscopy) whose probability P_A is exactly complimentary to this, i.e.

$$P_A = 1 - P_x.$$

It is evident that light elements (e.g. C, O, S) show a high probability of Auger electron emission and a negligible characteristic X-ray emission yield, and conversely for heavier elements. This has many important practical consequences and influences the range of applications and type of equipment used for either technique. Thus, some contamination of the specimen by carbon (e.g. through electron beam interactions) can be tolerated in the electron microprobe apparatus.

The developments in electron microprobe analysis have been covered by several international conferences held every three years since 1956, on X-ray optics and microanalysis. Proceedings of a conference on quantitative microanalysis were edited by Heinrich (1968). Recent advances were reviewed by Duncomb (1969). A brief review of Reuter (1971) gave particular emphasis to surface analysis applications.

3.2.7 Electron spectroscopy for chemical analysis (ESCA)

This makes use of the emission of photo electrons at characteristic energies due to the excitation of a sample by X-rays.

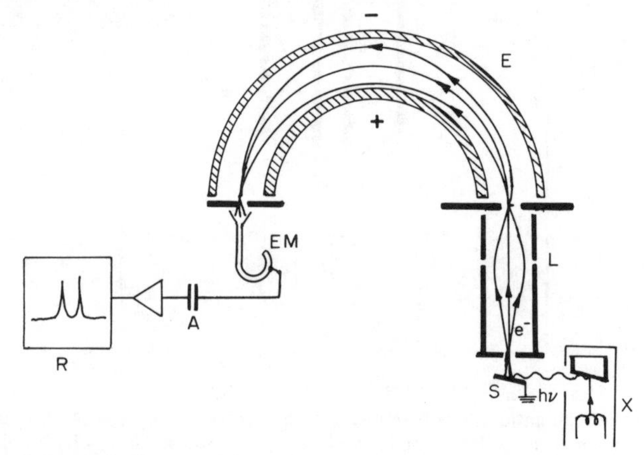

Fig. 3.16 Schematic of ESCA instrument.
E = hemispherical energy analyser
S = sample
EM = electron multiplier
R = X–Y recorder.
L = lens system
X = X-ray generator
A = amplifier and rate meter

Typically an X-ray tube with a suitable target, e.g. to generate aluminium $K\alpha$ (1500 eV) X-rays is used in an arrangement shown schematically in Fig. 3.16. Some of the photo-electrons emitted are characteristic of the sample and are energy analysed and counted. The energy levels associated with the sample atoms may be identified from standard X-ray tables, chemical shifts can be attributed to changes in the electron density of the outer orbitals involved in chemical bonding. These chemical shifts effects can give useful information on the nature of the chemical bonds. Additional energy peaks due to Auger electrons also occur and give useful additional information.

Although the primary X-rays penetrate well below the surface, the electrons leaving the solid without appreciable energy loss are confined to those released by atoms near to the surface so that information is confined to a depth of say 10 Å. An advantage of ESCA is that little damage is produced by the primary X-ray irradiation, but since the cross-section for photoelectron emission is small, long counting times are required.

The method of ESCA analysis has been pioneered by Siegbahn et al. (1967, 1972) whose publications give much information on the apparatus and techniques used and the many possible applications.

3.2.8 Photoelectron spectroscopy

Similar to ESCA (sometimes ESCA is assumed to include this technique) in which the primary ionization is due to u.v. light (rather than X-rays) and the characteristic energy levels of emitted photoelectrons are analysed by an electron spectrometer. The energy spectrum gives information on the binding energies of the outer electrons and most applications have been with gaseous samples rather than surfaces of solids.

The technique and its applications have recently been reviewed by Turner (1971).

3.2.9 Ion scattering spectrometer

A monoenergetic beam of ions (all of the same m/e) is directed at the surface. The ions backscattered through a certain angle pass through a high resolution energy analyser as shown in Fig. 3.17 and are recorded as an energy spectrum characteristic of both the primary ions and the surface atoms. The collision process can be pictured as a simple two

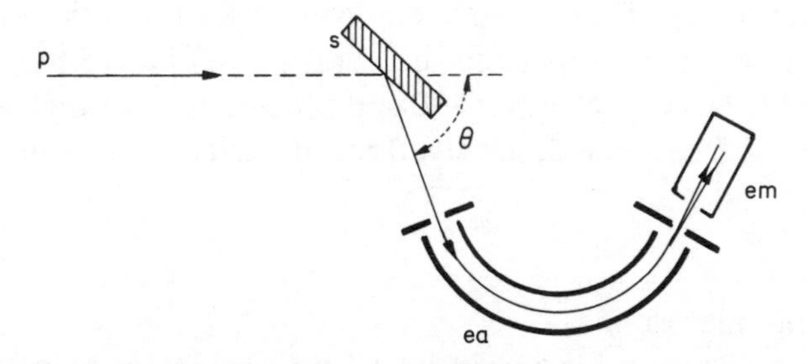

Fig. 3.17 Ion scattering spectrometer (following Smith, 1971).
p = primary ion beam ea = energy analyser; electrostatic deflection 127° cylindrical condenser type
s = specimen em = electron multiplier type ion detector
θ = scattering angle

body elastic interaction between the incident ion of mass M_1 with the single lattice atom of mass M_2 in which the remainder of the host lattice plays little or no part. Consequently, the backscattered primary ions have an energy maximum which depends only on the two masses M_1 and M_2 and the scattering angle. The energy ratio E_0/E_1 of primary to scattered ion is obtained from conservation of energy and momentum considerations for a scattering angle θ,

$$E_1/E_0 = M_1^2/(M_1 + M_2)^2\{\cos\theta + (M_2^2/M_1^2 - \sin^2\theta)^{1/2}\}^2$$

With $\theta = 90°$ as the scattering angle, this reduces simply to

$$E_1/E_0 = (M_2 - M_1)/(M_2 + M_1).$$

At incident energies greater than 10 KeV, it is found that the peak maximum corresponds well with the predicted position, but there is a broadening of the peak at the low energy side. This is attributed to collisions with atoms below the surface and energy losses when the scattered ion passes through the surface layers. When bombarding with a relatively high energy beam and in the case of single crystal samples, orientation of the crystal to the beam in a channelling or blocking direction will give rise to sharp peaks. The appearance of these help to bring out impurities present at the surface interstitially.

Operating at primary energies less than 2 keV, a charge exchange process between those primary ions which have penetrated below the immediate surface causes their neutralization, so that relatively sharp peaks without energy tails are obtained without the need for any special crystalline alignment or even with polycrystalline samples.

The primary bombarding ions are typically a noble gas to prevent chemical interactions with the surface. The mass M_1 of these ions is selected to suit the mass of the ion at the surface to be detected according to the energy relationship given above. The peak height is related to the relative quantities of the constituents. However, some observation effects may occur making interpretation of the results more difficult. The same beam may be used at higher current densities to sputter-etch some of the sample surface and obtain a depth profile analysis of each successive atomic layer.

Armour and Carter (1972) reviewed the application of ion backscattering techniques over the whole energy range. There have been several recent studies indicating the particular advantages of using low energy ion beams, e.g. Smith (1971), Goff (1972), Ball et al. (1972). Carter and Colligon (1968) and Nelson (1968) discussed the various applications of ion beams to analysis of surfaces including the effect of surface scattering.

3.2.10 Ion scattering microscope

The surface of a crystalline solid is bombarded with protons or light ions. In the instrument shown in Fig. 3.18(a) the backscattered ions are observed on a fluorescent screen or recorded photographically. Facilities are provided to orient the crystal surface in relation to the primary ion beam to obtain the typical blocking pattern generated by backscattering from preferred directions. The blocking pattern may be related to the crystal lattice parameters and crystal structure. The pattern may also help to identify special effects due to radiation damage, ion implantation and epitaxial deposits. The technique can usefully be combined with energy analysers of the backscattered beam as in the ion scattering spectrometer mentioned above, when operated in the higher energy mode.

The performance obtained with simple ion scattering microscopes was reviewed by Nelson (1967) and Livesey (1968). Experiments at much higher primary ion beam energies were reviewed by Behrisch (1968) and Barrett (1968).

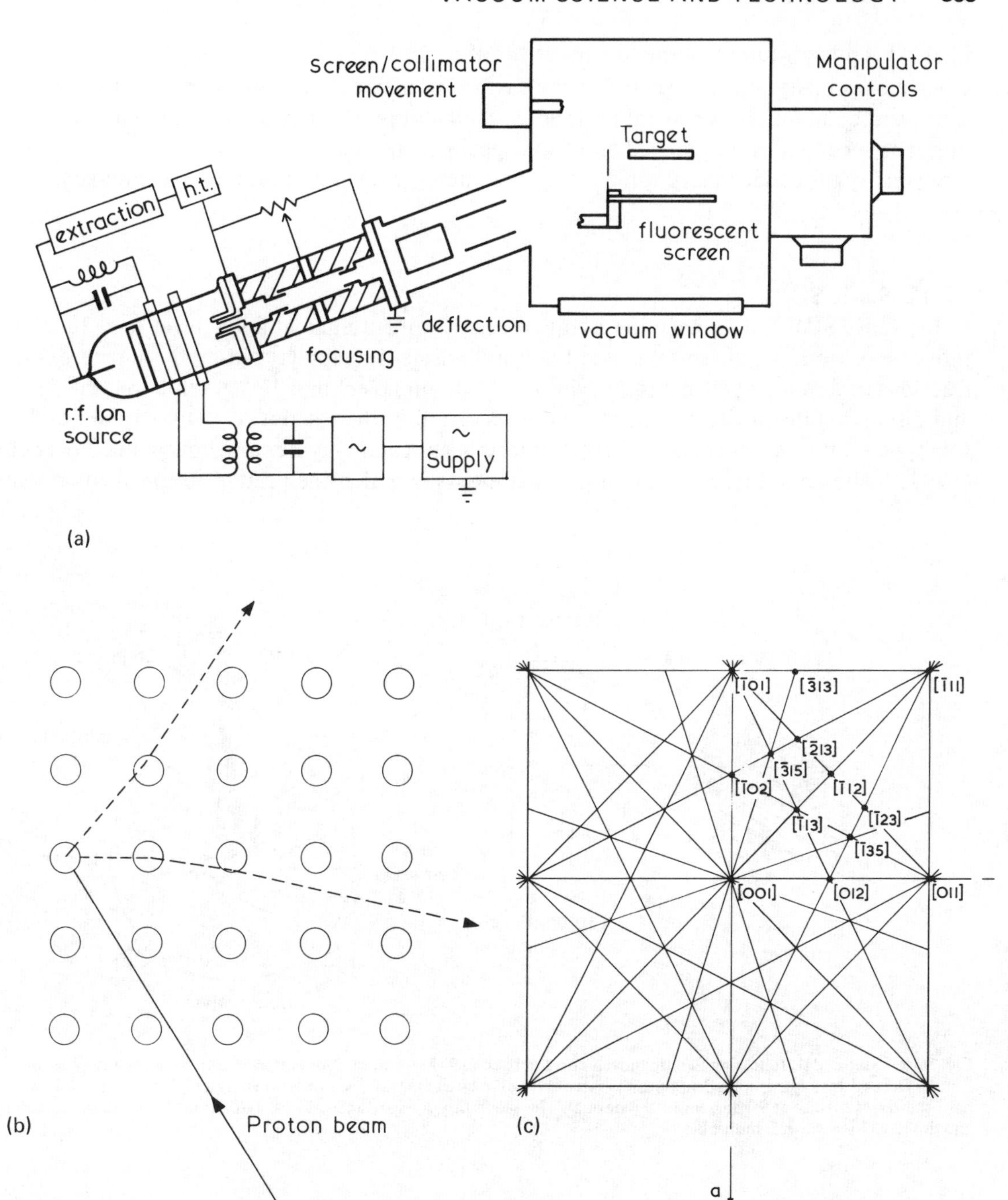

Fig. 3.18 (a) Ion scattering microscope. Ions from the r.f. ion source are focused on to the crystalline target. The back-scattered ions show up the blocking pattern on the fluorescent screen which may be viewed through the window or photograph. (b) Diagram to show the blocking phenomenon by the atomic rays in a crystal lattice. (c) Blocking pattern as produced from a cubic crystal.

3.2.11 Secondary ion mass spectrometer (SIMS)

A beam of primary ions is directed at the surface resulting in the sputtering of surface atoms and compounds as neutral particles as well as a percentage of positive and negative ions. The secondary ions emitted from the specimen are mass analysed.

Several types of secondary ion mass spectrometer (SIM) instruments have emerged.

3.2.12 Static SIM

In the static SIM, the primary ion bombardment is maintained at a low level to reduce the rate of erosion from the surface over the bombardment region (typically 0·1 cm^2 area) to fractions of a monolayer (detection limit 10^{-6} monolayers or 10^{-14} g) while recording and displaying the positive or negative ion spectrum in the usual way. An instrument of this type with mass separation of the primary beam following Benninghoven and Loebach (1971) is shown in Fig. 3.19. Simpler instruments in which the primary beam from an ion

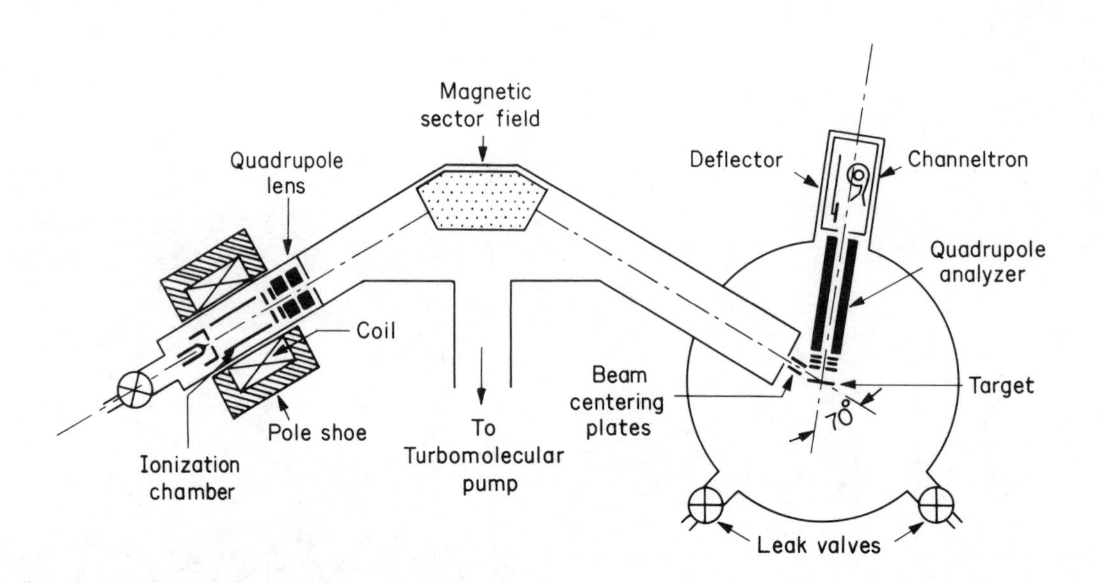

Fig. 3.19 Static SIM, following Benninghoven and Loebach (1971). Primary ions are generated by a plasma type ion source, focused by a quadrupole lens and separated by a 60 degree magnetic sector field analyser (R = 10 cm) to bombard the target surface at 70 degrees to the normal. The secondary ions are analysed by a quadrupole mass spectrometer and detected by a particle multiplier.

gun is directly accelerated onto the target surface have also been successfully used. To obtain a depth profile, layers of material are first removed by continued sputtering at a suitable rate. The performance of such instruments was reviewed by Werner (1969), Benninghoven (1970, 1971) Fogel (1972) and Rüdenauer (1971).

3.2.13 Ion imaging SIM

The ion imaging SIM instrument was developed by Castaing and Slodzian (1965) shown in Fig. 3.20. The secondary ions are mass selected by tuning the mass separator to a particular element. An ion and electron optical image obtained from these secondary ions indicates the spatial distribution of the selected element over the surface bombarded by the primary ion beam.

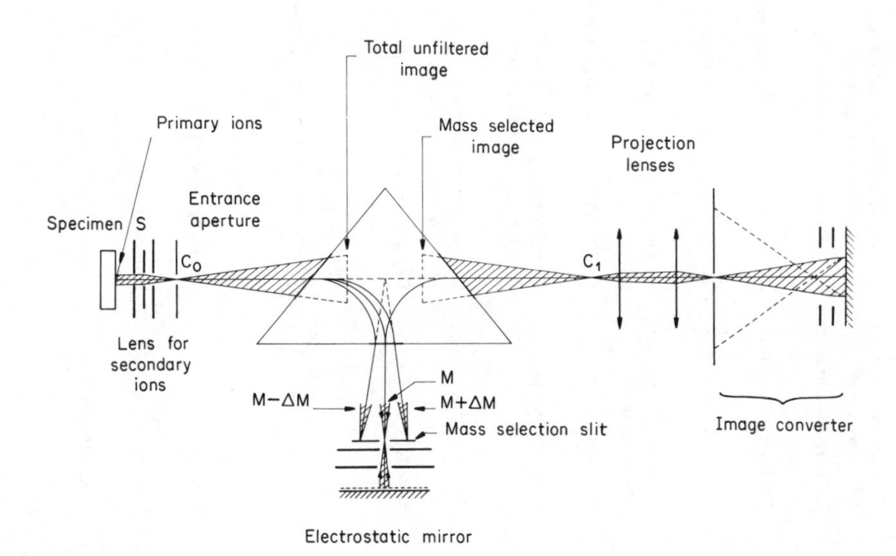

Fig. 3.20 Ion imaging SIM, following Rouberol et al. (1968). A beam of primary ions is focused on to the specimen. The secondary ions are extracted and mass separated by a double magnetic sector field. The ions of the selected mass are reflected by an electrostatic mirror passing again through the sector field and are focused on to a secondary electron emission conversion surface. The secondary electrons produce a magnified image on a fluorescent screen corresponding to the ion emission from the specimen surface.

3.2.14 Secondary ion microprobe

A secondary ion microprobe as developed by Liebl (1967) is shown in Fig. 3.21. In this case the mass selected primary ion beam is focused to a fine spot and scanned over the specimen surface. The secondary ions are mass separated and when tuned to a particular mass, the output records the spatial distribution, e.g. by cathode ray tube display, in synchronism with the primary beam scanning. Recent results obtained with such an instrument showing the advantage in some cases of using a chemically active primary ion beam (e.g. oxygen) were reviewed by Anderson (1969).

TABLE 3.3 Principal characteristics of the vacuum instruments for the analysis of surfaces

Instrument	Method of excitation	Emission or reflection	Analysis	Remarks
Ellipsometry	Polarized light	Polarized light	Change in angles of polarization	Non-destructive; sensitive to changes in surface down to fractional monolayers
LEED	Electrons (< 10 keV)	Mainly elastically scattered electrons	Diffraction pattern	Crystal structure near surface
HEED	Electrons (> 10 keV)	Mainly elastically scattered electrons	Diffraction pattern	Reflection at glancing incidence for surface and thin films
THEED	Electrons (> 20 keV)	Mainly elastically scattered electrons	Diffraction pattern	Reflection at glancing incidence for surface and thin films
Scanning electron diffraction	Electrons (> 10 keV)	Mainly elastically scattered electrons	Electron diffraction pattern scanned	Scanned pattern. Improved evaluation
Auger electron analyser (AES)	Electrons	Auger electrons	Energy	Characteristic levels for surface layers. Good yield light elements
Scanning electron microscope (SEM)	Fine focus electrons scanning surface	Scattered and secondary electrons	Intensity variation	Surface topography (confirmed with X-ray detectors as micro-probe characteristic levels)
Electron microprobe	Electrons	Characteristic X-rays	Wavelengths	Characteristic X-rays of both surface and deeper layers good yield for heavy elements
ESCA	X-rays (often from demountable X-ray tube)	Electrons	Energy	Constitution chemical binding near surface effects
Photoelectron spectroscopy	U.V. light (usually from windowless discharge tube)	Electrons	Energy	Constitution and chemical structure mainly gaseous samples
Ion scattering spectrometer	Ions mainly < 10 keV	Ions (mainly of primary species)	Energy	Characteristic of colliding masses
	> 10 keV	Ions (mainly of primary species)	Energy	Blocking and channelling effects in single crystals orientation, crystal structure and impurities
Ion scattering microscope	Ions (10 keV)	Ions	Intensity pattern	Blocking and channelling effects in single crystals orientation, crystal structure and impurities
SIMS	Ions	Sputtered secondary ions	Mass	Constitution of sputtered layer – sensitive to fractional mono-layers
Ion imaging SIMS	Ions	Sputtered secondary ions	Mass	Secondary ions produce ion optical image indicating their origins on surface
Ion microprobe	Ions scanned over surface	Sputtered secondary ions	Mass	Similar but by scanning primary beam over surface

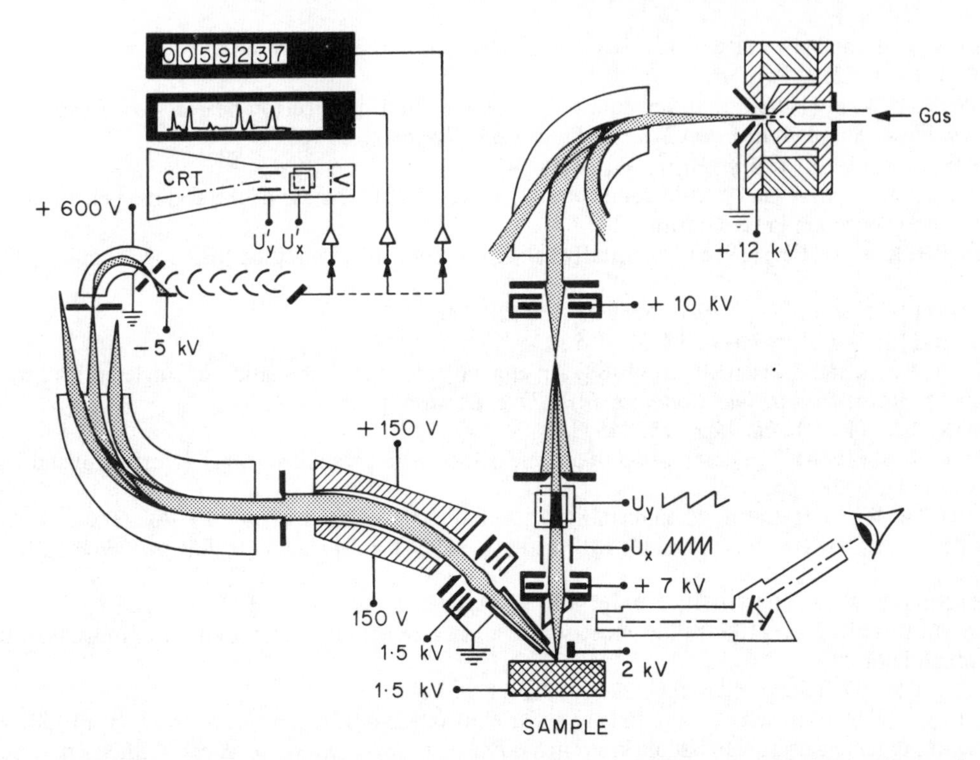

Fig. 3.21 Ion microprobe, following Liebl (1967). The primary ion beam is mass selected and focused on to the sample surface with provision for two-dimensional electrostatic scanning. The secondary ions are extracted and mass selected by a double focusing arrangement (an electrostatic analyser followed by a 90 degree magnetic sector field). The selected ions fall on a secondary electron conversion dynode and are detected after electron multiplication with cathode ray tube, recorder or digital display.

REFERENCES

ANDERSON, C. A. (1969) *J. Mass Spectr. Ion Phys.*, **2**, 61-74.
ARMOUR, D. G., and CARTER, G. (1972) *J. Phys. E. Sci. Instrum.*, **5**, 2-8.
BALL, D. J., BUCK, T. M., MACNAIR, D., and WHEATLEY, G. H. (1972) *Surface Science*, **30**, 69-90.
BARRETT, C. S., MUELLER, R. M., and WHITE, W. (1968) *J. Appl. Phys.*, **39**, 4695-4700.
BASHARA, N. M., BUCKMAN, A. B., and HALL, A. C. (1969) *Surface Science*, **16**, 462.
BEHRISCH, R. (1968) *Canad. J. Phys.*, **46**, 527-535.
BENNINGHOVEN, A. (1971) *Surface Science*, **28**, 541-572.
BENNINGHOVEN, A., and LOEBACH, E. (1971) *Review Sci. Instrum.*, **42**, 49-52.
BOOTSMA, G. A., and MEYER, F. (1969) *Surface Science*, **14**, 52-76.
CARTER, G., and COLLIGON, J. S. (1968) 'Ion bombardment of solids' (Heinemann Ed. Books Ltd., London).
CASTAING, R., and SLODZIAN, G. (1965) 'Proc 1st Int. Conf. Electron and Ion Beam Science and Technology' (J. Wiley, New York), p. 780-98.
CHANG, C. C. (1971) *Surface Science*, **25**, 53-79.
CREWE, A. V., EGGENBERGER, D. N., WALL, J., and WALTER, L. M. (1968) *Rev. Sci. Instr.*, **39**, 576-583.
DUNCOMB, P. (1969) *J. Sci. Instr.*, **2**, 553-560.
ESTRUP, E. J., and MCRAE, E. G. (1971) *Surface Science*, **25**, 1-52.
FOGEL, Ya., M. (1972) *Int. J. Mass Spectr. Ion Phys.*, **9**, 109-125.

GALLON, T. E., and MATTHEW, J. A. D. (1972) *Rev. Phys. Tech.*, **3**, 31-64.
GOFF, R. F. (1972) *J. Vac. Sci. Tech.*, **9**, 1.
GRIGSON, C. W. B. (1968) Studies of thin polycrystalline films by electron beams, *Adv. Electron & Electron Phys.*, Suppl. 4, *Electron Beam & Laser Beam Technology*, 187.
HARRIS, L. A. (1968) *J. App. Phys.*, **39**, 1419-1427.
HEARLE, J. W. S., SPARROW, J. T., and CROSS, P. M. (1972) 'The use of the scanning electron microscope' (Pergamon Press, Oxford), 284 pp.
HEINRICH, K. F. J. (Ed.) (1968) Quantitative electron probe microanalysis, NBS Special Publ. 298 (Oct.).
KAMMLOTT, G. W. (1971) *Surface Science*, **25**, 120-146.
LIEBL, H. (1967) *J. App. Phys.*, **38**, 5277-5283.
LIVESEY, R. G., and BUTCHER, G. (1968) An apparatus for the examination of single crystals by scattered protons, *4th Int. Vac. Congress. Inst. Phys. London*, pt. 1, 203-206.
NELSON, R. S. (1967) *Phil. Mag.*, **15**, 845-853.
NELSON, R. S. (1968) 'The observation of atomic collisions in crystalline solids' (North Holland Publ. Co., Amsterdam), 281 pp.
NISHIMURA, H., FUJIWARA, T., and OKANO, J. (1971) *Mass Spectroscopy*, **19**, 205-212.
OATLEY, C. W., NIXON, W. C., and PEASE, R. F. W. (1965) *Advance Electr. Electron Phys.*, **21**, 181-247.
PALMBERG, P. W., and RHODIN, T. N. (1968) *J. App. Phys.*, **39**, 2425-2432.
PASSAGLIA et al. (Eds.) (1964) Ellipsometry in the measurement of surfaces and thin films, Nat. Bur. Stds. Miscl. Publ. 256.
REUTER, W. (1971) *Surface Science*, **25**, 80-119.
ROUBEROL, J. M., GUERNET, J., DESCHAMPS, P., DAGNOT, J. P., and GUYON, J. M. (1968) Micro analyser using secondary ion emission (in French), *Proc. 5th Internat. Symp. X-ray & Microanalysis, Tubingen, Germany* (Springer-Verlag, Berlin, 1969), 311-318.
RUDENAUER, F. G. (1971) *Int. J. Mass Spectr. Ion Phys.*, **6**, 309-323 (see also (1972) **9**, 106-107).
SEWELL, P. B., MITCHELL, D. F., and COHEN, M. (1969) *Dev. Applied Spectroscopy*, **7A** (Plenum Press, New York), 61-79. Ed. E. L. Grove and A. J. Perkins.
SIEGBAHN, K., NORDLING, C., et al. (1967) *Nova Acta Reg. Soc. Sci. Upsaliensis, Ser.* IV, **20**.
SIEGBAHN, K. (1972) 'Electron Spectroscopy' (Ed. D. A. Shirley), Conf. Asilomar, Pac. Grove Calif. (North Holland Publ. Co.), p. 15-52.
SMITH, D. P. (1971) *Surface Science*, **25**, 171-191.
SOCHA, A. J. (1971) *Surface Science*, **25**, 147-170.
TAYLOR, N. J. (1969) *Rev. Sci. Instrum.*, **40**, 792-804.
THORNTON, P. R. (1968) 'Scanning electron microscopy' (Chapman & Hall, London), 312 pp.
TURNER, D. W., BAKER, C., BAKER, A. D., and BRUNDLE, C. R. (1970) 'Molecular photo electron spectroscopy' (John Wiley & Sons, New York), 396 pp.
WEBER, R. E., and PERIA, W. T. (1967) *J. Appl. Phys.*, **38**, 4355-4358.
WERNER, H. W. (1969) *Dev. Applied Spectroscopy*, **7A** (Plenum Press, New York), 239-266. Ed. E. L. Grove and A. J. Perkins.
VRAKKING, J. J., and MEYER, F. (1971) *Appl. Phys. Letters*, **18**, 226-228.

3.3 Ion impact sputtering - particle emission related to apparatus design and thin film growth

3.3.1 Introduction

When the gas pressure in an envelope is reduced to a fraction of a Torr and a cold cathode discharge is excited between metal electrodes inserted in the envelope the cathode electrode begins to emit atoms and atomic clusters, i.e. 'sputter'.

The cathode is bombarded by ions and energetic particles in the discharge which remove the target material and it is deposited on the envelope walls or in a coating system on adjacent receivers. Cathodes, particularly of the noble metals, form metallic deposits rapidly when the gas pressure is sufficiently reduced for the applied voltage exciting the glow discharge to be above 500 V, i.e. the glow discharge is in the *abnormal regime.* This 'critical' voltage is an apparent effect because the onset of sputtering can be detected (in the absence of gas – target particle collisions) at ion energies down to a few electron volts. When gas is present at a sufficient pressure the liberated particles, although initially energetic, lose energy by collision with gas molecules and cool to the gas temperature. The liberated particles then diffuse in the gas and can be returned by collision to the cathode.

Sputtering occurs in glow discharge lamps operated in the low pressure region (~10 Torr) but the resistance offered by the gas to particle flow reduces the deposition rate on the envelope to a negligible quantity. This lead to the erroneous early belief that sputtering did not occur in the *normal* glow discharge regime, but one can observe a sputtered film on discharge lamps operated in this regime at domestic mains voltage after extended use.

Film deposition from cathode sputtering is a phenomenon which accompanies the passage of a high voltage discharge between electrodes in a gas at about 10^{-1} Torr or less with film growth rates of a few atoms per second. Thus as soon as studies commenced in the last century of electrical discharges in low pressure gases sputtering was observed and turned to practical use for film formation. The ease with which many metals can be sputtered to form thin films made early (and some contemporary) workers neglect to apply existing knowledge of particle emission and transport when designing deposition systems. The situation was not helped by the uncertainty which existed until the last decade about the cause or causes of sputtering. Development of our understanding of the phenomenon (and our capacity to design sputtering systems to obtain desired emission and condensation conditions) is related to a technological ability to provide ion beams and ultra-high vacuum. With these targets can be sputtered under conditions that residual gas molecules neither influence the target emission nor the film growth by adsorption or transit collisions.

We are concerned here with the application of ion impact sputtering to thin film deposition, but ion/target impact produces other phenomena which have practical application and reference to some typical uses will help to enhance comprehension of target and receiver effects which can occur during thin film growth. These uses are: structural and profile etching of target surfaces; target analysis by mass spectrometry of secondary positive and negative ejected ions; and the gettering and ion pumping of gases in targets and sputtered deposits.

A wide range of substances can be sputtered to form coatings of: metals, alloys, semiconductors, metal oxides, carbides and complex mixtures of inorganic substances such as cermets. Techniques used in the preparation of mixed component or compound films are as follows:

(a) When sputtering from an alloy or mixed component target there is an induction period, which depends on preferred component emission and diffusion in the target, after which the target components are emitted in proportion to their bulk content, but their reformation as a film will depend on their individual transport rate, loss by pumping, condensation coefficient and mobility in the deposit at the substrate temperature.

(b) Films of mixed substances can be prepared by sputtering simultaneously from several targets on to a receiver or multi-layers prepared by sputtering from targets in sequence followed by heat-treatment if component diffusion is possible.

(c) Active gas may be added to the deposition atmosphere to react with a growing metal film to form a compound or restore a gas component released from a target and lost by pumping – the process is termed *reactive sputtering.*

(d) Generally organic compounds such as high molecular weight polymers release atoms and fragmented species when ion bombarded but some, e.g. PTFE, form condensates of low molecular weight material.

Many deposition systems have been devised for preparing the foregoing range of thin film materials on different types and shapes of substrate. Also, attention has been given to the sputtering of dielectrics and the prevention of film contamination by impurity gases. However, so extensive is the published literature that it is beyond the scope of this review to discuss all of these topics in detail. Therefore, an attempt has been made to analyse the conditions existing in typical sputtering systems so that the dependence of film structure and properties on the mode of deposition can be appreciated. It will be shown that the conditions of film growth can vary considerably with the design of the sputtering system. We shall commence by considering how target material is released and then discuss briefly some generalized forms of sputtering system. On this basis it will be possible to treat the transport and condensation of sputtered material in systems where gas or energetic particle collisions at surfaces or with sputtered atoms in space affect film growth and composition.

3.3.2 Physical sputtering

Chemical and physical processes: target material can be released by reacting with active species in an ionized gas to form volatile compounds, e.g. metal hydrides, halides. Such compounds may be unstable and dissociate outside of the active discharge zone to deposit target material. The process has been termed *chemical sputtering.* However, the principal source of target emission by energetic ion impact, even where such active gas is present, is *physical sputtering* which arises from energy released in the target.

Target effects: Bombardment of a solid by energetic ions can result in:

(a) Sputtering of energetic neutral target species with small amounts of secondary positive ions ($\sim$1%) and negative ions ($\sim 10^{-3}$%)

(b) structural damage

(c) ion implantation

(d) backscattering of energetic ions and neutral atoms, and

(e) the emission of secondary electrons and electromagnetic radiation.

Most of the these effects are of importance when sputtering is used for film deposition as the sputtering yield (atoms/ion) may change as the target surface is damaged and implanted with gas. Also, the backscattered and emitted species may reach the substrate with the sputtered particles and influence the film growth.

Energy exchange mechanisms: An incident ion transferring momentum and energy to target atoms can eject atoms by direct or *primary* collisions or set in motion *secondary* collisions which lead to atom ejection. The mechanism which is dominant depends on: the mass, energy and related cross-section and incidence angle of the ion and the mass and structural arrangement of the target atoms. As energies of only a few electron volts are needed for atom release from a target surface secondary collisions are mainly responsible for sputtering when incident ions have energies in the keV region. The maximum energy which is transferred by a particle of mass M_1 and energy E_1 on impact with a particle M_2 is

$$E_{max} = E_I \frac{M_1 M_2}{(M_1 + M_2)^2}$$

and

$$\alpha = \frac{E_{max}}{E_1 - E_0}$$

when the energy of the impacting ions is much higher than that of the target atoms (E_0) the accommodation coefficient $\alpha = E_{max}/E_1$.

If the target is a single crystal the structural order and close atomic packing in certain directions may produce correlated collisions. Wehner (1955, 1956) was the first to observe the formation of sputtered deposits showing patterns which arose from nearest neighbour collisions. Silsbee (1957) proposed that under certain conditions of ion/atom collision subsequent collisions between target atoms could be focused into the atomic lines with maximum energy transfer between like atoms. However, the impinging ions can damage the lattice and form interstitials which, before complete randomization of atom ejection, can modify the emission to make it appear of different crystal origin (Anderson and Wehner, 1960). Raising the target temperature can remove such damage by annealing thereby sharpening the deposit pattern. Thompson (1959) using a gold foil target with preferred orientation of the crystallites found that protons in the MeV range, with sufficient energy to sputter Au atoms from the far side of the foil, ejected the atoms preferentially in the close packed directions.

Sputtering from either single crystal or polycrystalline materials is observed in the transmitted direction only for very thin targets and for film deposition one is concerned with recoil or back-sputtering from massive targets.

If protons or helium ions approach a solid they are likely to penetrate its structure and if the target contains heavy atoms the penetrating ions will transfer little energy on each collision. Protons can be *channelled* between close packed atom rows in a crystal and transmitted or reflected to form by their distribution *blocking* patterns of the crystal structure. The sputtering yield for protons incident on heavy metals is low but Behrisch and Weissman (1969) have shown that 100 keV protons backscattered in a target can directly contribute to sputtering by colliding with atoms in the direction of removal. Thus the sputtering yield is comprised of two components $S = S_I + S_B$ where S_I is that for the incident beam which can create collision cascades and S_B is a form of 'transmission' sputtering from the backscattered ions. Of course if the target is an organic substance, e.g. hydrocarbon polymer, then atom and ion masses do not greatly differ and maximum energy transfer is presumably possible for H^+ incident on H.

The incident positive ions may be embedded in the target (ion implantation) and at high concentration implanted gas atoms can precipitate to form bubbles. At equilibrium gas atoms are released by sputtering as fast as implanted, but protons may diffuse into the target. If the gas were released only by removal of the target material then its implanted density would be equal to $1/S$. However, the measured value is usually less, because collision cascades desorb gas.

Ion energy, mass and cross-section: The number of collisions an ion makes in a given target depth depends on its collision cross-section σ so that the sputtering yield S (atoms/ion) is approximately given by

$$S = K\sigma \frac{M_1 M_T}{(M_1 + M_T)^2} E_1,$$

where K is a proportionality constant. The larger the ion mass m_I the greater will be σ and more collisions will occur in a given target depth with more energy transferred at each collision as m_I approaches m_T in value. This will both enhance S and raise the ion energy at which the sputtering yield tends to a maximum and roughly constant value S_{max}. The collision cross-section decreases as the ion energy rises and fewer collisions occur but more energy is transferred on each impact. Thus in the target depth (~100 Å) within which recoil cascades lead to sputtering the total amount of energy transferred by an ion tends to be constant when it is sufficiently energetic to penetrate deeply. When the energy of Ar^+ ions is above about 30 keV a constant amount of energy tends to be deposited in the region where recoils lead to sputtering in metal targets and the yield becomes independent of ion energy.

When energetic Hg^+ ions impact on solids their cross-section and mass (200 a.m.u.) can limit their penetration to the recoil zone effective for sputtering and the sputtering yield only reaches a maximum when the ions have energies of 100 keV or above. For sputtering efficiency in thin film deposition systems one should use heavy inert gas ions, e.g. argon with energies in the 1 to 10 keV range raising if necessary the ion current density to increase the sputtering rate.

Critical energy: Onset of sputtering is observed when incident ions reach a critical energy E_c of typically 20 to 30 eV for argon ions on metals. The minimum energy which must be transferred by an incident ion to a target atom for onset of sputtering will be related to its sublimation energy $E_{sub} = \alpha E_c$. However, as the target surface may be rough on an atomic scale, or roughened by ion etching, the critical energy required to liberate material should be less than that needed for sublimation from a smooth surface; atoms in a rough surface have greater surface energy than those in a plane. Gas implantation and structural damage can also affect the energy required for atom ejection as well as the nature of the collision cascades. The critical energy for sputtering *is not an ideal quantity.*

Collision depth and ion incidence angle: As stated above target atoms are released either by a direct collision between an impinging ion and a surface atom or by penetration of the ion in the solid producing internal collisions with recoils which reach the surface. Release of target material will be enhanced if the ion incidence angle is raised, because more energy will be transferred to the surface layer within which cascade collisions can reach the surface with sufficient energy for sputtering. The depth of this layer is usually about 100 Å and the sputtering yield S (atoms/ion) will tend to be proportional to $d/\cos\theta$, where θ is

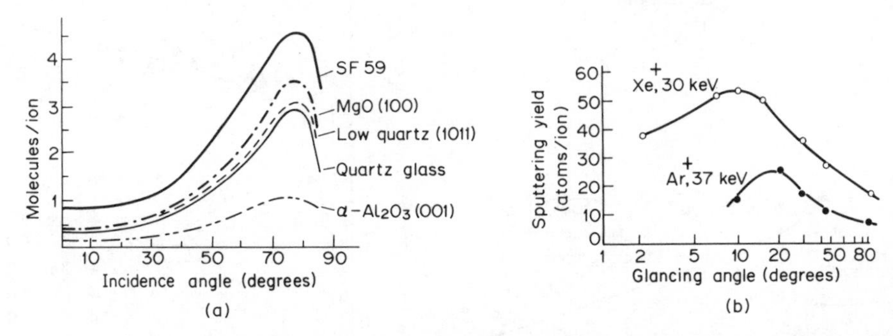

Fig. 3.22 (a) Angular dependence of the sputtering coefficients of insulating materials, after Bach, 1970. (b) Sputtering yield of copper as a function of Xe^+ and Ar^+ ion glancing angles ($90°$-θ) showing effect of ion mass after Cheney and Pitkin, 1965.

the ion incidence angle. However, as ions are reflected at high incidence by surface electric fields and can be neutralized by extracting electrons from the surface without energy exchange S reaches a maximum at $\theta \simeq 60$–$70°$.

As the energy and mass of an ion are raised both S and its maximum value S_θ are increased because the more energetic ions are not so easily reflected and the surface penetration is less for heavy ions. Bach (1970) has given values of S (molecules per ion)* as a function of θ for a number of silicate glasses and dielectrics bombarded by argon ions at 5·6 keV Fig. 3.22(a) and Cheney and Pitkin (1965) have reported the yield for copper for 37 keV Ar^+ and 30 keV Xe^+ ions as a function of glancing incidence $(90 - \theta)$ as shown in Fig. 3.22(b).

Angle of preferred emission: As the ion incidence angle θ to the target normal is increased the emission may show a maximum yield in a forward direction partly related to the incidence angle and passage of ions. Betz et al. (1970) used different gas ions incident at $\theta = 60°$ on several metals and measured the angle of preferred emission (to the surface normal). Their results are plotted in Fig. 3.23(a) against the total yield values S measured by Almén and Bruce (1961) for 45 keV Kr^{84} normally incident on the same metals. Betz et al. also

* Silica and metal oxide glasses are polymeric and must emit atoms or polymer fragments.

plotted the quantity $E_{sub}/\sqrt{M_T}$ (termed reduce sublimation energy) as a function of the angle of preferred emission Fig. 3.23(b). They deduced from these figures that:

(a) the lower the sputtering yield the more forward emission is preferred and
(b) the results were independent of the mass of the Ar^+ and Xe^+ ions used and of their energy in the 10 to 40 keV range.

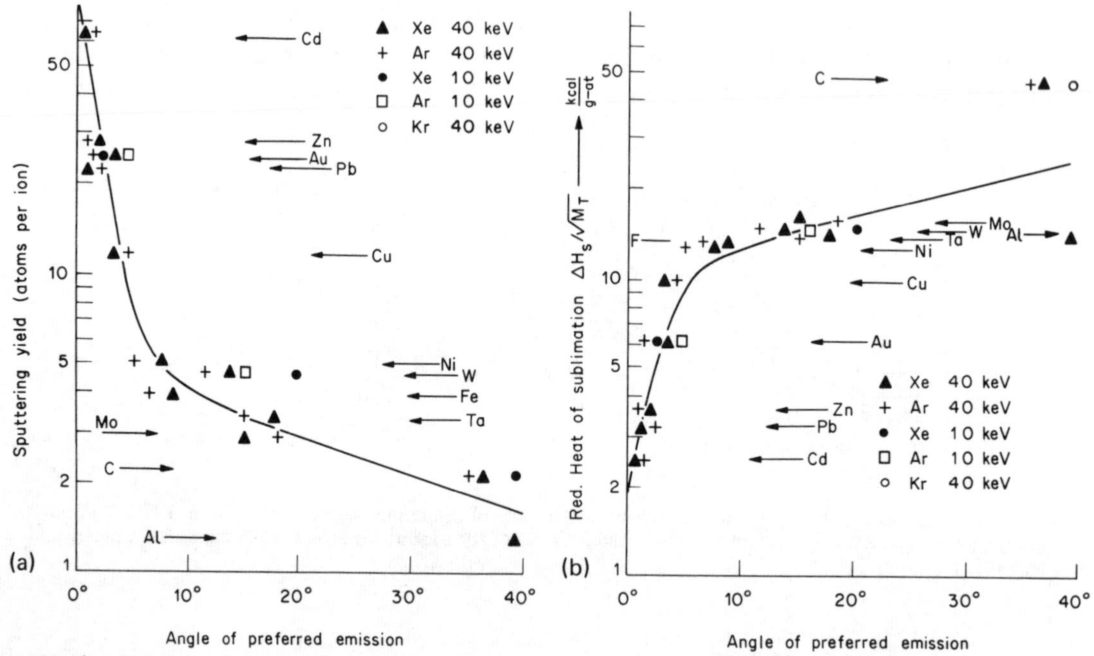

Fig. 3.23 Angle of preferred emission for ion incidence of 60° to the target normal measured by Betz et al. (1970) and plotted by them as a function of (a) the total sputtering yield S reported by Almén and Bruce (1961) and (b) the angle of preferred emission as a function of the reduced heat of sublimation.

Thus as a simple generalization one can say that polycrystalline metals which can be easily volatilized thermally will when sputtered, at normal temperature, tend to emit atoms with negligible relation to the ion incidence angle. This is because collision cascades with a large number of interacting target atoms can still transfer sufficient energy at the surface to release material. Obviously the greater the scope of the collision cascade the less dependence momentum exchange at the surface has on the initial direction of the incident ion. Energy liberation under these conditions tends to approach locally the random conditions otherwise produced uniformly by thermal agitation.

Ions impinging at normal incidence on poly- and non-crystalline target materials, independent of yield or sublimation energy, should tend to produce random emission because the collision cascade have greater depth of origin without initial asymmetry in direction. However, targets with partially ordered structures, e.g. worked metals with fibre structures, can give emissions which do not obey the cosine law and differ with respect to the direction of mechanical working.

Effect of temperature on sputtering yield: A problem in discussing the influence of temperature on target yield is the difficulty of dealing with the wide variation that exists in the chosen sputtering conditions. Target structures can range from almost amorphous to the single crystal, and the ions employed have a mass and energy varying from that of protons to mercury and with energies from a few eV up to many tens of keV. However, one can roughly identify two types of sputtering region when raising a target temperature above ambient, but the effects of the first can critically depend on the target structure and the extent of ion damage. When the temperature of a single target is raised there can be an enhancement both in yield and definition of the ejection pattern because ion damage disturbing focused collisions is removed by annealing. It has been shown (see Nelson, 1968) that preferred ejection can occur from deeply focused collisions and limited atomic encounters near the target surface therefore annealing can have varied importance depending on length of the collision sequence. Also some crystals are very easily damaged and annealing is essential to obtain ejection patterns as observed with germanium single crystals. However, single crystal patterns have been obtained with gold at below normal temperature.

With polycrystalline materials the position can be more complicated because it will depend on the size of and damage in the crystallites and the extent of focused collisions. Reports have appeared which indicate that in the annealing region the sputtering rate of copper can decline or with silver remain unchanged. One assumes that when focused collisions are of less importance annealing of damage removes energy, previously transferred to the target surface, which reduces the sublimation energy. However, with Mo, W and Ta the yield increases linearly in the range 350 to 1000°K for 2·5–10 keV Ar^+. The conditions of sputtering from polycrystalline targets in what I have termed the annealing region is still not clear but yield changes may depend on what initially predominated, localized order, or damage.

If one raises the temperature of a single crystal above the annealing region, the ejection pattern begins to become diffuse and the yield can rise rapidly. The yield of polycrystalline metals also rises when above the annealing region and presumably the emission distribution would tend to obey the cosine law even for targets of high sublimation energy who have an emission dependence on ion incidence angle at normal temperature. However, for both single and polycrystalline targets the average energy of the emitted particles tends to decrease (from >1 eV to <1 eV) this is because emission is now from thermal hot spots* or *thermal spikes*, as discussed by Seitz and Koeler (1956) and Nelson (1968). Thompson and Nelson (1962) found a small fraction of atoms ejected from gold at 500°C (bombarded with 43 keV Xe^+ ions) which had an average energy of 0·23 eV (1750°K) lasting about 10^{-11} s. At lower temperature the ejected atoms would have had average energies about 10 eV with cascades lasting for shorter periods. Nelson gives sputtering weight losses for several metal targets bombarded by 43 keV Xe^+ which show an exponential rise in yield with temperature. I have given the temperature region for a rapid rise in yield and added the

* When sputtering volatile substances, e.g. Cd or Zn, thermal evaporation can occur if ion bombardment unduly raises the bulk target temperature.

temperature at which v.p. = 10^{-2} Torr, i.e. the temperature at which bulk thermal evaporation becomes easily observable.

TABLE 3.4 Temperature region for rapid sputtering of some metal targets with 43 keV Xe^+

Target	Bi	Zn	Ag	Cu	Au	Ge
Rapid increase in S(°C)	100–200	250	650	800	600–900	600–1000
V.P. = 10^{-2} Torr (°C)	698	343*	1047	1273	1465	1251
m.p.	271	419	961	1083	1063	959

* Sublimes.

These results indicate that yield enhancement with temperature is not a mere addition of a bulk sublimation component but arises from thermal spikes. Nelson states that 'The components of sputtered material which results from focusing collisions and other high energy events should vary only slightly with temperature, decreasing by a few per cent just below the melting point'.

Reflection and ejection of energetic particles: When one bombards a massive target with ions whose mass are comparable to, or greater than, those of the target, appreciable energy can be transferred as the ions penetrate the target. This can produce recoil collision sequences which arrive at the surface with enough energy to eject surface atoms. Compton and Lamar (1933) determined the momentum transferred to a target by an impacting ion by measuring the deflection of a pendulum whose bob was a metal cathode.* They found the accommodation coefficient α for argon ions (35 and 125 eV) on molybdenum was 0·8 and on aluminium unity; in earlier work they measured 0·35 to 0·55 for helium ions incident on molybdenum.

If the ions were incident normal to the target surface and their respective masses were M_1 and M_T then the maximum energy transferred to a target atom by the head-on collision of an ion with energy E_I† would be related to the accommodation coefficient by

$$\alpha = E_{max}/E_I = 4\,M_I M_T/(M_I + M_T)^2 \text{ when } M_T > M_I, \text{ when } M_T < M_I, \text{ then } \alpha = 1.$$

In the above cases the ions were energetic enough to make more than one collision with surface atoms but a non-penetrating specie such as a thermal He atom incident on Mo had a value of $\alpha = 0{\cdot}1$ which was almost the calculated average value for He atoms rebounding and scattered between 0 to 90° to the surface normal.

Since the early work the backscattering of ions and neutrals has been extensively studied and has been used for surface analysis by determining the energy distribution of backscattered particles, for a review of the subject see Armour and Carter (1970). Shown in Fig. 3.24 is the scattered ion energy spectrum for 19 keV N^+ ions incident on stainless steel measured in the writer's laboratory by Livesey (1972). Peaks from ions scattered by

* Allowance was made for a radiometer force due to gas molecules leaving the bombarded region at a higher temperature.
† As an ion can only cool to the target temperature the energy E_I is strictly that above the thermal value but the latter is negligible (~0·03 eV) compared with that of an energetic ion.

Fe and O atoms in the surface can be identified. Energetic neutrals from gas ions can become implanted in sputtered deposits and this is one of the pumping modes in Penning pumps and a source of gas contamination in thin film deposition systems.

The gases typically used in deposition work are argon for sputtering pure films and reactive gases such as oxygen and nitrogen in reactive sputtering. These have atomic and molecular weights in the region 16 to 40 and energetic rebounds can be expected when sputtering heavy metals such as Ta and less energetic rebounds with aluminium or metal oxide covered targets.

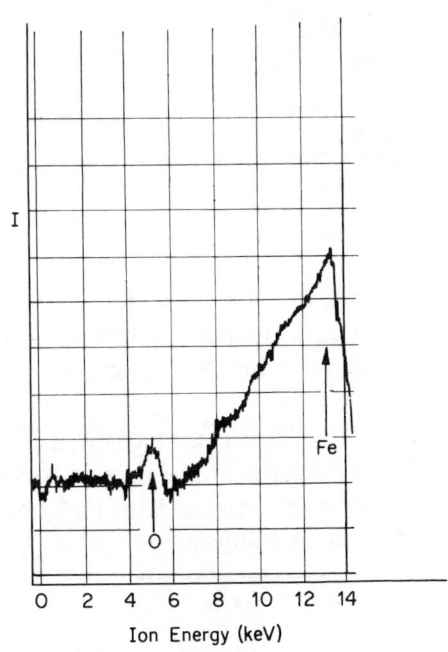

Fig. 3.24 Scattered ion energy spectrum for 19 keV N^+ ions incident on stainless steel, after Livesey, 1972.

Sputtered particles can be energetically ejected by impinging ions and from recoil collisions. Wehner (1960) studied the forces on the vanes of a torsion balance which formed the target probes in a mercury plasma operated at 2×10^{-4} Torr so that sputtered particles could freely escape from the target. He examined 22 metals using Hg^+ ions in the energy range 20 to 300 eV and concluded that energy accommodation was complete ($\alpha = 1$) on a clean metal surface. With metal oxide on the target surface α was less than unity, e.g. zirconium oxide gave $\alpha = 0{\cdot}8$ in the range 20 to 300 eV. With metal targets a force on the vanes was not detected until the energy had reached the value for the onset of sputtering and the force was attributed to a reaction from the release of energetic target particles.

When a sputtered particle leaves a target it creates a reaction force depending on the kinetic energy transferred from an ion or a collision cascade. If one assumes atom ejection is from a structureless or polycrystalline target and obeys the cosine law then the force F on an area A is $\frac{1}{2}PA$. P would be the equilibrium pressure of sputtered particles if they were released in an enclosed target within which their surface bombardment rate equalled their sputtering rate.

As $P = \frac{1}{3} nmv^{-2}$, where n is the equilibrium density of sputtered particles, m their mass, and $\bar{v}$ their average velocity,* we obtain for the reaction force $F = \frac{1}{6} Anmv^{-2}$. The sputtering rate is given by SAN^+ where S is the yield (atoms/ion) and N^+ the ion impact rate per unit area. The atomic flux leaving unit surface area SN^+ would under equilibrium conditions be equal to $n\bar{v}/4$ the sputtered particle impact rate. Substituting $4SN^+ = n\bar{v}$ we obtain $F = \frac{2}{3} ASN^+mv$ and as $AN^+ = I^+/ze$, where I^+ is the total ion current and z the number of electronic charges per ion, this gives

$$F = \frac{2}{3} \frac{I^+ Sm\bar{v}}{ze}$$

As $e = 1{\cdot}6 \times 10^{-19}$ C, $m = M$(atomic mass number) $\times 1{\cdot}67 \times 10^{-27}$ kg and $\bar{v}$ is in m/s.

$$F = \frac{2}{3} \frac{10^{-8} I^+ SM\bar{v}}{z} \text{ Newtons.}$$

The relation of I^+ to the measured ion current which includes a secondary electron contribution is discussed further on. As electrons have a small mass their effect on the reactive force can generally be neglected, but their component of the measured current may need to be allowed for.

Energy spread of sputtered particles: Wehner has measured F and determined the average velocity and kinetic energy of the sputtered atoms using a torsion balance whose vanes formed target probes in an Hg-plasma. Mean values for Hg^+ ions of 100 to 300 eV energy incident on three different metals are as follows:

Target	$\bar{v}$(m/s)	eV
Cu	$3{\cdot}1 \times 10^3$	3·2
Ag	3×10^3	5
Au	$2{\cdot}2 \times 10^3$	5

As the energy of the incident ions is increased into the keV region the energy of the ejected atoms can rise by a few more eV. This has been shown by energy analysis of sputtered neutrals and secondary ions which reach typical energies of 10 eV for argon ions in the 1–10 keV range as used in thin film sputtering. The degree to which energy is retained and transferred to a growing film surface by backscattered ions, neutrals and ejected particles depends on their transit path, the nature of the gas and its pressure. If the target reaches a high temperature atoms can be ejected mainly from collision cascades which have the character of *thermal hot spots.* The average energy of sputtered atoms will then decrease (<1 eV) but their equivalent temperature will be higher than that of the bulk cathode.

Shown in Fig. 3.25 is the spectrum obtained by Macdonald et al. (1970) for 10 kV Ar^+-ions ejecting *neutral* material from polycrystalline Cu. Macdonald et al. have also plotted

* In this simple treatment ejection velocities are assumed constant.

results obtained by other workers for neutral sputtered copper, and the emitted particles have a mean energy of ~10 eV.

The ejected material may be in atomic or cluster form depending on the nature of the target and the ion/target interaction. Woodyard and Burleigh Cooper (1964) sputtered polycrystalline copper with 0–100 eV ions extracted from a low pressure magnetically

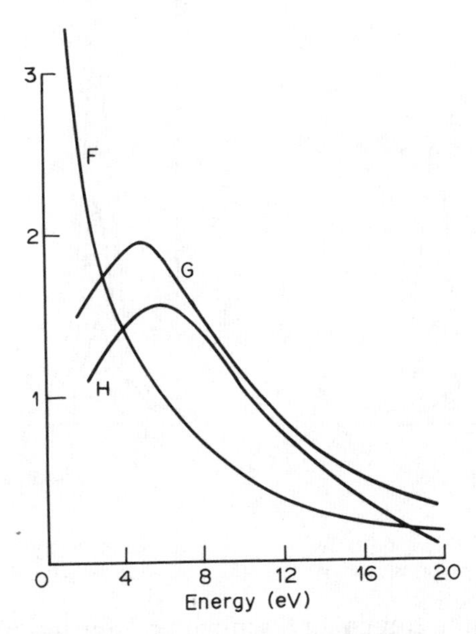

Fig. 3.25 Energy spectra of neutral sputtered copper collated by Macdonald et al., 1970. F, Farmery and Thompson (1968); G, Stuart and Wehner (1964); H, Macdonald et al. (1970), 10 keV Ar^+.

confined argon discharge. The sputtering source resembled a 'triode' system used for film deposition except that the neutral sputtered atoms entered, via an orifice, the electron impact (Nier type) ion source of a mass spectrometer. The Ar^+-ions used were produced by electron impact with the Ar^{++} yield kept to a negligible level by using a low electron energy.

Plotted in Figs. 3.26(a) and (b) are the ion collector current/voltage scans for sputtered and ionized isotopes of Cu and Cu_2 mass analysed in a magnetic sector field instrument. (The writer notes that the energy spread now arises from both that of the neutral sputtered particles and their place of extraction from the Nier source when ionized.) For 100 eV Ar^+ ions the Cu_2 : Cu atomic ratio was 0·05 : 1 with Cu appearing with a target voltage of −19 V and Cu_2 with −50 V; Cu_3 was not detected.* It is shown on page 394 that studies of metal sputtering in cold cathode discharges give evidence of appreciable emission of atomic clusters which disintegrate in traversing the ionized gas, but the writer has suggested that clusters can also be sputtered surface contaminants.

* Mass discrimination increases in voltage scanning instruments as the accelerating voltage decreases, for high values of m/e, because the ion energy spread can become comparable to the accelerated energy.

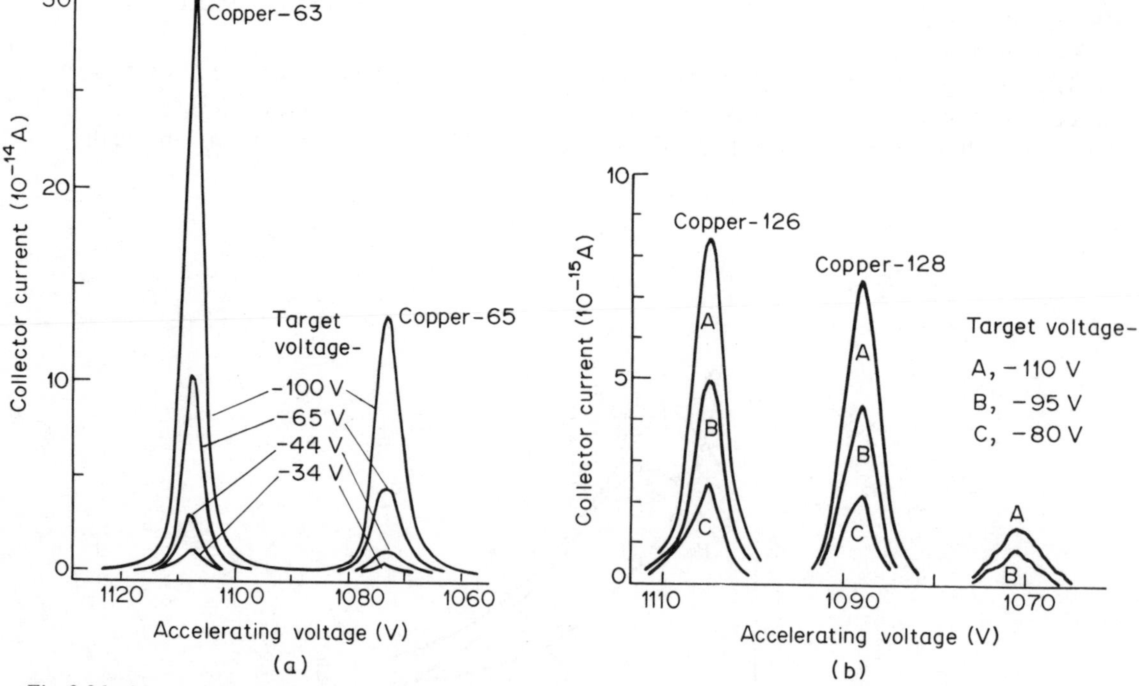

Fig. 3.26 Mass spectra of neutral (a) Cu and (b) Cu_2 isotopes sputtered in an argon arc excited by electron impact. The accelerating voltage scans show energy spread of extracted ions. After Woodyard and Burleigh Cooper, 1964.

Summary of effects: From the foregoing discussion of sputtering mechanisms several things have emerged of importance when film deposition is being considered:

(a) Sputtered particles have higher energies than those of thermally evaporated substances and their atomic and molecular cluster distribution for a given element, compound or mixture can differ from that of vacuum evaporation.

(b) Atomic emission with respect to direction tends to be random or preferred depending on: the ion energy and incidence angle and the target sublimation energy and atomic structure.

(c) The sputtering yield is enhanced by increasing the ion incidence angle.

(d) The sputtering yield rises rapidly with temperature above the ion damage-annealing range but before thermal sublimation becomes appreciable.

(e) A neutralized ion may rebound from a target with appreciable energy because of incomplete energy accommodation.

When thin films are prepared by vacuum evaporation we must for a given substrate material and condensation temperature consider the influence on film structure and composition of the condensation rate, vapour incidence and gas sorption. However, in sputtering we must also allow for the effect on film growth of the high kinetic energy of the atoms condensing to form a film whilst exposed to reflected ion beam components (and other energetic particles if the substrate is immersed in a plasma).

3.3.3 Basic types of sputtering system

Diode d.c. and a.c. systems: The simplest form of sputtering system (but by no means the simplest to analyse!) is the cold cathode diode arrangement for which several forms are outlined in Fig. 3.27. The writer showed that unwanted glow discharges could be prevented from parts of the cathode and lead-in electrode surfaces by placing anode shields so that secondary electrons were trapped before ionizing sufficient gas molecules to sustain the discharge. The electrode spacing to achieve this must be smaller than the Crookes or cathode dark space (d in Fig. 3.27). Diode systems may be constructed using alternating

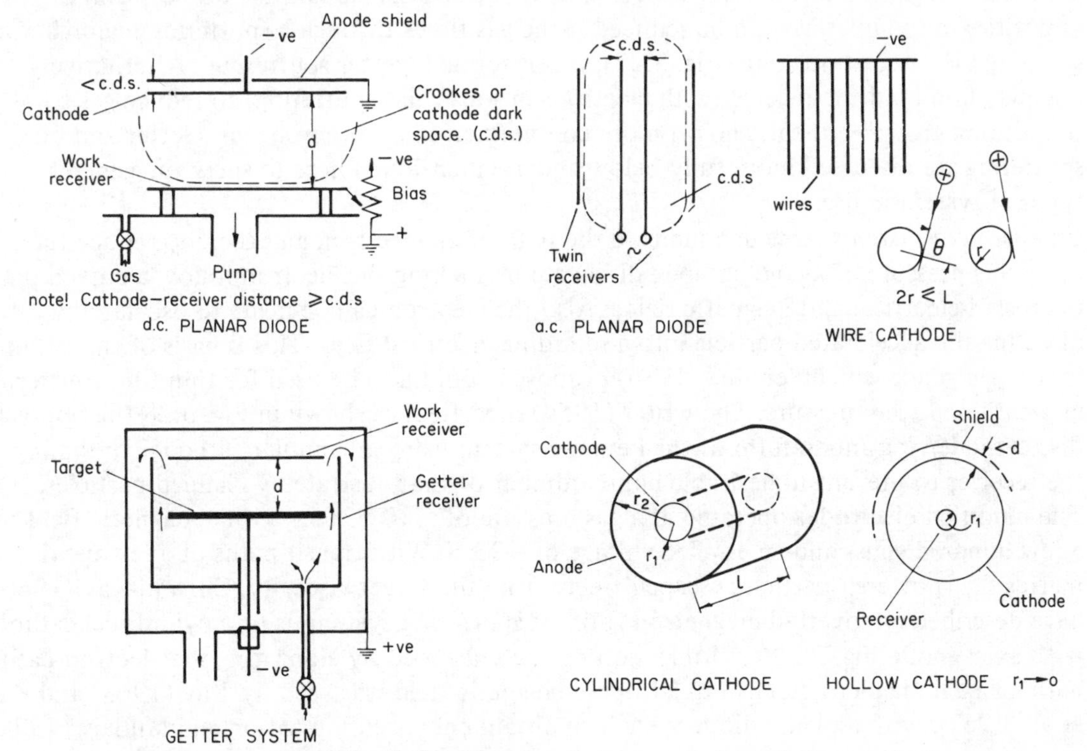

Fig. 3.27 Cold cathode glow discharge sputtering systems.

h.t. supplies and an a.c. planar type with an electrode spacing less than d to sustain the discharge is shown in Fig. 3.27 for coating two opposed substrates. Arrangements have been devised for coating both surfaces of a plane receiver using opposed target electrodes each with an earth backing shield which obstructs the electrons promoting ionization.

It will be shown further on that wire cathodes give an enhanced sputtering yield because a high proportion of the ions arrive obliquely to the wire surface. Also, when sputtering in a low pressure gas the deposition efficiency for a wire is greater than that of a massive target when the wire diameter is commensurate with or smaller than the mean free path L of the liberated particle in the gas. Elevated temperatures can be reached with wire cathodes and enhanced sputtering and thermal evaporation can occur. Wire targets are not commonly used because, although efficient in terms of emission yield their ratio of thermal flux to deposition rate at the receiver can be higher than that of a water-cooled bulk target for the same deposition rate. Also, it is not easy to provide uniform films with wire or mesh cathodes, because the localized target can produce undulations in the film thickness distribution. The enhanced yield may have preferred directions of emission which can only be effectively used if the target mesh is surrounded by receivers.

Sputtered films may absorb gases during growth either by implantation of energetic species in the glow discharge or by chemical reactions. Apart from care in choosing materials with a low degassing rate for construction of the sputtering apparatus, the sorption of impurities in the inlet gas can be reduced if the gas flows through a sputtering region before reaching the deposition zone (Fig. 3.27); this is termed 'getter sputtering'. Alternatively one may bombard the receiver with electrons or ions (bias sputtering) to reduce gas sorption during growth but this can promote contamination by other means. Getter and bias sputtering are discussed more fully below and are mentioned here to show the general types of system in use.

Crossed field systems: One can increase the path of an electron, and thereby reduce the operating pressure of a cold cathode discharge by making the electron follow a curved path in crossed electrical and magnetic fields. Also the electron can be made to oscillate by directing the accelerated particle into a retarding electrical field. This is basis of the Penning ionization gauge which Penning (1936) proposed could also be used for thin film sputtering at greatly reduced pressure. The writer (1954) modified, as shown in Fig. 3.28 the opposing disc cathode/ring anode form of the Penning system using an annular cathode containing the receiver to prevent undesirable bombardment of the substrate by charged particles. The modified electrodes operated at a gas pressure of $\sim 10^{-3}$ Torr with a magnetic field of a few hundred gauss and an applied voltage of ~3 kV. With transit paths of a few centimetres the sputtered particles escaped freely from the target viscosity. Gill and Kay (1965) have described an inverted magnetron form of sputtering system using a cylindrical cathode with axial anode Fig. 3.28(b). Ionization can be enhanced by elongating the electron paths with an axial magnetic field. A quadrupole magnetic field was used by Kay (1963) and Kay et al. (1967), with a planar diode system to obtain enhanced ionization and Mullaly (1971) has used a concave target and an anode ring, as in Fig. 3.28(a). The cusp shaped magnetic field is produced by passing current in opposite directions through the exciting coils. Using a coil with a magnetic flux density of 200 gauss, an applied voltage of ~0·5 kV, a gas pressure of 5×10^{-3} Torr and a discharge current of ~5 A deposition rates for copper in the region of 500 nm min^{-1} were obtained uniformly over a receiver radius of 75 mm.

A so called sputter-gun* has been developed which consists of a short cathode cylinder 75 mm diameter x 23 mm high with a disc anode mounted at one end of the cylinder and behind the anode a cylindrical permanent magnet. The gun has the electrical characteristics of a magnetron type discharge with the physical appearance of a hollow cathode. Operating at 0·42 kV 1·5 A in argon at 5×10^{-3} Torr the source could be used to deposit films on a rotating plane workholder.

* I am indebted to Mr P. J. Clarke of Sputtered Films, Incorp., Stony Brook, N.Y. 11790 for this information.

Plasma probe: One can produce a plasma by electron impact using a hot cathode emitter (Fig. 3.28(d)) or ionize a gas with an r.f. electric field. Positive ions can be extracted from the plasma to a target probe. This technique was used by Wehner (1955) for studying

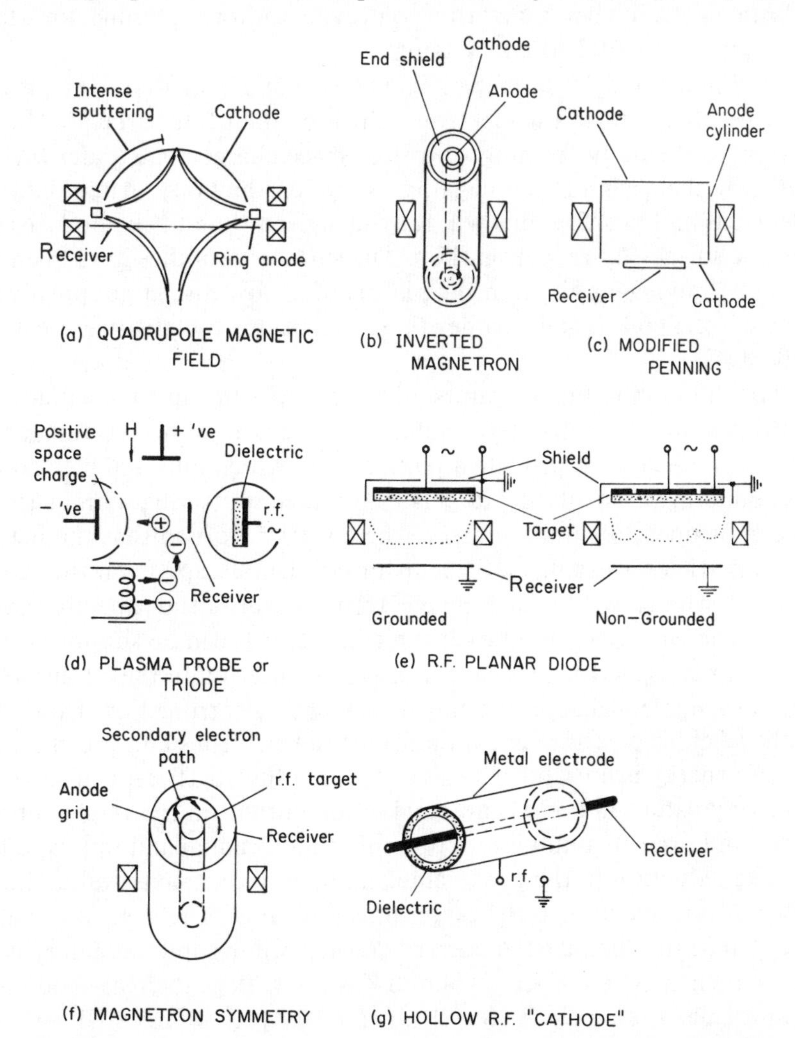

Fig. 3.28 High vacuum and plasma probe sputtering systems. To simplify the drawing the vacuum vessel, pump and gas inlet connections have been omitted from each system. Usually the magnetic field is provided by an external coil and is about 700 G. (a) Quadrupole magnetic field, after Mullaly, 1971 (see Kay, 1963; Kay et al., 1967); (b) Inverted magnetron, Gill and Kay, 1965 (see Penning, 1936); (c) Modified Penning, Holland, 1954; (d) Plasma probe: d.c. probe–Wehner, 1955; r.f. probe–Anderson et al., 1962; (e) R.F. planar diode: Davidse and Maissel, 1965 – grounded; Holland et al., 1968 – non-grounded; (f) R.F. magnetron symmetry, Rivlin, 1971; (g) Hollow r.f. 'cathode', Davidse and Whitaker, 1971.

sputtering by Hg^+ ions at reduced gas pressure. Triode deposition systems using hot cathodes and electron beam ionization were first introduced under the trade name of Plasma-Vac.* A later form used an auxiliary anode to extract the electron beam into the ionizing

* Consolidated Vacuum Corp.

region so that ion extraction had less disturbance on the electron flow. Developments in r.f. sputtering, as discussed below, first for use with dielectric targets and later as a general deposition tool, have tended to make plasma probe systems of less interest. Particularly as it is difficult with electron beam ionization to provide uniform plasmas for use with targets with dimensions above about 100 mm square.

R.F. sputtering: When a body is immersed in a plasma and insulated from ground it quickly acquires a negative charge from the electrons which bombard its surfaces. The less mobile positive ions flow to the body forming a positive space charge zone under the influence of the field created by the accumulated negative charge on the body. At equilibrium an equal number of electrons and positive ions will flow in a given time to the body to restore those removed from the surface by recombination. The surface of bodies have been cleaned by weak sputtering by immersing them in the plasma of a glow discharge (positive column) or that formed by a Tesla coil. Usually under these conditions the positive sheath voltage is only about 20–30 V.

More energetic ion bombardment can be obtained by using an r.f. oscillator with one or both of its output terminals connected to electrodes backing a desired target material. The r.f. electric field can be used to provide a plasma from which ions and electrons can be extracted to conducting or insulating targets to produce sputtering with charge equilibrium. R.F. plasmas can be operated at gas pressures below 10^{-2} Torr so that the mean free path of the sputtered particles is usually sufficient for efficient escape from the target. When the target is an insulator it can with a backing electrode comprise the capacitor coupling the r.f. supply to the plasma, and when the target is a conductor it can be supported on an insulating target to form a capacitor or a capacitor can be inserted in the r.f. input. By these means a negative charge collects on the target surface as electrons flow from the plasma when the applied field is positive. The mobility of the electrons and the charge accumulation ensures that their energy is not above a value at which the coefficient of secondary electron emission is one or greater. As there is no conduction current in the target capacitor the negative charge could rise to a maximum potential that is equal and opposite to the applied peak voltage. The reduction in the peak value and its resultant mean value then depends on the rate of flow of positive ions to the target plus their rate of release of secondary electrons. Let us assume that in the time Δt the current density of ions and secondary electrons I is an average value then as $\Delta V C_T = I \Delta t$, where ΔV is the voltage decrease and C_T is the target capacity per unit area. Taking $C_T = 10^{-7}$ F/m^2 $I = 10$ A/m^2 we have $10/10^{-7} = 10^8$ V/s. If the duration of the ion bombardment period was about 10^{-7} sec and the peak voltage ~ 1 kV then the surface potential would only decrease by 1%. Normally one uses an r.f. supply operating at 13·56 MHz.

When a charged particle is in an alternating field of high frequency its displacement in a given direction during acceleration and deceleration by the reversing field may be insufficient for it to reach the electrode or confining surfaces and the particle oscillates in the field.* If

* There is a $\pi/2$ phase difference between the alternating electric field and the alternating velocity of the electron. Electrons without this phase relation would be accelerated out of the plasma. Of course the internal electrical field in the plasma can vary from local space charge effects, e.g. injection of secondary electrons from surfaces or extraction of charged particles.

the electric field E_p (volts/m) is that due to the peak applied potential and its frequency (Hz) is f then the maximum velocity reached by a particle with a charge ze is

$$v_{max} = \frac{zE_p}{M2\pi f}$$

As $e = 1{\cdot}6 \,.\, 10^{-19}$ C and $M = m$ (atomic mass number) x $1{\cdot}67 \,.\, 10^{-27}$ kg (H rest mass)

$$v_{max} = 1{\cdot}5 \,.\, 10^7 \frac{zE_p}{mf}$$

If $E = 10^3$ volts/m; $M = 40$ a.m.u. for Ar^+ and $z = 1$ then with $f = 13{\cdot}56$ MHz we have $v_{max} \simeq 30$ m/s.

When the oscillating particle is an electron, rest mass $M = 9{\cdot}1 \,.\, 10^{-31}$ kg, then under the same field conditions $v_{max} \simeq 2{\cdot}10^6$ m/s. The kinetic energy $\frac{1}{2}mv_{max}^2$ for the argon ion would be $2{\cdot}10^{-4}$ eV and for the electron 11 eV.

If the electrons are to ionize gas molecules and produce a plasma they require to reach an average energy of about 10 eV and the gas pressure must be reduced so that the mean free path L_e of an electron is not smaller than the electric field displacement required to reach this energy. During the acceleration and deceleration in a given direction the electron is displaced $2a$ where a is the amplitude of the oscillation given by

$$a = \frac{zE_p}{M4\pi^2 f^2}$$

Under the foregoing field and frequency conditions the value of a for an argon ion is 3×10^{-7} m, whereas that for the electron is $\sim 0{\cdot}02$ m. The latter is roughly the mean free path of an electron in a gas such as argon at a pressure in the 10^{-1}–10^{-2} Torr region. When a plasma is formed secondary electrons released from the envelope walls and target are accelerated across the positive ion sheaths and enter the plasma. *Thus there is an additional supply of electrons which can promote ionization in a similar way to that occurring in a cold cathode discharge.* Also, a magnetic field can be applied to the plasma to prevent electrons from escaping thereby enhancing the degree of ionization.

These effects will reduce the field strength necessary to produce a plasma and r.f. discharges can be operated in the 10^{-3} Torr region or lower with most of the applied voltage dropped across the positive space charge sheath in front of the target.

Although r.f. sputtering had been observed some 40 years ago Anderson et al. (1962) were the first to propose a system which could be used for deposition of sputtered dielectric materials. They employed a plasma probe system as described above with an r.f. potential applied to the backing electrode of the probe. One may use a single r.f. supply to both generate and extract the positive ions and Davidse and Maissel (1965) achieved this with a planar target, shown in Fig. 3.28(e) grounding the remaining electrode. If one grounds the r.f. electrode sputtering of earthed fixtures is limited if there is not an accumulated negative charge to create a field for positive ion flow. Also a large area of earthed metal fixtures will reduce the ion current density I and this is related to the sheath voltage V by $I \alpha V^{3/2}$. However, insulating oxide films on grounded fixtures and insulating substrates resting on grounded receivers can accumulate a negative charge under an applied r.f. potential similar to that of the target. For these reasons and *to localize* the plasma the writer and his co-workers developed the non-grounded system (Holland et al. 1968) shown in Fig. 3.28(e).

A characteristic of r.f. sputtering systems is the intense electron bombardment of the receiver which can arise from secondary electrons released from the target being accelerated across the positive sheath. Holland et al. (1968) stated that electron bombardment could be reduced by arranging the magnetic field to bend the electron paths. Rivlin (1971) has described a magnetron symmetry (Fig. 3.28(f)) in which the magnetic field bends the paths of the secondary electrons so that they are captured on a grid. With this system he has been able to deposit films on to plastics which are easily degraded by electron bombardment.

If one applies an r.f. potential to a metal cylinder containing a dielectric target also in the shape of a cylinder a hollow r.f. 'cathode' discharge can be created for sputtering on to an axial receiver (Davidse and Whitaker 1971). Of course one may use a metal target cylinder capacitively coupled to the r.f. supply and obtain metal sputtering at reduced pressures. This type of discharge is the same as that obtained in an r.f. ion source using a glass tubular envelope with external electrodes. It is also of the kind in which early workers first studied r.f. sputtering.

The foregoing systems have been classified in accordance with the mode of ion production and extraction. However, to understand the nature of the films they produced, the system must be analysed in terms of the condition of sputtered particle transport and the type and energy of the particles bombarding the target and the receiver. This has partly been done schematically in Fig. 3.29 where it is shown that for Penning, plasma probe or r.f. systems the product (Pd) of pressure and transport distance is some ten to a hundred times less than that of cold cathode discharges. We shall discuss in the next section the influence these changed deposition conditions can have on film growth.

Ion beam: If the energetic ions required for sputtering are formed in a separate cell and extracted and accelerated into a target vessel, one may choose the vacuum conditions in the target vessel independently, from those in the ion source. For research into the mechanism of sputtering it may be desirable to use an ion source with a limited energy spread for the extracted ions so that a monoenergetic beam can be formed. Also to prevent sputtering of adsorbed gases from the target an ultra high vacuum may be required in the vessel. Several kinds of ion beam system have been developed for deposition purposes and the following are only examples of those reported. Heisig et al. (1965) used a duoplasmatron ion source and multiple target sputtering of NiCr and SiO_2 in sequence. Chopra and Randlett (1966) have described a system using a duoplasmatron ion source giving a final target beam (8 mm diameter) of 50 mA Ar^+ at 2 kV. With the beam and receiver at 45° incidence to the target and a receiver distance of 80 mm the deposition rate was 40 nm/min for silver; SiO_2 and TiO_2 were also sputtered. Gaydon (1967) has described an ion beam arrangement in which ions are formed in a magnetically confined arc sustained between a hot

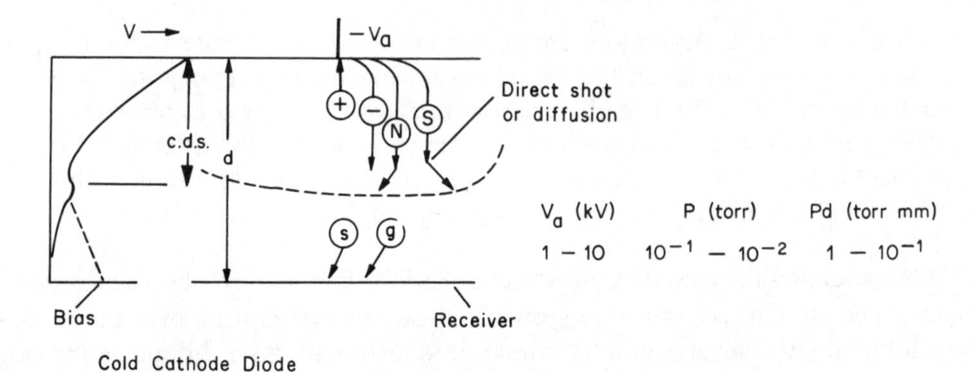

Note in practical systems d ⩾ c.d.s.

d

$-V_a$
or
r.f.

Receiver
bombardment

Positive
sheath

Plasma

Plasma (Probe, r.f. Penning)

0·5 – 5 10^{-2} – 10^{-3} 10^{-1} – 10^{-2}

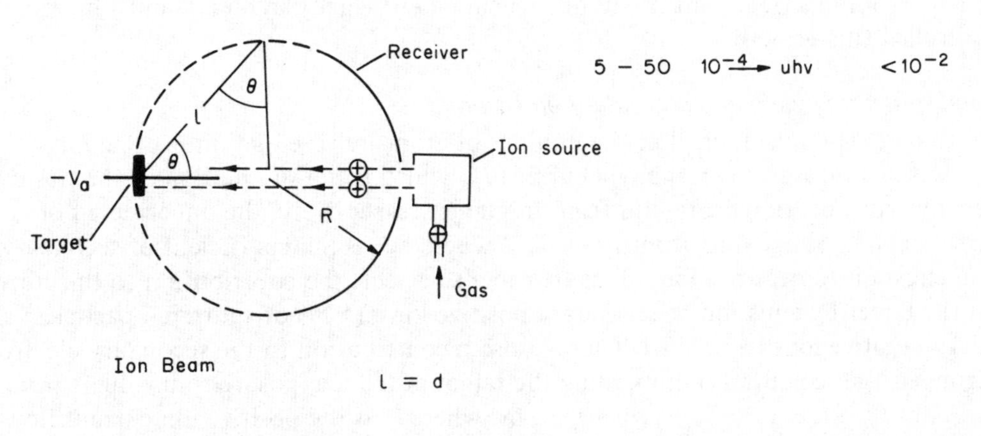

Key

(g) gas molecule
(+) gas ion
(−) secondary electron
(N) energetic neutralized ion
(s) sputtered species

Fig. 3.29 Variations in thin film deposition and growth conditions arising from gas pressure and target/receiver distance.

cathode and an anode. Argon ions are extracted through the anode and strike a target in a high vacuum vessel. The target faces a spherical work-holder carrying the components to be coated (as in Fig. 3.29). If emission from the target follows the cosine law, uniform deposits grow on the work-holder when the target is on the circumference. Gaydou deposited Cu at 34·5 nm/min, Al at 27·3 nm/min and Ta at 12 nm/min with a target work-holder normal distance of 380 mm and a current of 1·4 to 2·0 A at 1·5 kV.

3.3.4 Factors influencing film growth and deposition rate in the two deposition regimes

Precise calculation of the thin film growth rate and its distribution over a receiver surface for a given target/receiver geometry would need information on the sputtering conditions. These are: the target sputtering yield and its angular dependence for the given ion energy, impact rate and incidence angle used; the mass distribution of the emitted species; their transport velocities or diffusion coefficients if gas is present; and the condensation and resputtering coefficients at the receiver. Such rigorous calculations are unlikely because apart from incomplete sputtering data, such as the sputtering yields for the components in a mixed target, the deposition conditions are usually complex and incompletely analysed. Thus in cold cathode or plasma extraction systems the ions can be single and multiply charged and have a wide energy spread; also energetic neutral atoms can contribute to the sputtering.

Even so we shall discuss film growth in specified systems making simple assumptions regarding target emission and atom transport and show how one can estimate the film growth rate and its distribution and the efficiency with which a material is sputtered and deposited. From this treatment it will become apparent what parameters must be monitored for controlled film growth.

Measured sputtering yield and secondary emission

The sputtering yield S is defined as the number of atoms released per impacting ion. However, as one measures externally a current I_m which is the summation of the positive and negative currents flowing to and from the target respectively, the uncorrected or measured yield S_m is equal to atoms/$(1 + \delta_+^-)$ where δ_+^- is the coefficient of secondary electron emission for positive ions. This treatment neglects the contribution to the current flow at the target by emission of secondary positive ions (~1% of sputtered particles) and secondary negative ions (~10^{-3}%). There is also a contribution to the secondary electron yield from energetic neutrals bombarding the target or from exposure to u.v. light from a plasma. Thus $I_m = i_+ + i_+^- + i_n^- + i_r^- + i_{s^-} - i_{s^+}$, where i_+ is the positive ion current flowing to the target, i_+^-, i_n^-, and i_r^- are the coefficients of secondary electron emission for incident positive ions, neutrals and photons respectively and i_{s^-} and i_{s^+} the currents for secondary negative and positive ions. When an ion beam is used secondary electrons from the edges of a collimator can flow to the target in a field free space and under the influence of the positive space charge of the beam.

Measurement of sputtering yield requires an ion beam with negligible energy spread containing ions of known charge and free from energetic neutrals. Also, as shown below the measured yield can be reduced if gas is continually adsorbed on the target and ion energy is used in its removal.

Secondary electron emission: The greatest contribution to the secondary current flow is usually from electron emission but the value of δ_+^- depends on many factors, δ_+^- for single charged ions can decrease as gas is adsorbed on the target, particularly for oxygen chemisorbed on metals, but for multiply charged species, e.g. He^{++}, δ_+^- can be enhanced by gas coverage.

The secondary electron emission for a given ion/target pair will depend on the mass, charge, energy and incidence angle of the ion as it is related to ion neutralization and kinetic collision mechanisms. If the ion incidence angle θ is increased so that collisions in the target release electrons nearer the surface then fewer of the low energy electrons are absorbed in the target and the emission follows the relation $\delta_{+\theta}^- = \text{constant } \delta_+^- \sec\theta$. The equation has been found to hold closely for several ion/target pairs with equation constants in the region of 0·8 to 5.

Generally δ_+^- ranges from 0·1 to 2 for gas ions, such as Ar^+, Ar^{++}, impinging on metals with energies from 1 to 20 keV. If a target surface changes in composition as sputtering proceeds, e.g. by removal of a surface oxide or the partial pressure of an active gas in the system changes, then S_m would change if δ_+^- varied appreciably. (Of course the yield S would be different for the gas covered surface because gas as well as target atoms and their combinations are now sputtered.)

Gas density and temperature: For deposition purposes we usually control the ion energy and I_m to obtain a given sputtering rate and a related deposition rate. If we keep the measured target current constant as the temperature of a glow discharge or plasma varies then we shall tend to maintain constant gas density. For electron impact ionization the positive ion yield depends on the number n of gas molecules per unit volume. As $P = nkT$ it would be incorrect to maintain constant gas pressure in the discharge as the gas temperature rose as this would diminish the molecular number density and therefore the target current. The relation between the 'pressure' measured by a vacuum instrument and the conditions in the plasma depend on the operating mechanism of the gauge (e.g. ionization) and the vacuum conditions, i.e. whether the pressure is uniform or determined by thermal transpiration. When the gas mean free path is smaller than the gauge pipe diameter and the vessel gas is at temperature T_1 and the gauge T_2 the pressure is constant and the gas densities $n_1/n_2 = T_2/T_1$. When the mean free path is the larger thermal transpiration gives $n_1/n_2 = \sqrt{T_2/T_1}$.

Aston's equations for Pd and I

When sputtering in a cold cathode discharge one can estimate the cathode dark space depth d and potential V_c (which for a limited positive column length approaches the applied voltage) using Aston's empirical relations. In terms of pressure P and current I these are

$$d = \frac{A}{P} + \frac{B}{I^{1/2}}$$

and

$$V_c = E + FI^{1/2}/P.$$

Values for the equation constants A, B, E and F for various metal/gas combinations are given in works on glow discharges but the data can vary widely because of measuring difficulties and imprecisely known conditions. The validity of the equations have been investigated by Westwood and Boynton (1972) using a 125 mm diameter tantalum cathode in argon. Visual measurements were made of the dark space depth d working in a pressure range 12 to 49 mTorr with currents of 20 to 90 mA and voltages from 2 to 5 kV. Confirming the equation for d they determined $A = 84 \pm 1$ (cm mTorr) and expressing B in current density terms $B = 3{\cdot}1 \pm 0{\cdot}3$ (cm mA$^{1/2}$).

Plots of $I^{1/2}$ as a function of applied voltage at given pressures produced a linear curve but with a distinct change in slope at high current densities which could have arisen from a decrease in gas density with rise in gas temperature as discussed above. The modified equation $V_c = (G + F_1 I^{1/2})/P$ gave good agreement at low current densities with $G = 23{\cdot}3 \pm 3$ (kV mTorr) and $F_1 \simeq 100 + 0{\cdot}8P$ (kV mTorr mA$^{-1/2}$ cm) for currents below 21 mA (1·6 A m^{-2}).

Sputtering yield values

Some typical sputtering yield values for metals and dielectrics are given in Table 3.5 for given ions and impinging energies. Compound and polymer targets can be fragmented when sputtered. For example metal oxides may emit both free metal and oxide groups as shown by secondary ion analysis. The sputtering of fluorides and some sulphides can be almost completely dissociative.

TABLE 3.5 Sputtering yields for target materials with normal incident ions

Target	Ion	keV	Sm*	Ref.
Au	Ar^+	1	4	Weijsenfeld (1966)
Ag	Ne^+	5	5·5	Grønlund and Moore (1960)
Ag	Ar^+	5	7	Pitkin et al. (1960)
Cu	Ar^+	1	2·75	Weijsenfeld (1966)
Ni	Ar^+	1	2·25	Weijsenfeld (1966)
Al	Ar^+	1	2	Weijsenfeld (1966)
Fe	Ar^+	1	1·25	Weijsenfeld (1966)
Mo	Ar^+	1	1·2	Weijsenfeld (1966)
Si	Ar^+, Ar^{++}	1	1·8	Wolsky and Zdanuk (1961)
Ge	Ar^+	5·6	2·47	Bach (1968)
SiO_2	Xe^+	50	1·7 max	Hines and Waller (1961)
SiO_2	Xe^+	20	0·7	Hines and Waller (1961)
SiO_2	Ar^+	150	1·9	Schroeder et al. (1971)
SiO_2 quartz	Ar^+	5·6	0·36†	Bach (1968)
Cer-Vit	Xe^+	80	3·7	Schroeder et al. (1971)
Al_2O_3	Ar^+	5·6	0·18†	Bach (1968)
LiF	Ar^+	5·6	1·38†	Bach (1968)
MgO	Ar^+	5·6	0·40†	Bach (1968)
PbS	Ar^+	5·6	1·61†	Bach (1968)
ZnS	Ar^+	5·6	1·30†	Bach (1968)

* atoms/$(1 + \delta_+)$. † molecules/$(1 + \delta_+)$.

The measured sputtering yield S_m can be expressed in terms of the mass sputtered W(kg), the target atomic mass (a.m.u.) and the current I_m (amps) flowing for t (s), i.e. the charge (C) that has flowed to the cathode

$$S_m = \frac{10^8 W}{MI_m t} \text{ atoms}/(1 + \delta_{+}^{-}) \quad (1)$$

Although it is usually assumed that the sputtering rate is proportional to measured current, this should not be taken for granted. When ions are extracted from plasma and cold cathode discharges their energy spread and charge distribution depends on the positive sheath conditions, as discussed directly below, and changes in either for a fixed extraction voltage could show an apparent dependence of S_m on current, but the writer has not seen this reported.

Random emission – direct shot transport: Let us consider sputtering from a metal target for which $S_m = 2$, which is the average yield for a metal bombarded by an Ar^+ ion of 1 keV energy. If the measured current density is 10 A/m^2 (1 mA/cm^2) then as $e = 1{\cdot}6 \,.\, 10^{-19}$, C the sputtering rate is $10 \times 2/1{\cdot}6 \,.\, 10^{-19} = 1{\cdot}25 \times 10^{20}$ atoms m^{-2} s^{-1}.

If all the sputtered atoms condensed on the opposing receiver of a planar (semi-infinite) diode system then, with $2{\cdot}5 \times 10^{19}$ atoms/m^2 in a monolayer for an assumed atomic diameter of 2×10^{-10} m, a deposit would grow at five monolayers per second or with a depth of a nanometer per second for a closely packed structure. Such a growth rate would produce an opaque metal coating in less than a minute. In practice the emitted atoms will arrive at the receiver with reduced intensity, because either the target is of finite size and emission is within a large solid angle or diffusion occurs with edge losses when gas is present.

In this example we have assumed the conditions of a monoenergetic ion beam but ions extracted by a cathode in a glow discharge can have energies ranging from a few eV up to that gained from passing through the cathode dark space. Such discharges also contain multiply charged ions, e.g. Ar^{++}, with the smallest collision cross-sections. Such ions can traverse the cathode dark space gaining maximum energy, e.g. for Ar^{++} $2\,eV_c = \frac{1}{2}mv^2$.

For an amorphous or polycrystalline target bombarded by ions in the keV energy region for which emission tends to be random we can consider a system, as shown in Fig. 3.29 in which an ion beam strikes a small target in a vessel under high vacuum where $l < L$ the mean free path. Assuming emission following the cosine law for a small plane target the sputtered film on a receiver will have the surface density n

$$n = \frac{N \cos\theta \cos\phi}{\pi l^2} \text{ atoms/m}^2 \quad (2)$$

where θ and ϕ are the atom incidence angles at target and receiver, l is the transit distance and N are the number of atoms ejected given by $N = S_m It/1{\cdot}6 \times 10^{-19}$. If the source is at the base of a sphere of radius R as in Fig. 3.29 then $\theta = \phi$ and we have

$$n = \frac{N\cos^2\theta}{\pi l^2} = \frac{N}{4\pi R^2} \quad (3)$$

showing that a uniform film would be deposited on the inside of the sphere. Taking the sputtering rate for a target, with $S_m = 2$ and a current density of 10 A/m^2, as calculated above, we have for a target of 5×10^{-4} m^2 area with $R = 50$ mm

$$\frac{n}{t} = \frac{1{\cdot}25 \times 10^{20} \times 5 \times 10^{-4}}{4\pi \times 2{\cdot}5 \times 10^{-3}} = 2 \times 10^{18} \text{ atoms m}^{-2} \text{ s}^{-1}$$

The growth rate in this example has fallen to about a tenth of a monolayer per second and if for economic reasons (or to reduce the amount of gas absorbed by the condensing film) the growth rate must be enhanced this will require increasing the current flowing to the target with more localized heating. Although the system does not suffer from loss of sputtered material by back-diffusion to the target, as in a low pressure diode system, the heat energy liberated at the target could be a problem if an alloy is to be sputtered and thermal diffusion in the target is undesirable. Obviously for a given cooling arrangement there is an upper limit for which heat energy can be removed from unit target area. For this reason higher deposition rates should be attainable with r.f. planar or cylindrical systems where the ratio of the area of the receiver to that of the target is smaller, i.e. providing the transit path of the liberated atom is less than its mean free path in the gas.

Effect of target topography on yield

Wire targets: Wire cathodes give a higher sputtering yield than massive electrodes and Güntherschulze (1926) attributed this to fewer sputtered atoms diffusing back to the wire as its diameter became comparable to their mean free path in the gas, i.e. ~0·5 mm at 10^{-1} Torr. Berghaus (1936) reported that the sputtering yield of a target was increased by using wire cathodes of increasingly smaller diameter. He claimed that the yield for copper rose to a maximum for wire diameters of 0·3 mm or less and the maximum for small diameters rose further as the gas pressure was reduced from 0·5 to 0·17 Torr. Berghaus attributed the effect to the higher temperature reached by thinner wires. Fetz (1942) studied the sputtering of Mo by Hg^+-ions at $1{\cdot}5 \,.\, 10^{-3}$ Torr and correcting for the wire temperature showed that the yield was intrinsically increased as the wire became thinner. Apart from heating and gas diffusion effects there is a rise in sputtering yield when using a wire target because many of the ions impinge obliquely and S rises with the incidence angle.

Roughened targets: One might expect that a massive target with a rough surface would give a higher sputtering rate for ions normally incident to the electrode plane because locally the ions have incidence angles greater than normal. However, the deposition rate is not higher at the receiver and may be less than that for a smooth target surface. This is because part of the sputtered material is trapped in the cavities in the rough surface. *Also, if the direction of emission is preferred it will tend to be in the trapping direction.* When targets with components of different yield are sputtered the low yield regions can form protuberances on the surface which act as trapping regions.

Holland and Priestland (1972) have given the deposition rate of silica r.f. sputtered in argon from targets (066 grade translucent Vitreosil) 75 mm diameter with different surface finishes as shown in Table 3.6. The table shows that a rough worked surface gives reduced yield which increases with erosion, whereas a smooth surface gives the highest yield which reduces with time. The micrographs showed that with erosion the rough surfaces were smoothed but all targets including the polished developed a surface density of defect cavities and protuberances.

TABLE 3.6 Deposition rates when r.f. sputtering in argon silica targets of different surface finish, after Holland and Priestland (1972)

Target		Deposition rate (Å/min)	
Surface finish	Initial centre line average (μm)	Initial	After 4 μm removed
Polished	0·05 –	68 –	61 63
Fine ground	0·6 –	61 –	59 57
As cut	1·6 –	55 –	60 61
Rough ground	0·9 –	56 –	58 58

Influence of residual gases

Mass ratio and mean free path: If the gas pressure in the deposition system is such that particles ejected from the cathode collide with gas molecules then sputtered material can arrive at the receiver by direct shot and diffusion processes. The magnitude of either depends on the mass and energy of the target particles and gas molecules and the gas pressure for a given transit path.

If the ejected particles are of large mass in a gas of light mass (e.g. Ag/H_2) then because of the high energy of sputtered atoms (~10 eV) they may arrive at the receiver mainly by direct shot even though their mean free path is less than the transit distance.

Using the relation

$$E = E_s \frac{4M_1M_2}{(M_1 + M_2)^2},$$

where E_s is the energy of ejection we can estimate the maximum energy transferred per collision. For (M)Ag = 108, $(M)H_2$ = 2 we have $E/E_s = 0{\cdot}07$ and for (M)Cu = 64, (M)Ar = 40 we have $E/E_s = 0{\cdot}95$.

If the system pressure were between 10^{-2} to 10^{-1} Torr as used for d.c. cold cathode discharges then the mean free path of argon atoms in argon at thermal energies is 5 to 0·5 mm. Copper atoms with a transit distance of 50 mm in argon would make between ten to one hundred collisions in travelling to the receiver in this pressure range.* One would therefore expect the sputtered atoms to cool to the gas temperature after two or three collisions and

* Dushman (1949) has given the molecular diameters ($\times 10^{-10}$ m) of H_2, O_2 and Ar as 2·75, 3·64 and 3·67 respectively as found from kinetic theory. The mean free paths of O_2 and Ar should be similar. It is assumed that copper would have the same mean free path in argon as an argon atom.

then diffuse to the receiver. Güntherschulze (1942) found normal diffusion for copper sputtered in argon in a low pressure glow discharge at 1 kV applied voltage ($P \simeq 0{\cdot}1$ Torr).

The mean free path of H_2 in H_2 is about 1 mm at 10^{-1} Torr, i.e. twice the value for N_2. This value would be reduced for Ag in H_2 because of the greater cross-section of silver atoms which would make some fifty collisions in travelling 50 mm to a receiver. However, as the maximum energy lost per collision is only 7% silver atoms and clusters can arrive at a substrate mainly by direct shot. Güntherschulze has observed that at 3·3 kV ($P_{H_2} \simeq 0{\cdot}05$ Torr) silver was deposited in accordance with an emission tending to obey the cosine law of evaporation. If one assumes 25 collisions were made in transit the energy remaining would be $E_s(1 - 0{\cdot}07)^{25} = 0{\cdot}16\, E_s$, i.e. the atoms would retain a tenth or more of their original energy. If this was initially ~10 eV they would still arrive with an energy thirty or more times that of the gas.*

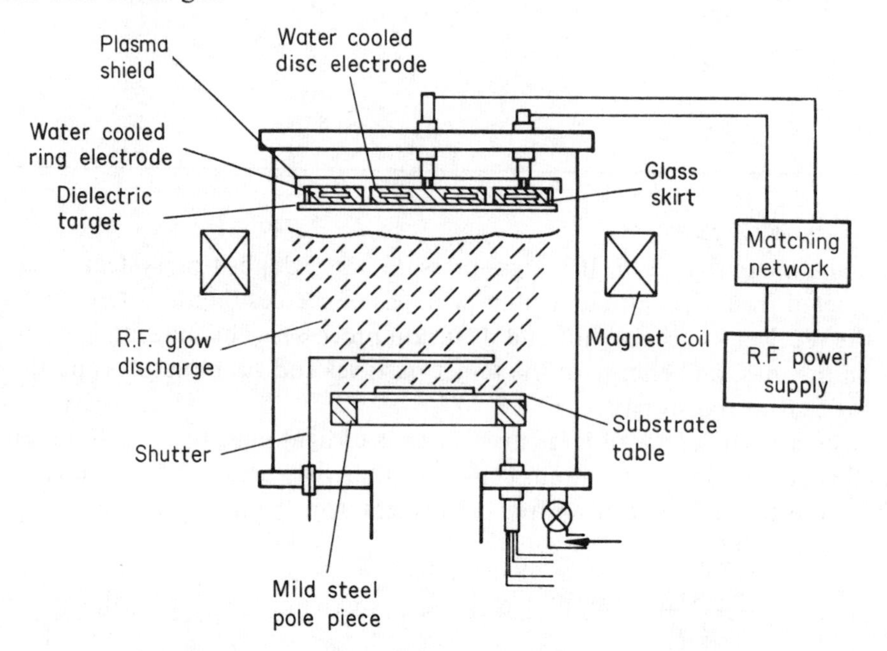

Fig. 3.30 Balanced r.f. supply with fully insulated lead in system and disc and annulus target backing electrodes, after Holland, Putner and Jackson, 1968.

Film thickness distribution: The writer and his colleagues have found a similar change from direct shot to diffusion when r.f. sputtering silica in argon in the pressure region 10^{-3} to 10^{-2} Torr when using the system shown in Fig. 3.30 with target/receiver distances of about 25 mm. The normalized thickness distribution of a silica deposit from a 0·2 m diameter target is plotted in Fig. 3.31 for two different argon pressures. That obtained at $1{\cdot}3 \,.\, 10^{-3}$ Torr agrees with a curve computed assuming cosine emission and weighting each elementary emission zone, annular element, to allow for variations in sputtering rate over the target surface. With the non-grounded system there is a higher sputtering yield at the junction of

* 1 eV is equivalent to a temperature of 7740°K.

the disc and annular electrode due to the curvature of the space charge sheath and consequent oblique ion incidence; a similar effect occurs at the edge of a disc target. When the gas pressure is raised to 4 x 10^{-3} Torr the direct influence of the higher erosion rate regions on the film thickness distribution is reduced as diffusion occurs in the gas. The basic particles

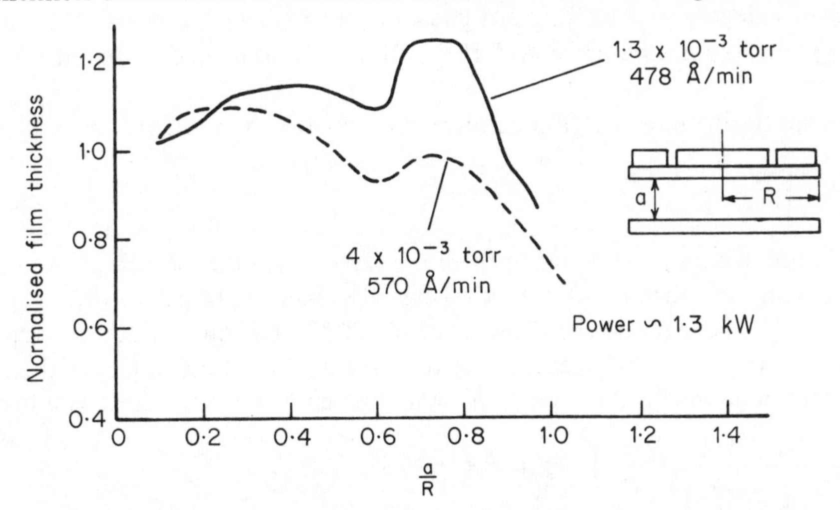

Fig. 3.31 Normalized thickness distribution of silica film for two argon working pressures in r.f. sputtering system shown in Fig. 3.30.

emitted from a silica target could be Si, SiO, SiO_2, O, O_2 and these have mass ratios M_s/M_{Ar}: Si(28) = 0·7, SiO (44) = 1·1, SiO_2(60) = 1·5. An SiO molecule should lose most of its energy to an argon atom on the first head on collision and the maximum energy transfer ratios for Si and SiO_2 in argon are 0·97 and 0·96 respectively which also indicate a rapid energy transference. An Si- atom would tend to bounce back towards the target on direct impact with the heavier gas atom but a SiO_2 molecule would have some persistence of velocity in the substrate direction as the argon would bounce towards the receiver. These effects are unlikely to influence greatly the diffusion of silica fragments in argon but obviously if the target contained atoms with large differences in their mass, ejection velocity and collision cross-section in the gas this would determine their transport conditions to the receiver. Thus if recapture occurred on the target *the film composition could vary at distances from the target although at equilibrium the target components were sputtered proportionately to their bulk concentration.* Losses by diffusion to the envelope would also affect the component distribution.

Diffusion in planar and cylindrical systems

Planar diode: If sputtering was to occur in a semi-infinite system – such as a parallel plane diode of spacing d – unbounded by an envelope with complete diffusion in a gas then the deposition rate Q in kg m^{-2} s^{-1} for a given ion current density and energy would be given by

$$Q = \frac{C_T - C_R}{d}\left(\frac{P_0 D_0}{P}\right)$$

where C_T and C_R are the concentrations at the target and receiver in kg m^{-3}; D_0 is the diffusion coefficient m^2/s at a pressure P_0; and P is the working pressure. In a diode system d the target/receiver distance would be slightly larger than the Crookes or cathode dark space. As P increases to a tenth of a torr in value the dark space contracts but not to the same inverse degree so that Pd is not constant but rises with pressure. At a pressure above ~0·5 Torr Q falls to a negligible value.* If the film condensation coefficient is unity then $C_R = 0$.

When the sputtering rate and film condensation coefficient are constant we have

$$Q = \frac{\text{const}}{Pd}$$

Note if diffusion occurs some distance from the target after the ejected atoms have cooled to the gas temperature then d will effectively be less than the target/receiver gap.
Cylindrical diode: If the diode consisted as in Fig. 3.27 of a rod cathode of radius r_1 and length l on the axis of a cylindrical receiver of radius r_2 and of equal length then the deposition rate under diffusion conditions would be given by, neglecting end losses,

$$Q = \frac{C_T - C_R}{r_2 \log_e \left(\frac{r_2}{r_1}\right)} \left(\frac{D_0 P_0}{P}\right) \text{ kg m}^{-2}\text{ s}^{-1},$$

and if C_T and C_R are constant we can write as before.

$$Q = \frac{\text{constant}}{r_2 \log_e \left(\frac{r_2}{r_1}\right)}.$$

With copper targets at 1 kV Güntherschulze (1942) operated: *a planar diode* system with a target 166 mm diameter and $d = 40$ mm deriving a system constant = $0{\cdot}58 \times 10^{-3}$ and for a *cylindrical diode* of $l = 0{\cdot}8$ m, $r_1 = 35$ mm and $r_2 = 83$ mm a constant = $1{\cdot}03 \times 10^{-3}$. If the condensation coefficient is unity or the same value in both systems then the constants should have been equal but due to the finite size of the planar system more sputtered materials (~40%) was lost by diffusion to the system walls.

Stirling and Westwood have studied the sputtering of Al (S. and W. 1970) and Ni and Fe (S. and W. 1971) in argon using a cylindrical diode with a central cathode measuring the deposition rate with an h.f. crystal microbalance and the concentration gradient of sputtered atoms in the argon discharge by optical absorption. In the gas pressure range studied the sputtered components would have cooled by collision to the gas temperature near to the cathode where the optical absorbance should have been a maximum, but maximum absorbance was observed near the edge of the cathode dark space as shown in Fig. 3.32(a) for sputtered aluminium. The displaced maximum was attributed to a large

* Generally when $Pd \leq 0{\cdot}1$ Torr mm all sputtered atoms reach the receiver.

fraction of the sputtered material being molecules or atom clusters which were fragmented to atoms as they passed through the ionized gas. The atom density distribution, in terms of optical absorbance, showed a maximum position moving towards the target with rise in current density (Fig. 3.32(b)) but the maximum position was almost independent of applied voltage (Fig. 3.32(a)).

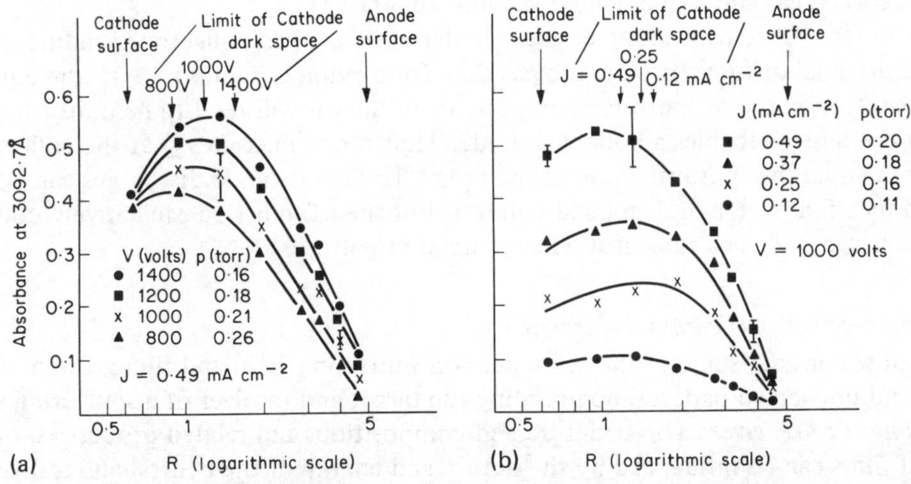

Fig. 3.32 The absorbance at 3092·7 Å arising from ground state Al – atoms in argon atmosphere of cylindrical diode, after Stirling and Westwood, 1970: (a) For different applied voltages; (b) For different current densities. (Note; the nett sputtering rate in terms of V and I should change by a factor of about four for the curves in (a) and under two for those in (b). If displaced Al – density maximum arose from ejection of surface contaminants, as suggested by writer, their content should decrease with higher sputtering rate and dissociation occur earlier in transit at higher pressures.)

Calculated deposition rates using the atom density gradients derived from the optical absorbance curves gave values much lower than those determined directly with the h.f. microbalance and this also indicated the presence of aluminium in a state which was not excited by optical absorption. However, the yield of metal atom clusters which would be needed to explain these experimental results appear unduly high (see for example earlier discussion on emission from copper).* As the experiments were made in a high vacuum vessel evacuated by an oil diffusion pump with water-cooled baffle (ultimate pressure 10^{-6} Torr) it is possible that some clusters sputtered were *chemisorbed compounds* from back-streaming and desorbed gas, e.g. hydrocarbons, H_2O, CO_2 and H_2. The inlet gas was dried at −72°C, but this would not trap gases released in the vessel. It has been shown above that for a given rate of surface contamination one can reduce the surface coverage and relative effect on the sputtering yield by raising the ion impact rate, i.e. current density, and this could explain the results in Fig. 3.32(b). Taking aluminium as an example Table 3.5 shows S_m = 2 atoms/(1 + δ_{+}^{-}) for 1 keV Ar^+ falling to 0·18 atoms/(1 + δ_{+}^{-}) for Al_2O_3 sputtered with 5·6 keV Ar^+. At 1 A m^{-2} the impact rate of Ar^+ ions is comparable to the impact rate of water vapour with a partial pressure of 10^{-6} Torr and this is an appreciable component of high vacuum vessels. Thus assuming a high reaction rate for the water vapour

* Herzog et al. (1973) have observed appreciable ion cluster emission from Al and Si using high current density ion beams of 10^3 A m^{-2} in the 4 to 12 keV range.

the aluminium would be sputtered from an oxide coating yielding metal atoms and oxide molecules and fragments. (Experiments on secondary ion mass spectrometry have shown that introducing oxygen into an argon ion beam increases the yield of positive metal ions from a target, these would of course be returned to the cathode in a glow discharge.) Extremely high sputtering and deposition rates are needed to prepare aluminium films free from sorbed gas and oxide in *high* vacuum apparatus.

Target cylinder: If one uses a long target cylinder with an axial collector of radius r_1 and sputtered material diffuses through the gas then for a cylinder radius $r_2 \gg r_1$ the equilibrium concentration C_T of sputtered components in the gas will tend to be uniform because as $r_1 \to 0$ the target becomes a hollow cathode. Thus if one measures Q at the collector this will tend to equal the sputtering rate at the target. Hollow cathode discharges can be produced by r.f. as well as d.c. applied voltages, but the r.f. must be capacitively coupled to the discharge, e.g. via a capacitor when a metal target is used.

Target and receiver particle impingement

We shall now consider some of the influences on sputtering yield and film growth of charged and uncharged particles bombarding the target and receiver of a sputtering system.

Bombardment of receiver: The structure and composition, and related properties, of sputtered films can be influenced by the nature and energy of the neutral and ionized particles which bombard the receiver during film growth. Such energetic particles, apart from sputtered species, can be gas ions rebounding from the target after charge neutralization and, depending on the substrate potential, negative or positive gas ions or electrons. Energetic neutrals and ions can sputter the substrate and condensing film so that under some conditions the deposition rate is a nett quantity. As shown in Section 3.3.2 energetic neutrals can be formed by ions which arrive at glancing incidence to a target or which rebound from one of high atomic mass. Thus in a planar system argon ions should rebound with greater energy from a gold (197 a.m.u.), tantalum (181) or tungsten (194) target than from one of silicon (28) or aluminium (27). Likewise the energy of rebound will rise even for the latter targets if the gas atomic mass is reduced, e.g. He^+ ions are used, but this would give negligible sputtering rates.*

Energetic neutrals may not reach the receiver because they lose their excess energy by gas collisions if the sputtering is in the low pressure region ($\sim 10^{-1}$ Torr). Coburn (1970) has analysed the mass and energy of the particles arriving at the film receiver in a planar diode d.c. system. He observed ionized gas and target species and found argon ions arrived at the film substrate with zero energy unless a bias potential (as discussed below) was applied to the substrate. Positive ions that are reflected (or emitted) from a cathode would be retarded in cold cathode or plasma probe systems by the positive space charge sheath.

* Energetic neutrals are formed in sputter-ion pumps when the atomic mass of the target is high compared with that of the gas ions. Thus by replacing a titanium cathode by a tantalum one it is possible for rebounding argon neutrals to retain enough energy to become implanted in the sputtered film on the anode electrode. Helium is trapped when rebounding from titanium. Generally when the ratio of the target to the gas atomic mass is about three or higher energetic neutrals are trapped in the growing deposit.

When an r.f. or d.c. plasma probe system is employed with a target/receiver spacing of a few centimetres and a working pressure in the 10^{-3} Torr region the energetic neutrals will reach the film receiver. Also, if the receiver is an insulator it may accumulate a negative surface charge which accelerates positive ions to the film surface. When an r.f. sputtering system is used of the type shown in Fig. 3.30 in either the grounded or non-grounded mode there is a positive space charge sheath in front of the target and another above the work surface. The voltage drop across a sheath, its current density and the sputtering yield must

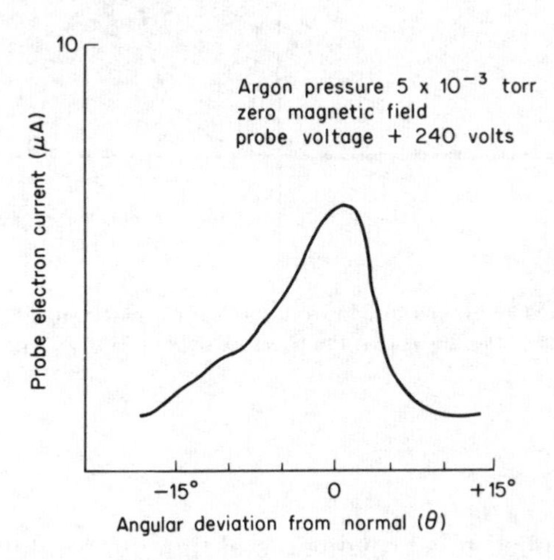

Fig. 3.33 Electron current measured by a Faraday cage probe below aperture in work-table. Maximum current measured when probe aperture receives electrons arriving at normal incidence to work plane, after Holland and Priestland, 1972.

combined give a greater sputtering rate at the target than at the receiver otherwise deposition would not occur. Thus when a film is deposited it is also exposed to positive ions extracted from the plasma across the receiver sheath (typical voltage 50–100 V). Also, secondary electrons released from the target reach the receiver. These electrons are accelerated across the first dark space sheath (typically by about 0·5 to 1 kV) and decelerated (50–100 V) across the second.

Thus superimposed on the plasma electrons (which are in the 25–50 eV range) are electrons with energies of several hundred electron volts. (Negative ions can form in the plasma and have been detected by a probe.) As the accelerated secondary electrons reaching the receiver have energies for which $\delta \gtrsim 1$ more electrons are released from an insulating substrate than arrive from the target and this would tend to further diminish the negative surface charge on a low capacitance receiver. Holland and Priestland (1972) have mounted a Faraday cage probe below a hole in the substrate table of a non-grounded r.f. diode and determined the electron current at the table potential (Fig. 3.33) and by biasing the probe the plasma potential with respect to ground as shown in Fig. 3.34.

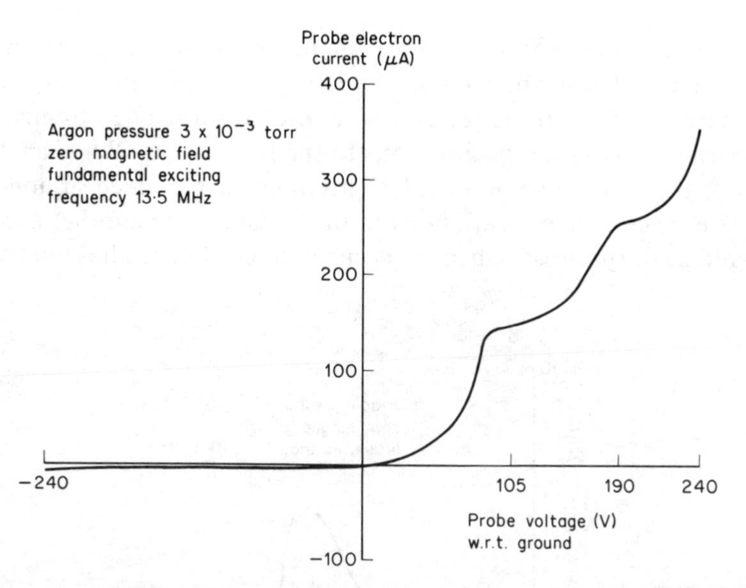

Fig. 3.34 Probe current measured below aperture in work table of r.f. sputtering plant, after Holland and Priestland, 1972. The first 'knee' in the probe curve shows that the plasma would be 100 V+ with respect to a grounded work holder.

Bias sputtering

A bias voltage may be applied to a growing metal film deliberately to enhance particle bombardment of a given sign. An r.f. bias voltage applied to a work-holder carrying a dielectric substrate permits enhanced positive ion bombardment of the substrate and growing deposit. The bias voltage in a non-grounded system can be controlled by altering the balance of the r.f. supply with respect to ground.

Negative bias: Bias sputtering by positive ion bombardment of a receiver held at a negative potential with respect to a plasma, has been proposed for removing adsorbed gas from a film surface during growth. Although the surface coverage of adsorbed gas is reduced by the technique the impinging ions can knock adsorbed atoms into the film and be implanted themselves. Whilst the penetration depth may only be a nanometer or so for ions of low energy, their burial is completed by subsequent film growth. Such an effect produces a uniform depth distribution of implanted ions in a film. Of course both the deposit and implanted ions can be sputtered by the impinging ions, and if all ions were trapped on impact the concentration of gas atoms in the film would reach at equilibrium the value $1/S_F$ where S_F is the number of film atoms released per incident ion.

Mattox and Kominiak (1971) report that sputtered Au-films with atomic ratios of He : Au of up to 0·4 can be produced in helium d.c. diode discharges. High energy neutrals reflected from the cathode enter the films and gas content reaches a maximum at about −80 V bias; raising the receiver temperature from room to 300°C reduced the atomic ratio to 0·08 at zero bias. (Güntherschulze early contended that sputtering was non-existent in helium and arose from impurities in the gas. The Au-films prepared by Mattox and Kominiak were sputtered slowly over several hours, but this does not affect the observation of helium absorption.)

Winters and Kay (1972) studied film deposition in N_2–Ar mixtures and noted that bias sputtering *reduced* the nitrogen contents of sputtered W- and Ni-films but raised that of sputtered Au-films. The increase for gold was from N_2^+ implantation, whereas enhanced implantation in tungsten and nickel was offset by resputtering of previously sorbed nitrogen.

Mitchell and Maddison (1971) have measured the gas sorbed by sputtered and evaporated gold films. With zero bias the sputtered Au-films had about two atomic percent of Ar in Au attributed to target reflected energetic neutrals entering the film. This was two orders higher than the gas content of films evaporated under similar pressure conditions. Nitrogen in the residual gas was sorbed by both films but least by the sputtered coating and fell with increasing negative bias; the nitrogen partial pressure was 3×10^{-8} Torr. It has occurred to the writer that the lower sorption of N_2 by the sputtered films arose from their higher Ar content, such effects are observed with ion pumps if an implanted gas replaces another.

Positive bias: If the bias potential is reversed (i.e. made positive) with respect to ground then negative ions and electrons will more energetically flow to the receiver. Adsorbed gas may be effectively removed by electron bombardment without the problem of gas implantation but another and equally serious effect can occur. Thus electron bombardment can induce cross-linking and retention of adsorbed organic molecules if they are present in the sputtering atmosphere. Negative oxygen ions may enhance film oxidation, because with a condensed metal coating a positive bias voltage in an oxygen plasma will form an oxide layer and the process has been termed *plasma anodization.*

It, therefore, follows that one must use bias sputtering with caution and if either a negative or positive bias improves a film property, apparently because of improved purity, this may not indicate that *it is pure* but is less contaminated!

Film properties: These effects are shown by a study of the influence of substrate bias on d.c. sputtering of tantalum in a 10% oxygen/90% argon mixture by Westwood and Wilcox (1971). They found that using a target voltage of $-3{\cdot}5$ kV the oxygen content of Ta-films was reduced from 35 at %, for non-biased coatings, to 10 at % for a -100 V bias. However, using a positive bias of $+100$ V the oxygen content fell to 6 at %. They mention that a negative bias reduced the oxygen content by raising the sputtering rate and thereby the gettering action as well as directly removing gas from the receiver surface. Of course electron and ion interactions with chemisorbed gas will be different and it is not possible to say which would in general use be more effective. The electron current density was several times that of the positive ion and they noted that the Ta-film resistivities for the positive bias coatings were not the low values expected for the reduced oxygen content. It was concluded that carbon from electron cracked pump oil had entered the film.

A negative bias potential applied to a work-holder before deposition will sputter-clean the receiver and subsequently improve film adhesion. Substrate material can be back-sputtered on to the target so that on commencing film deposition a mixed coating is initially obtained which improves film adhesion. Vossen and O'Neill (1968) have used an r.f. bias to clean a dielectric substrate in a d.c. diode system.

Boron carbide and boron nitride have been sputtered by C. R. D. Priestland using a negative bias for pre-cleaning and during deposition. The coefficient of friction of the boron carbide films in contact was 0·2 after burnishing which was that obtained by Mordike (1960) using a dynamic friction measurement on the bulk material. Boron nitride films had a coefficient of friction of 0·17. The negative bias permitted thicker deposits to be prepared before the coatings crazed under internal stresses; boron nitride films up to about 3 μm thick were deposited. Using the negative bias no significant change was found in the coefficient of friction but the film abrasion resistance was enhanced.

As the substrate is heated by bias enhanced bombardment a deposited film may undergo other physical changes such as annealing during growth.

Substrate bombardment may be undesirable even when it is not enhanced by biasing. Thus Goldstein and Bellina (1970) report that Ta-films could not be deposited on PTFE (Teflon) in a diode sputtering system (3 kV, argon, 0·02 Torr) with a cathode/anode spacing of 6·35 cm unless a stainless steel mesh screen was interposed between the substrate and receiver. Presumably the screen reduced electron bombardment of the PTFE. An explanation of the effect was not given but PTFE will thermally evaporate at 300°C and sputter under ion bombardment emitting molecular fragments.

Effects of adsorbed gas on S yield: Active gases, e.g. H_2, N_2, CO_2, H_2O and hydrocarbon gases can be desorbed from vacuum system materials, backstream from the pumps or be present in inert gases admitted to the system. Some metals form nitrides by reacting with nitrogen or oxide layers by reacting with oxygen or water vapour. Contaminant or reaction layers on a target surface can be sputtered as atomic groups or dissociated by ion bombardment. Thus part of the ion energy released in the target is dissipated in liberating surface compounds or active gas which, if part of the residual atmosphere, returns to react with the target. The sputtering yield will be apparently reduced if active gas molecules are chemisorbed on a target at a rate commensurate with the impingement rate of the inert gas ions used for sputtering. Let us consider O_2 molecules incident on a clean metal surface for which we assume that the sticking coefficient is unity and that impacting singly charged ions have a current density of 1 A/m^2. Then we can find the O_2 pressure at which the gas molecules will arrive at the same number rate per unit area as the ions. At 1 Torr and 25°C there are $n = 3{\cdot}2 \times 10^{22}$ O_2 molecules per m^3 with an average velocity of $\bar{v} = 4{\cdot}4 \times 10^2$ m/s; $e = 1{\cdot}6 \times 10^{-19}$ C. The ion impingement rate $1/1{\cdot}6 \times 10^{-19} = 6{\cdot}6 \times 10^{18}$ ions m^{-2} s^{-1}. Assuming the gas is in a random state the impact rate per unit area at 1 Torr is given by $n\bar{v}/4 = 3{\cdot}5 \times 10^{24}$ molecules m^{-2} s^{-1}. Thus reducing the O_2 pressure to 2×10^{-6} Torr will give a molecular impact rate of 7×10^{18} molecules m^{-2} s^{-1} comparable to that of the ions.

Depending on the value of the sticking coefficient – this simple calculation shows the importance of even trace impurities in gases admitted to plasma sputtering systems or released from the vacuum system. Oxygen molecules may dissociate on chemisorption forming twice the number of bonds for impact severance. Generally when the O_2 content of a plasma rises to about one per cent in an inert gas at a pressure of 10^{-3} Torr the deposition rate of a metal falls for current densities of $\sim$1 A/m^2. Initially there is a compensating effect as the oxygen reacts with the metal deposit increasing the mass deposited.

Reactive and compound sputtering

Oxygen or nitrogen can be added to the inert gas atmosphere of a diode or plasma sputtering plant to promote the formation respectively of metal oxide or nitride films by sputtering of metal compounds formed on the target surface or by reaction with free metal at the receiver. Holland (1956) has reviewed early work in his laboratory and elsewhere on *reactive sputtering* as the technique is termed.

The active gas concentration must be the minimum required for compound formation as the sputtering yield is reduced by gas adsorption as discussed above. Helwig (1952) was one of the earliest to observe the fall in deposition rate as the O_2 pressure was raised when sputtering Cd to form CdO-films. A partial pressure of 10^{-4} Torr of O_2 or N_2 is usually adequate for complete reaction with an active metal film. Zinc sulphide films have been prepared by admitting H_2S to the plasma when sputtering from a zinc target (Helwig and Koenig 1955). Frazer and Melchior (1972) were able to prepare CdS-films in a diode system with dark resistivities of 10^7 Ω cm. Using a CdS target and a pressure of 0·05 Torr they needed to add H_2S to argon to ensure sulphur replacement in the film.

With the development of r.f. sputtering it became possible to prepare films directly from compound and dielectric targets although for some substances it was still necessary to add a reactive component to the inert gas plasma. SiO_2-films have been prepared by reactively sputtering Si in $Ar^+ + O_2$ and by r.f. sputtering SiO_2 in argon but some added O_2 may be necessary in the latter case. Metal oxides and particularly silica and sodium free silicate glasses have been successfully r.f. sputtered to form thin films. Many carbides, nitrides and compounds based on Cd, Se, Te and In have been r.f. sputtered but without composition analysis one must not assume that a compound stoichiometrically reforms on a receiver.

Rawlins (1971) has r.f. sputtered ZnS-films using a ZnS target in argon from which H_2, O_2 and H_2O had been removed and found that the deposits resembled those prepared by evaporation of ZnS in ultra high vacuum.

Unvala and Pearmain (1970) state that they have produced stoichiometric layers of GaAs by ion beam sputtering of a GaAs-target using 9 keV Ar^+-ions in a vessel at 10^{-4} Torr.

Atomic hydrogen formed in a plasma can react with some metals to form volatile hydrides. Free hydrogen or that released by a water vapour reaction can be sorbed by a metal target or condensing film. Chemical reactions may also occur with carbon oxide gases which are often present in a vacuum atmosphere. We shall now consider the vacuum conditions which are required if pure metal coatings are to be sputtered.

3.3.5 Vacuum system design and partial pressure monitoring

The nature and partial pressure of the gas components in the sputtering atmosphere will be determined by the composition of the inlet gas (used for providing the energetic ions), the gas desorption characteristics of the constructional materials and target and the type and disposition of the pumps. As most metals, with the probable exception of gold but including the noble metal such as platinum and palladium, chemisorb oxygen and absorb hydrogen, it is essential that the partial pressures and related surface impingement rates of these active gases and water vapour should be low when pure films are to be deposited. If growth rate of the sputtered film is between one to two monoatomic layers per second then a partial pressure of 10^{-6} Torr of oxygen at room temperature would give a similar molecular arrival rate. Water vapour, a common contaminant of tank gas and desorption component from materials, will often react more readily with a metal than pure oxygen and after reaction a deposit can contain oxide, hydroxyl and hydrogen impurities.

Assuming a sticking coefficient of 0·1 for gases reacting with the film then for a gas impurity level less than one part per million the total pressure of active gas in the target/receiver vessel must not exceed the ultra high vacuum value of 10^{-11} Torr with film growth rates of one to two monoatomic layers per second. However, inert gas ions could still enter the film after rebounding from the target, as previously discussed, and the deposit may require heat-treatment during growth to remove such gas.

Getter-sputtering

One can reduce film contamination in two ways by careful design of the vacuum system and by the use of the sputtered deposit to getter those gases which would contaminate the film. Berghaus and Burkhardt (1943) were the first to propose getter sputtering and the technique was reintroduced more recently by Theuerer and Hauser (1965). Preferably one arranges the sputtering system so that gas passes over a sputtered deposit before entering the main coating region.* Holland and Priestland (1972) have achieved this in an r.f. system using a single target, as shown in Fig. 3.27.† One target side was used for film preparation and the other for depositing a getter coating in a separate vessel into which inert gas flowed before entering the film making region. Of course the target may contain gaseous impurities which are desorbed and retrapped in the deposit unless exhaustion is adequate.

Vacuum pumps and systems: Oil vapour and hydrocarbons emitted from gaskets and pumps can be degraded contaminating the deposited film and the film deposition region in a getter system must be isolated from these emissions.

A vacuum system can be made to ultra high vacuum standards and degassed by baking but generally for simplicity one must employ conventional components and positive displacement pumps. Metal gaskets can be used for sealing infrequently dismantled components but an elastomer seal is usually necessary on the vessel closure plate for ease of operation. If the sputtering pressure is in the 10^{-3} Torr range the system can be evacuated with a LN_2-trapped oil diffusion pump which for backstreaming oil molecules can have an ultimate pressure less than 10^{-9} Torr. When the operating pressure is $>5 \cdot 10^{-3}$ Torr the pumping speed (litres/second) of a diffusion pump falls and the throughput (torr litre/second) tends to a constant value. To obtain a stable pressure for a given flow of gas into the system, via

* Of course gettering occurs in any sputtering systems when films are deposited on the vessel walls and the substrate. Early operators often used a shield enclosing the coating zone to prevent undue film deposition on the vessel and thereby inadvertently produced a type of getter sputtering arrangement before its use was fully appreciated!

† The School of Engineering Science, University College of North Wales at Bangor, have shown at the 1970 Physics Exhibition a double vessel sputtering system in which gas is purified in one vessel before passing into the deposition chamber.

an adjusted leak, one must throttle the pump with a high vacuum valve to obtain a pressure at which the throughput has a unique value. One can raise the pumping speed in the low pressure region by using a *vapour booster* or ejector pump in place of a diffusion pump. Vapour stream pumps with combined high vacuum and ejector stages have been developed for pre-exhaustion of sputtering systems and operation at low pressures with high gas loads; Holland (1956) has reviewed his work on this type of system.

Visser (1971) recently has discussed pumping systems for sputtering plant and compared turbomolecular and diffusion pumps. He states that diffusion pumps have low throughputs in the sputtering pressure region but appears to be unaware of ejector type pumps whose stages are adapted for high pressure use.

Axial flow compressors or turbomolecular pumps are used for exhausting sputtering systems and they are simple to operate although very costly. There is a common belief that, unlike oil pumps, backstreaming does not occur with such pumps. The lubricated bearings on turbomolecular pumps are on the backing side and emitted oil components are swept away from the vessel. Hydrocarbon lubricants are fragmented emitting components of light molecular mass when their molecules are exposed to abrasion by friction and these components can enter the vacuum vessel when the pump is idle. Thus LN_2 trapping may be necessary if the pump is not baked to degas it after each operation.

The pumping speed of a turbomolecular pump tends to be constant for most gases but the compression ratio, determining back-flow of gas, is less for the light gases and can fall to typically 100 for H_2. If a plasma pressure were 10^{-3} Torr the partial pressure of H_2 could be 10^{-4} Torr if the backing pump, e.g. rotary mechanical had a H_2 partial pressure of 10^{-2} Torr from a high blank-off pressure or low pumping speed for H_2. *Because of the fragmentation of lubricant fluids which release volatiles it is always wise to use a trapped rotary pump* and a trap which is always in operation. An alumina sorbent has been found the most effective in the writer's laboratory.

The choice of pumping speed for evacuation depends on the chief source of gas impurity in the sputtering vessel. If the inlet gas is contaminated then raising the evacuation rate may not change the partial pressure of the inlet impurity at a given total operating pressure, because the pumping speeds of the different gas species are increased by similar order. In that case one must purify the gas before entry using if necessary getter sputtering and a well throttled pump. If the impurity is mainly from desorption a high evacuation rate is necessary to reduce its partial pressure. It is essential to use a residual gas analyser to determine the composition of a sputtering atmosphere to determine the best pumping technique.

REFERENCES

ANDERSON, G. S., MAYER, WM, N., and WEHNER, G. K. (1962) *J. Appl. Phys.*, **33**, 2991–2992.
ANDERSON, G. S., and WEHNER, G. K. (1960) *J. Appl. Phys.*, **31**, 2305–2313.
ALMÉN, O., and BRUCE, G. (1961) *Nucl. Inst. Methods*, **11**, 257–278.
BACH, H. (1968) *Naturwissenschaften*, **55**, 439–440.
BACH, H. (1970) *Non-Crystalline Solids*, **3**, 1-32.
BAEDE, A. P. M., JUNGMANN, W. F., and LOS, J. (1971) *Physica*, **54**, 459–467.

BEHRISCH, R., and WEISSMANN, R. (1969) *Phys. Letts,* **30A**, 506–507.
BERGHAUS, B., and BURKHARDT, W. (1943) German Pat. 736130.
BERGHAUS, B. (1936) Brit. Pat. 486629.
BETZ, G., DOBROZEMSKY, R., and VIEHBÖCK, F. P. (1970) *Nederlands Tijdschrift V. Vacuumtechniek,* **8**, 203–206.
CHENEY, K. B., and PITKIN, E. T. (1965) *J. Appl. Phys.,* **36**, 3542–3544.
CHOPRA, K. L., and RANDLETT (1966) 'A duoplasmatron ion source for vacuum sputtering of thin films', Tech. No. TR-96. Ledgement Lab., Kennecott Copper Co.
COBURN, J. W. (1970) *Rev. Sci. Instrum,* **41**, 1219–1223.
COMPTON, K. T., and LAMAR, E. S. (1933) *Phys. Rev.,* **44**, 338–344.
DAVIDSE, P. D., and MAISSEL, L. T. (1965) *Third Inter. Vacuum Congress, Stuttgart,* Vol. 2, Part III, 651–655.
DAVIDSE, P. D., and WHITAKER, H. L. (1971) Brit. Pat. No. 1242492.
DUSHMAN, S. (1949) 'Scientific foundations of vacuum technique' (J. Wiley, New York).
FARMERY, B., and THOMPSON, M. W. (1968) *Phil. Mag.,* **18**, 415–424.
FETZ, H. (1942) *Z. Phys.,* **119**, 590–595.
FRAZER, D. B., and MELCHIOR, H. (1972) *J. Appl. Phys.,* **43**, 3120–3127.
GILL, W. D., and KAY, E. (1965) *Rev. Sci. Instrum.,* **36**, 277–282.
GOLDSTEIN, R. M., and BELLINA, J. J. JUN. (1970) *Rev. Sci. Instrum.,* **41**, 1110.
GAYDOU, F. P. (1967) *Vacuum,* **17**, 325–327.
GRØNLUND, F., and MOORE, W. J. (1960) *J. Chem. Phys.,* **32**, 1540–1545.
GÜNTHERSCHULZE, A. (1926) *Z. Phys.,* **38**, 575–583.
GÜNTHERSCHULZE, A. (1942) *Z. Phys.,* **119**, 79–91.
HEISIG, U., SCHILLER, S., and BEISTER, G. (1965) Microminiaturization, *Proc. IFAC/IFIP Symp., Munich,* 547–565.
HELWIG, G. (1952) *Z. Phys.,* **132**, 621–632.
HELWIG, G., and KOENIG, H. (1955) *Z. Angew. Phys.,* **7**, 323.
HERZOG, R. F. K., POSCHENRIEDER, W. P., and SATKIEWICZ, F. G. (1973) *Radiation Eff.,* **18**. 199–205.
HINES, R. L., and WALLOR, R. (1961) *J. Appl. Phys.,* **32**, 202–204.
HOLLAND, L. (1954) Brit. Pat. No. 736512.
HOLLAND. L. (1956) 'The vacuum deposition of thin films' (Chapman and Hall Ltd., London).
HOLLAND, L., PUTNER, T. I., and JACKSON, G. N. (1968) *J. Phys. E, J. Sci. Instrum.,* **1**, 32–34.
HOLLAND, L., and PRIESTLAND, C. R. D. (1972) *Vacuum.,* **22**, 133–149.
KAY, E. (1963) *J. Appl. Phys.,* **34**, 760–769.
KAY, E., CAMPBELL, and POENISCH. A. P. (1967) US Pat. No. 3,325,394.
LIVESEY, R. G. (1972) *Vacuum.* **22**, 595–597.
MACDONALD, R. J., OSTRY, D., ZWANGOBANI, E., and DENNIS, E. (1970) *Nederlands Tijdschrift V. Vacuumtechniek,* **8**, 207–212.
MATTOX, D. M., and KOMINIAK, G. J. (1971) *J. Vac. Sci. Technol.,* **8**, 194–198.
MITCHELL, I. V., and MADDISON, R. C. (1971) *Vacuum,* **21**, 591–595.
MORDIKE, B. L. (1960) *Wear,* **3**, 374–377.
MULLALY, J. R. (1971) *Res/Dev.,* 40–44.
PENNING, F. M. (1936) U.S. Pat. 2,146,025.
PITKIN, E. T., MACGREGOR, M. A., SALEMME, V., and BIERGE, R. (1960) 'Investigation of the interaction of high velocity-ions with metallic surfaces', ARL Tr-60-299 OTS, Dept. Commerce, Washington D.C.
RAWLINS, T. G. R. (1971) 'Sputtered ZnS films on silicon', SRDE Rep. No. 71006.
RIVLIN, J. (1971) 'Sputtering on plastics: definition of the state of the art today', *Trans. Conf. and School – The Elements, Techniques and Applications of Sputtering, Brighton* (M.R.C. Ltd.), 53–57.
SCHROEDER, J. B., DIESELMAN, H. D., and DOUGLASS, J. W. (1971) *Appl. Opt.,* **10**, 295–299.

SEITZ, F., and KOELER, J. S. (1956) *Solid State Phys.*, **2**, 307.
SILSBEE, R. H. (1957) *J. Appl. Phys.*, **28**, 1246.
STIRLING, A. J., and WESTWOOD, W. D. (1970) *J. Appl. Phys.*, **41**, 742–748.
STIRLING, A. J., and WESTWOOD, W. D. (1971) *Appl. Phys., J. Phys. D.*, **4**, 246–252.
STUART, R. V., and WEHNER, G. K. (1964) *J. Appl. Phys.*, **35**, 1819–1824.
THEURER, H. C., and HAUSER, J. J. (1965) *Trans. Metall. Soc. AIME.*, **233**, 588–591.
THOMPSON, M. W. (1959) *Phil. Mag.*, **4**, 139–141.
THOMPSON, M. W., and NELSON, R. S. (1962) *Phil. Mag.*, **7**, 2015.
UNVALA, B. A., and PEARMAIN, K. (1970) *J. Mat. Sci.*, **5**, 1016–1018.
VISSER, J. (1971) Vacuum systems for sputtering, *Trans. Conf. and School. – The Elements, Technique and Applications of Sputtering, Brighton* (M.R.C. Ltd.) 105–130.
VOSSEN, J. L., and O'NEILL, Jr., J. J. (1968) *R.C.A. Rev.*, **29**, 566–581.
WEHNER, G. K. (1955) *J. Appl. Phys.*, **26**, 1056–1057.
WEHNER, G. K. (1956) *Phys. Rev.*, **102**, 690–704.
WEHNER, G. K. (1960) *J. Appl. Phys.*, **31**, 1392–1397.
WEHNER, G. K., and ROSENBERG, D. (1960) *J. Appl. Phys.*, **31**, 177–179.
WEHNER, G. K. (1961) *Trans. 8th Vac. Sympos. and 2nd Inter. Congr.*, 239–244.
WEIJSENFELD, C. H. (1966) 'Yield energy and angular distribution of sputtered atoms', Thesis, Rijksuniversiteit Te Utrecht.
WESTWOOD, W. D., and BOYNTON, R. (1972) *J. Appl. Phys.*, **43**, 2691–2697.
WESTWOOD, W. D., and WILCOX, P. S. (1971) *J. Appl. Phys.*, **42**, 4055–4062.
WINTERS, H. F., and KAY, E. (1972) *J. Appl. Phys.*, **43**, 794–799.
WOODYARD, J. R., and BURLEIGH COOPER, C. (1964) *J. Appl. Phys.*, **35**, 1107–1117.

BIBLIOGRAPHY

Reference works on ion impact sputtering and related effects

BEHRISCH, R. (1964) 'Sputtering of solid bodies under ion bombardment' (in German), *Ergebnisse der Exakten Naturwissenschaften*, **35** (Springer Verlag).
CARTER, G., and COLLIGON, J. S. (1968) 'Ion bombardment of solids' (Heinemann Educational Books, London).
NELSON, R. S. (1968) 'The observation of atomic collisions in crystalline solids' (North-Holland Publishing Co., Amsterdam).
KAMINSKY, M. (1965) 'Atomic and ionic impact phenomena on metal surfaces' (Springer-Verlag, Berlin).
PALMER, D. W., ed. (1970) 'Atomic collision phenomena in solids', *Proc. Intern. Conf. Sussex Univ. 1969* (North-Holland Publishing Co., Amsterdam).
THOMPSON, M. W. (1969) 'Defects and radiation damage in metals' (Cambridge University Press).
TRILLAT, J. J., ed. (1962) 'Ionic bombardment theory and applications', Int. Sympos. Nat. Scient. Res. Center, Bellevue (Gordon and Breach Sci. Publ., New York).

Reference works on sputtering for thin film deposition

ANON. (1966) 'Symposium on the deposition of thin films by sputtering', Co-sponsored by Univ. Rochester and Consolidated Vacuum Corpt., Rochester, New York.
AMER. SOC. METALS (1964) 'Thin films' Seminar Oct. (1963) (Chapman & Hall, London).
ANDERSON, J. C., ed. (1966) 'The use of thin films in physical investigations' (Academic Press, London).
BERRY, R. W., HALL, P. M., and HARRIS, M. T. (1968) 'Thin film technology' (New Jersey, Van Nostrand).
CHOPRA, K. L. (1969) 'Thin film phenomena' (McGraw-Hill, New York).
HOLLAND, L. (1956) 'Vacuum deposition of thin films' (Chapman & Hall, London) (6th printing 1966 with up-dated ref. list).

HOLLAND, L. (1965) Chapter 4, Vacuum deposition apparatus and techniques, 'Thin film microelectronics' (Chapman & Hall, London).

KAY, E. (1962) 'Impact evaporation and thin film growth in a glow discharge', *Advances in Electronics*, **17**, 245–321.

MAISSEL, L. I., and GLANG, R., ed. (1970) 'Handbook on thin film technology' (McGraw-Hill, New York).

Papers and reviews

'The physics of thin films' Volumes 1 to 6 (Academic Press); the transactions of the American Vacuum Society and the International Vacuum Congresses (IUVSTA).

The journals: *Vacuum, Thin Solid Films, J. Electrochemical Society, J. Vacuum Science and Technology, Le Vide, Hochvakuum, Solid State Technology and Research/Development*, publish papers in these fields.

INDEXES

Manufacturers Index

Equipment Index

Subject Index

INDEX TO ADVERTISERS

Advertising agents:
T. G. Scott & Son Ltd
1 Clements Inn
London WC2A 2ED